Lineability

The Search for Linearity
in Mathematics

MONOGRAPHS AND RESEARCH NOTES IN MATHEMATICS

Series Editors

John A. Burns
Thomas J. Tucker
Miklos Bona
Michael Ruzhansky

Published Titles

Application of Fuzzy Logic to Social Choice Theory, John N. Mordeson, Davender S. Malik
and Terry D. Clark

*Blow-up Patterns for Higher-Order: Nonlinear Parabolic, Hyperbolic Dispersion and
Schrödinger Equations*, Victor A. Galaktionov, Enzo L. Mitidieri, and Stanislav Pohozaev

Cremona Groups and Icosahedron, Ivan Cheltsov and Constantin Shramov

Difference Equations: Theory, Applications and Advanced Topics, Third Edition,
Ronald E. Mickens

Dictionary of Inequalities, Second Edition, Peter Bullen

Iterative Optimization in Inverse Problems, Charles L. Byrne

Line Integral Methods for Conservative Problems, Luigi Brugnano and Felice Iaverno

Lineability: The Search for Linearity in Mathematics, Richard M. Aron,
Luis Bernal González, Daniel M. Pellegrino, and Juan B. Seoane Sepúlveda

Modeling and Inverse Problems in the Presence of Uncertainty, H. T. Banks, Shuhua Hu,
and W. Clayton Thompson

Monomial Algebras, Second Edition, Rafael H. Villarreal

*Nonlinear Functional Analysis in Banach Spaces and Banach Algebras: Fixed Point
Theory Under Weak Topology for Nonlinear Operators and Block Operator Matrices with
Applications,* Aref Jeribi and Bilel Krichen

*Partial Differential Equations with Variable Exponents: Variational Methods and Qualitative
Analysis*, Vicenţiu D. Rădulescu and Dušan D. Repovš

A Practical Guide to Geometric Regulation for Distributed Parameter Systems,
Eugenio Aulisa and David Gilliam

Signal Processing: A Mathematical Approach, Second Edition, Charles L. Byrne

Sinusoids: Theory and Technological Applications, Prem K. Kythe

Special Integrals of Gradshetyn and Ryzhik: the Proofs – Volume I, Victor H. Moll

Forthcoming Titles

Actions and Invariants of Algebraic Groups, Second Edition, Walter Ferrer Santos
and Alvaro Rittatore

Analytical Methods for Kolmogorov Equations, Second Edition, Luca Lorenzi

Complex Analysis: Conformal Inequalities and the Bierbach Conjecture, Prem K. Kythe

Forthcoming Titles (continued)

MONOGRAPHS AND RESEARCH NOTES IN MATHEMATICS

Lineability
The Search for Linearity in Mathematics

Richard M. Aron
Kent State University, Ohio, USA

Luis Bernal González
University of Sevilla, Spain

Daniel M. Pellegrino
Federal University of Paraiba, João Pessoa, Brazil

Juan B. Seoane Sepúlveda
Complutense University of Madrid, Spain

CRC Press
Taylor & Francis Group
Boca Raton   London   New York

CRC Press is an imprint of the
Taylor & Francis Group, an **informa** business

A CHAPMAN & HALL BOOK

CRC Press
Taylor & Francis Group
6000 Broken Sound Parkway NW, Suite 300
Boca Raton, FL 33487-2742

First issued in paperback 2018

© 2016 by Taylor & Francis Group, LLC
CRC Press is an imprint of Taylor & Francis Group, an Informa business

No claim to original U.S. Government works

ISBN-13: 978-1-4822-9909-0 (hbk)
ISBN-13: 978-1-138-89443-3 (pbk)

Visit the Taylor & Francis Web site at
http://www.taylorandfrancis.com

and the CRC Press Web site at
http://www.crcpress.com

Contents

Preface

Vector spaces and linear algebras are elegant mathematical structures which, at first glance, seem to be "forbidden" to families of "strange" objects. In other words, one might believe that finding large vector spaces or linear algebras within such families is rather unlikely. The present book intends to show that this is not at all the case.

Just to give a first historical instance, Vladimir Gurariy (Figure 1) proved as early as 1966 [239] that the family of continuous functions on the closed unit interval $[0, 1]$ that are differentiable at *no* point contains, except for the null function, an infinite dimensional vector space. This is a rather surprising result, because, in fact, it is difficult to provide explicitly one of such functions, as constructed by Karl Weierstrass in 1872. But not everything is positive: Levine and Milman [287] had already shown in 1940 the non-existence of closed infinite dimensional vector spaces contained in the family of continuous bounded variation functions on $[0, 1]$, when this family is considered as a subspace of the space of continuous functions on $[0, 1]$ under the convergence uniform topology. More recently, the appearance of a survey paper on this topic [105] has prompted the appearance of this more specialized monograph. Therefore research on the conditions for the existence of large algebraical substructures is justified.

Although there are several interesting findings in the subject, obtained in the last third of the preceding century, they are very scattered and, in fact, that research was not officially initiated till the beginning of the current millennium. The main aim of this book is not only to bring the main results found to date nearer postgraduate students, but also nearer young and senior researchers who want to enter the subject. The monograph is mainly addressed to all these people.

Concerning prerequisites, some knowledge of certain mathematical areas is assumed in order to fruitfully read the book, namely, calculus of one and several real variables, Lebesgue integration, elementary set theory, rudiments of linear algebra, geometry and general topology and basic theory of Hilbert/Banach spaces and operators on them. The reader is also expected to handle basic notions of complex variables; especially, a certain familiarity with holomorphic functions would be convenient.

Nevertheless, one of our aims has been to make the book, as far as possible, self-contained. This is why a section *What one needs to know* has been included

Preface

FIGURE 1: Vladimir I. Gurariy (1935–2005).

at the beginning of each chapter, except in Chapter 6. We hope that such a section will be useful in order to put at the disposal of the reader a number of tools which are specific for each chapter. Some of these tools are basic, but others might be unknown at an elementary level.

The book has been divided into eight chapters. We start with an unnumbered one called *Preliminary Notions and Tools*, whose title is rather explanatory. In this chapter, the basic concepts about existence of linear structures (namely, lineability, dense-lineability, spaceability and algebrability), which were officially introduced at the beginning of the 21st century, are presented. Roughly speaking, the last four properties mean, respectively, existence of large vector spaces, of dense vector spaces, of large closed vector spaces and of large algebras within a given subset of vector space X (where X is endowed, in addition, with a topological structure if some topological property is required). Prior to this, a number of definitions and results from elementary set theory are referenced for the convenience of the reader: cardinal numbers and order relation among them, the Cantor-Bernstein-Schroeder theorem, Zorn's lemma, the cardinal of the set of natural numbers and of the set of real numbers, the Continuum Hypothesis and arithmetic with cardinals. Moreover, we have included a few elementary notions about linear algebra (such as those of vector subspace, linear dependence and independence, linear span of a subset of a vector space, Hamel basis and norm on a vector space) and a selection of the most important results, including the fact that two Hamel bases of a vector space have the same cardinal. We have considered convenient to include a formulation and a proof of the important Baire's theorem, because this theorem is the main starting point to establish existence results in the context of complete metric spaces and, in particular, in normed spaces. In order to study the existence of more richer substructures, the notion of (linear) algebra is also recalled, as well as those of unital algebra, homomorphism between algebras, generator system of an algebra and free generated algebra.

Chapter 1 deals with lineability properties of families of functions defined on some subset (mostly, an interval) of the real line. We begin with the mentioned continuous nowhere-differentiable functions, and provide a historical report of them. This is followed by a linear study of the family of differentiable functions that are not monotone at any interval, after which the set of infinite differentiable functions which are not representable by the Taylor series at any point is considered. We also study families of functions satisfying some of the following properties, or even more restrictive ones: are continuous at no point; are surjective in any interval; attain their maximum at one point; are continuous and fill in some portion of a high dimensional space.

The main concern of *Chapter 2* is the lineability of special families of holomorphic (or analytic) functions defined on some domain of the complex plane. A topic that has attracted the attention of many mathematicians in the area is the existence of holomorphic mappings that cannot be extended beyond the boundary of the domain. After establishing the topological genericity of this phenomenon, its lineability is analyzed, even for subspaces of functions enjoying richer properties. Special attention is devoted to the case in which the domain is the open unit disc. Other families studied in this chapter are the set of holomorphic functions in the whole complex plane tending rapidly to zero, when z approaches infinity along large sets, and the set of holomorphic functions sending each curve tending to the boundary onto a dense subset of the complex plane. In the final section we introduce the so-called Gevrey differentiable functions, which are placed between the analytic functions and the infinitely differentiable functions. Then we exhibit several results about the lineability of its complementary set.

Chapter 3 is focused on spaces of sequences (not necessarily real valued or complex valued) and spaces of integrable functions. Lineability and spaceability properties are studied in families of sequences (functions, resp.) that are summable (integrable, resp.) with respect to an order but not with respect to another(s) order(s). More specialized questions are also considered for instance, the lineability of the family of X-valued sequences is analyzed, with X a Banach space, satisfying that the sequence of their images under a given mapping $X \to Y$ satisfying appropriate conditions does not belong to certain Y-valued sequences. Turning to a classical question, we also treat operators non-attaining the norm, in the specific case in which the operators act between sequence spaces, as well as the classical (lack of) relationship between Riemann integrable functions and Lebesgue integrable functions.

Since 1929 when Birkhoff constructed his surprising entire function whose translates formed a dense subset of the space of entire functions, the volume of this kind of universality results has increased exponentially up to the present, so as to became a solid theory. A universal (or hypercyclic) vector is, roughly speaking, a very erratic or chaotic element under the action of a linear operator. That the phenomenon of universality is topologically generic is well known, and in *Chapter 4* we deal with its genericity from an algebraic point of view. The dense-lineability property, which is general and known from Herrero (1991), as well as spaceability and algebrability properties of the family of universal vectors are established and proved in this chapter. We have considered, in addition, several related phenomena such as supercyclicity, frequently hypercyclicity, universality of Taylor series and distributional chaos.

Chapter 5 treats the linear structure of the set of zeros of a polynomial defined on a (real or complex) Banach space, with focus on the infinite dimensional case. The systematic study of the zeros in the complex case dates from the 1950s, and nowadays this investigation can be approached with tools of algebraic geometry, complex analysis, and functional analysis. At the end of the last century it was proved the lineability of the set of zeros of every homogeneous polynomial on an infinite dimensional complex Banach space, and this fact will be registered in this chapter. We will also see how the situation is radically different in the real case. Moreover, concrete examples of Banach spaces will be furnished where additional information about the structure of the set of zeros may be given.

In *Chapter 6* we discuss, under the point of view of lineability, a number of disconnected topics that we also have considered interesting and that, in spite of the fact that they might have been included in some of the previous chapters, they are more specialized and hence they may be overlooked in a first reading. Namely, the lineability of the following families of vectors is studied: series in classical Banach spaces, Dirichlet series, continuous functions having unbounded sequence of partial Fourier sums, functionals attaining the norm, continuous functions vanishing at infinitely many points, sequences with finitely many zeros, and Sierpiński-Zygmund functions, among others.

Finally, *Chapter 7* offers an account of the general techniques in order to discover lineability, in its diverse degrees. It happens that there are only a bunch of these techniques at our disposal up to date. They have mostly been discovered by extracting the core ideas of the proofs of the several concrete findings in lineability. We will establish and prove here a number of sufficient conditions for dense-lineability, spaceability and algebrability. Several applications, as well as some negative examples, will be given. The final section is devoted to a brief account about a few cardinal invariants and the recently introduced concept of additivity, which may shed light in order to find lineability in some settings.

A collection of exercises is proposed in the second to last section of each chapter. The goal of these exercises is twofold, namely, the assimilation of the results given in the chapter through examples and the acquisition of a number of techniques in order to prove new results (or results stated, but not proved, in the chapter). Most exercises appear accompanied by a hint to direct the interested reader.

The book does not intend to be exhaustive with respect to the contents developed in it, but rather a source of information about lineability. With this in mind, and after having established the pertinent concepts and basic tools in each chapter, we have selected a number of assertions with their corresponding proofs, with the warning that these assertions are not necessarily the most general known up to date. Nevertheless, comments and references on improvements of such assertions and on new results are put either in every particular section or in the special final section *Notes and Remarks* of each chapter, except in Chapters 5 and 6, due to the fact that they are shorter and, thus, we have opted for including all remarks in their main text. As for the results whose proofs do not appear in the text, the interested reader can find them in the pertinent references.

The authors apologize for the more than likely omission of (probably many) relevant lineability results. According to the aim and scope of this book and taking into account the rapid development of this theory during the last years, it would be virtually impossible to gather the whole findings in the topic. Of course, we take responsibility for all flaws, including any mathematical errors, and if some result has not been included, this by no means indicates that it did not deserve to be mentioned. Most likely, the omission was due to a lack of space, and if the result does not even appear in the Notes and Remarks section this simply means that the authors were not aware of it.

There are many people who deserve our gratitude in connection with writing this book. A number of colleagues, family, and friends have read the manuscript or parts of it, made a number of useful suggestions improving the text, and provided interesting discussions and constructive remarks. In alphabetical order, our warmest thanks to Gustavo Araújo, Antonio Bonilla, Geraldo Botelho, Alberto Conejero, Vinícius Fávaro, Mar Fenoy, José Luis Gámez, Gustavo Muñoz, Marina Murillo, Tony Kleverson Nogueira, Diogo Diniz Pereira da Silva e Silva, Magnolia Inés Rodrigo, and Víctor Sánchez.

We are grateful as well to José Antonio Prado-Bassas, who has helped in the elaboration of a number of pictures. Our thanks also go to Bob Stern and Sarfraz Khan as editors-in-chief of Taylor and Francis, London, for their valuable assistance and quite positive reception of the project. The help received from the staff during the production process is invaluable; we are especially indebted to Marcus Fontaine for his effective assistance with the use of LaTeX.

We acknowledge the support of Plan Andaluz de Investigación de la Junta de Andalucía FQM-127 Grant P08-FQM-03543, of Spanish MEC Grant MTM2012-34847-C02-01, of Spanish MICINN Project MTM2011-22417, of MTM2012-34341 and of Conselho Nacional de Desenvolvimento Científico e Tecnológico-CNPq-Brazil, Grant 401735/2013-3-PVE-Linha 2.

The first author wants to dedicate this book to his family. The second author dedicates this book to the memory of his parents, Francisco and María Luisa. The third author dedicates this book to Noêmia and Ana Letícia. The fourth author dedicates this book to the memory of his uncle Juan A. Sepúlveda Aramburu.

Kent, Sevilla, João Pessoa and Madrid   Richard M. Aron
April 2015   Luis Bernal González
                                                                  Daniel Pellegrino
                                                       Juan B. Seoane Sepúlveda

Author Biographies

Richard M. Aron did his doctoral work under the supervision of Leopoldo Nachbin at the University of Rochester, receiving his degree in 1971. He has held positions at Trinity College Dublin (Ireland), and since 1986 he has been Professor of Mathematics at Kent State University (Kent, Ohio, USA). His areas of primary research interest include functional and nonlinear analysis, and his doctoral students—from Argentina, Brazil, Ireland, Spain and the US—have made significant contributions in these areas. Aron is on the Editorial Boards of RACSAM (Revista de la Real Academia de Ciencias Exactas, Físicas y Naturales. Serie A. Matemáticas) and the Mathematical Proceedings of the Royal Irish Academy. In addition, he is an Editor in Chief of the Journal of Mathematical Analysis and Applications.

Luis Bernal González graduated in 1980 at the Universidad de Sevilla (Spain). He obtained his Ph.D. degree in Mathematics from this university in 1984. Bernal has been a permanent faculty member at Sevilla since 1980 and was promoted to associate professor in 1987, and to full professor in 2010. He was an invited speaker at the International Congress on Hypercyclicity and Chaos for Linear Operators and Semigroups in Valencia (Spain) in 2009. His main interests are Complex Analysis, Operator Theory and, lately, the interdisciplinary subject of Lineability. Bernal has authored or coauthored more than 80 papers in these areas, many of them concerning the structure of the sets of the mathematical objects discovered. He has given talks at national and international conferences and is very active as a reviewer for several journals. His main hobbies—apart from Mathematics—are reading, hiking and cooking.

Daniel Marinho Pellegrino was born in Belo Horizonte, Brazil. He obtained a doctoral degree in Mathematical Analysis from Unicamp (State University of São Paulo) in 2002. Currently he holds an Associate Professorship position at Universidade Federal da Paraíba, Brazil. Pellegrino is also a researcher of the CNPq (National Council for Scientific and Technological Development) in Brazil. In 2012 he was elected affiliate member of the Brazilian Academy of Sciences and in 2014 a young fellow of the TWAS (The World Academy of Sciences).

Juan B. Seoane Sepúlveda received his first Ph.D. at the Universidad de Cádiz (Spain) jointly with Universität Karlsruhe (Germany) in 2005. His second Ph.D. was earned at Kent State University (Kent, Ohio, USA) in 2006. His main interests include Real and Complex Analysis, Operator Theory, Number Theory, Geometry of Banach spaces and Lineability. He has coauthored over one hundred papers up to this day. Seoane is currently a professor at Universidad Complutense de Madrid (Spain) and has delivered invited lectures at many international conferences and research institutes around the world. When he is not doing Mathematics he enjoys traveling, spending time with his family and friends, movies, visiting the local comic book store and hiking with his dogs. He is a fan of the Marx Brothers and Bob Hope.

List of Figures

Preliminary Notions and Tools

I   Cardinal numbers

The notion of cardinal numbers and some basic facts about cardinal arithmetic shall be of crucial importance in order to fully understand most of the notions in this monograph. Cardinal numbers are, in some sense, an extension of the natural numbers and cardinal arithmetic is the extension of the basic arithmetic from natural to cardinal numbers. Although our approach here is not intended to be, at all, exhaustive it will surely be sufficient for our purposes. Let us begin by providing the first definition (for any nonempty sets A and B), in which $card\,(A)$ stands for the cardinality of A.

- $card\,(A) \leq card\,(B)$ if there is an injection from A to B;

- $card\,(A) \geq card\,(B)$ if there is a surjection from A to B;

- $card\,(A) = card\,(B)$ if there is a bijection from A to B;

- $card\,(A) < card\,(B)$ if $card\,(A) \leq card\,(B)$ and there is no bijection from A to B;

- $card\,(A) > card\,(B)$ if $card\,(A) \geq card\,(B)$ and there is no bijection from A to B.

Clearly, if $card\,(A) \leq card\,(B)$ and $card\,(B) \leq card\,(C)$ then $card\,(A) \leq card\,(C)$. The following results show that the above definitions comply with what one would expect.

Remark I.1. *Notice that we have not defined $card\,(A)$ and $card\,(B)$ in a direct manner. However, we shall soon see that $card\,(A)$ behaves like an extension of the notion of natural number and the symbols $\leq, \geq, >, <$ and $=$ will behave naturally.*

Proposition I.2. *$card\,(A) \leq card\,(B)$ if and only if $card\,(B) \geq card\,(A)$.*

Proof. Let $f : A \to B$ be an injective map. Choose $a_0 \in A$ and define the surjective map $g : B \to A$ by $g\,(b) = f^{-1}\,(b)$ if $b \in f\,(A)$ and $g\,(b) = a_0$ otherwise. Conversely, if $g : B \to A$ is surjective, given $a \in A$ choose $b \in g^{-1}(a)$ and define the injective map $f : A \to B$ by $f(a) = b$. $\qquad\square$

The first main theorem in this Section may seem obvious. However, its depth can be noticed when one sees the subtleties of the arguments contained in its proof. It is known as the Cantor-Bernstein-Schroeder theorem and reads as follows.

Theorem I.3. *If* $card(A) \leq card(B)$ *and* $card(A) \geq card(B)$ *then* $card(A) = card(B)$.

Proof. Suppose that $f : A \rightarrow B$ and $g : B \rightarrow A$ are injections. Let $x \in A$. If $x \in g(B)$, we consider $g^{-1}(x) \in B$; if $g^{-1}(x) \in f(A)$, we consider $f^{-1}(g^{-1}(x))$, etc. There are three possibilities:

(a) The process continues indefinitely (in this case we say that $x \in A_\infty$).

(b) It ends with an element in $A \setminus g(B)$ (in this case we say that $x \in A_A$).

(c) The process ends with an element in $B \setminus f(A)$ (in this case we say that $x \in A_B$).

Thus, A is the disjoint union of A_∞, A_A and A_B. In a similar manner, B is the disjoint union of B_∞, B_B and B_A.

Notice that f maps A_∞ onto B_∞ and A_A onto B_A, whereas g maps B_B onto A_B. So, defining $h : A \rightarrow B$ by $h(x) = f(x)$ if $x \in A_\infty \cup A_A$ and $h(x) = g^{-1}(x)$ if $x \in A_B$, then h is a bijection. $\square$

From now on if A is a set, the symbol $card(A)$ will be called cardinal number (or simply cardinal). A cardinal number may be thought as an equivalence class of sets.

A cardinal number $card(A)$ is called:

• finite if A is finite; otherwise it is called infinite cardinal number;

• countable if it is finite or $card(A) = card(\mathbb{N})$.

For countable sets it is plain that $card(A)$ can be viewed as the number of elements of A. In particular if there is a bijection from A to $\{1, ..., n\}$ we write

$$card(A) = n.$$

Henceforth we write

$$card(\mathbb{N}) = \aleph_0 \text{ and } card(\mathbb{R}) = \mathfrak{c},$$

and we say that $\mathbb{N}$ has cardinality "aleph zero" and $\mathbb{R}$ has cardinality the continuum. For the sake of simplicity, cardinal numbers shall usually be denoted by letters. It is a simple exercise to verify that the set of all rational numbers, $\mathbb{Q}$, also has cardinality $\aleph_0$. It is convenient to define the cardinality of the empty set as zero.

The next result about cardinals (Theorem I.5) shows another nice property of cardinal numbers: any two cardinal numbers are comparable. The proof of this result, as well as several other proofs in this book, makes use of Zorn's lemma and for the sake of completeness we recall here its statement. First we need to recall some terminology.

- A partial order $\leq$ over a set P is a relation that satisfies the following for all a, b, c in P:

  (i) $a \leq a$;

  (ii) if $a \leq b$ and $b \leq a$ then $a = b$;

  (iii) if $a \leq b$ and $b \leq c$ then $a \leq c$.

- A total order in a set P is a partial order such that we have $a \leq b$ or $b \leq a$ for all $a, b \in P$.

- An upper bound of a subset S of some partially ordered set P is an element $p \in P$ such that $s \leq p$ for all $s \in S$.

- A maximal element of a partially ordered set P is an element of $m \in P$ such that if $m \leq p$ and $p \in P$, then $m = p$.

Zorn's lemma can be stated as follows.

Theorem I.4. *If P has a partial order and every subset $S \subset P$ with a total order has an upper bound in P, then the set P contains at least one maximal element.*

The proof of Zorn's lemma is out of the scope of this book, but we can say that it is equivalent to the Axiom of Choice: given any family of nonempty sets, their Cartesian product is a nonempty set. Next, we establish a theorem asserting the *total ordering of cardinals*.

Theorem I.5. *If α, β are cardinals then $\alpha \leq \beta$ or $\beta \leq \alpha$.*

Proof. Let A, B be two sets such that $card\,(A) = \alpha$ and $card\,(B) = \beta$. It suffices to prove that there is an injective function $f : A \to B$ or an injective function $f : B \to A$. Let us consider the set

$$\Psi = \{(C, D, g) : C \subset A, \ D \subset B \text{ and } g : C \to D \text{ a bijection}\}.$$

In Ψ we consider the natural order

$$(C_1, D_1, g_1) \leq (C_2, D_2, g_2)$$

if, and only if, $C_1 \subset C_2$, $D_1 \subset D_2$ and the restriction of g_2 to C_1 is precisely g_1. It is routine to verify that $(\Psi, \leq)$ satisfies the hypotheses of Zorn's lemma. Thus, let (A_m, B_m, f) be a maximal element in Ψ.

We will show that either $A_m = A$ or $B_m = B$. Suppose, by contradiction, that $A_m \neq A$ and $B_m \neq B$. Then there are $a \in A \setminus A_m$ and $b \in B \setminus B_m$ and we can define a suitable bijection

$$g : A_m \cup \{a\} \longrightarrow B_m \cup \{b\}$$

that obviously contradicts the maximality of (A_m, B_m, f). So we conclude that $A_m = A$ or $B_m = B$. If $A_m = A$ it is plain that $f_0 : A \to B$ given by $f_0(a) = f(a)$ is injective and, if $B_m = B$, it is obvious that $f_1 : B \to A$ given by $f_1(b) = f^{-1}(b)$ is injective. □

The next result shows how to "increase cardinalities." It shows us that we have infinitely many cardinal numbers. If A is any set then $\mathcal{P}(A)$ shall stand for the set of all subsets of A.

Proposition I.6. *For any set A, one has $card\,(A) < card\,(\mathcal{P}(A))$.*

Proof. Since $f : A \to \mathcal{P}(A)$ given by $f(a) = \{a\}$ is injective, we have $card\,(A) \leq card\,(\mathcal{P}(A))$. On the other hand if $g : A \to \mathcal{P}(A)$ is any map, note that $B := \{a \in A : a \notin g(a)\} \notin g(A)$. In fact, if $B = g(a_0)$ for some $a_0 \in A$, then we have two possibilities: $a_0 \in B$ and $a_0 \notin B$. If $a_0 \in B$ then $a_0 \notin g(a_0) = B$; and if $a_0 \notin B$ then $a_0 \in g(a_0) = B$. Since both possibilities lead us to a contradiction, we immediately conclude that $B \notin g(A)$ and g can not be a surjection. □

As one would expect, the smallest infinite cardinal number is $\aleph_0$, as stated in the next result.

Proposition I.7. *If α is an infinite cardinal number, then $\alpha \geq \aleph_0$.*

Proof. Let M be such that $card\,(M) = \alpha$. Choose $x_1 \in M$. Since M is not finite, there is $x_2 \in M \setminus \{x_1\}$, etc. Thus, we have a sequence (x_n) such that $x_i \neq x_j$ whenever $i \neq j$. So, the map from $\mathbb{N}$ to M corresponding j to x_j is an injection and therefore $\alpha \geq \aleph_0$. □

Let us recall that the Continuum Hypothesis asserts that there does not exist a set H such that

$$\aleph_0 < card\,(H) < \mathfrak{c}. \tag{0.1}$$

It is well known that the Continuum Hypothesis (CH) is independent from the standard axioms of Set Theory (usually denoted by ZFC, i.e., the Zermelo–Fraenkel set theory together with the Axiom of Choice). Thus, when one says that we are assuming the Continuum Hypothesis, it actually means that we are assuming (as an axiom) that there is no such set H satisfying (0.1). On the other hand, if we assume the negation of CH, it means that we assume (as an axiom) that there exists a set H satisfying (0.1).

II Cardinal arithmetic

As we already mentioned in the first section, the notion of cardinality is, in a very natural sense, an extension of the notion of natural numbers. Hence, a natural step that one would take is to define operations within cardinal numbers.

Definition II.1. Let α, β be cardinal numbers. We set:

(i) $\alpha + \beta := card\,(S)$, where $S = A \cup B$ with $\alpha = card\,(A)$, $\beta = card\,(B)$ and $A \cap B = \varnothing$.

(ii) $\alpha \cdot \beta := card\,(P)$, where $P = A \times B$, with $\alpha = card\,(A)$ and $\beta = card\,(B)$.

(iii) $\alpha^{\beta} := card\,(C)$, where C is any set of the form

$$C = \prod_{i \in I} A_i,$$

with $card\,(I) = \beta$ and $card\,(A_i) = \alpha$ for all $i \in I$. Equivalently, if $card\,(A) = \alpha$, then $\alpha^{\beta} = card\,(A^I)$, with

$$A^I = \{f : \ f \text{ is a function from } I \text{ to } A\}.$$

It is a simple task to verify that the above notions are consistent in the sense that they do not depend on the choices of the sets involved. It is also simple to verify that, for all cardinal numbers $\alpha, \beta, \gamma \geq 1$, one has:

- $\alpha + \beta = \beta + \alpha$,

- $(\alpha + \beta) + \gamma = \alpha + (\beta + \gamma)$,

- $\alpha \cdot \beta = \beta \cdot \alpha$,

- $(\alpha \cdot \beta) \cdot \gamma = \alpha \cdot (\beta \cdot \gamma)$,

- $\alpha \cdot (\beta + \gamma) = (\alpha \cdot \beta) + (\alpha \cdot \gamma)$,

- $(\alpha \cdot \beta)^{\gamma} = (\alpha^{\gamma}) \cdot (\beta^{\gamma})$,

- $\alpha^{\beta + \gamma} = (\alpha^{\beta}) \cdot (\alpha^{\gamma})$,

- $(\alpha^{\beta})^{\gamma} = \alpha^{\beta \cdot \gamma} = (\alpha^{\gamma})^{\beta}$.

Besides, the relation $\leq$ is compatible with the operations. More precisely, if $\alpha_1 \leq \alpha_2$ and $\beta_1 \leq \beta_2$, then

- $\alpha_1 + \beta_1 \leq \alpha_2 + \beta_2$,

- $\alpha_1 \cdot \beta_1 \leq \alpha_2 \cdot \beta_2,$

- $\alpha_1{}^{\beta_1} \leq \alpha_2^{\beta_2}.$

Also, it is simple to check that if α is not finite then

$$\alpha + n = \alpha$$

for all positive integers n. The following result is an extension of this last fact.

Theorem II.2. *Let* α, β *be cardinal numbers, with* $1 \leq \beta \leq \alpha$ *and* α *infinite. Then*

$$\alpha + \beta = \alpha.$$

Proof. It is straightforward that

$$\alpha \leq \alpha + \beta \leq \alpha + \alpha.$$

Thus, it suffices to assume that $\alpha = \beta$.

Now, suppose that $card\,(A) = \alpha$. Let Υ be the set of all bijections f from $X_f \times \{0,1\}$ to X_f, where (for each f) X_f is a subset of A.

Consider the partial order in Υ provided by

$$f \leq g \iff \begin{cases} X_f \subset X_g \\ f(x,y) = g(x,y) \text{ for all } (x,y) \in X_f \times \{0,1\}. \end{cases}$$

Notice that Υ is nonvoid. In fact we can find $Z \subset A$ such that $card\,(Z) = card\,(\mathbb{N})$ and it is clear that there is a bijection from $Z \times \{0,1\}$ to Z.

We shall now use Zorn's lemma to ensure the existence of a maximal element in Υ. Let $F := \{f_\lambda\}_{\lambda \in I}$ be a totally ordered subset of Υ. Consider the function f given by

$$f: \quad \bigcup_{\lambda \in I} X_{f_\lambda} \times \{0,1\} \quad \longrightarrow \quad \bigcup_{\lambda \in I} X_{f_\lambda}$$
$$(x,y) \quad \longmapsto \quad f_\lambda(x,y) \text{ if } x \in X_{f_\lambda}.$$

It is straightforward to check that f is well defined and f is an upper bound to F. Hence, from Zorn's lemma there exists a maximal element in Υ. Let us denote by $h: X \times \{0,1\} \to X$ this maximal element. Since h is bijective, we have

$$card\,(X) = card\,(X \times \{0,1\})$$
$$= card\,(X) + card\,(X).$$

We claim that

$$card\,(A \setminus X) < \infty.$$

Indeed, if it were not true then we would be able to find a subset $B \subset A \setminus X$ such that there is a bijection $t : B \times \{0,1\} \to B$. Therefore, the map $h_t : (X \cup B) \times \{0,1\} \to X \cup B$ given by

$$h_t(x,y) = h(x,y) \text{ if } x \in X,$$
$$h_t(x,y) = t(x,y) \text{ if } x \in B$$

would be a bijection and, since $h \leq h_t$ and $h \neq h_t$, this is a contradiction. We thus have $card(A \setminus X) < \infty$ and therefore X is not a finite set and

$$card(A) = card(A \setminus X) + card(X) = card(X).$$

Finally, we have

$$\alpha = card(A)$$
$$= card(X)$$
$$= card(X) + card(X)$$
$$= card(A) + card(A)$$
$$= \alpha + \alpha,$$

as required. $\qquad\square$

Corollary II.3. *If α, β, γ are infinite cardinal numbers, with $\alpha + \beta = \gamma$ and $\alpha < \gamma$, then $\beta = \gamma$.*

The following result is a multiplicative version of Theorem II.2.

Theorem II.4. *Let α, β be cardinal numbers, with $1 \leq \beta \leq \alpha$, and α infinite. Then*

$$\alpha \cdot \beta = \alpha.$$

Proof. Clearly, we have

$$\alpha \leq \alpha \cdot \beta \leq \alpha \cdot \alpha.$$

Thus, it suffices to assume that $\alpha = \beta$.

It is easy to show that

$$\aleph_0 \cdot \aleph_0 = \aleph_0.$$

Let A be a set with $card(A) = \alpha$. Let Υ be the set of all bijections f from $X_f \times X_f$ to X_f, where for each f, X_f is a subset of A. It is not difficult to see that Υ is nonvoid. Consider the partial order

$$f \leq g \iff \begin{cases} X_f \subset X_g \\ f(x,y) = g(x,y) \text{ for all } (x,y) \in X_f \times X_f \end{cases}$$

defined in Υ. Let $F \subset \Upsilon$ be a totally ordered set and note that

$$T : \left(\bigcup_{f \in F} X_f\right) \times \left(\bigcup_{f \in F} X_f\right) \longrightarrow \bigcup_{f \in F} X_f$$
$$(x,y) \longmapsto f(x,y) \text{ if } (x,y) \in X_f \times X_f$$

is well-defined, bijective, and $g \leq T$ for all g in F. Thus, we can invoke Zorn's lemma and so there is a maximal element $h : X \times X \to X$ in Υ. Note that

$$card\,(X) \cdot card\,(X) = card\,(X) \tag{0.2}$$

and it suffices to prove that $card\,(A) = card\,(X)$.

Since $X \subset A$, we have $card\,(X) \leq card\,(A)$. Thus, we just need to show that $card\,(X) < card\,(A)$ is not possible. If we had $card\,(X) < card\,(A)$ we would have $card\,(A) = card\,(A \smallsetminus X)$, since

$$card\,(X) < card\,(A) = card\,(X) + card\,(A \smallsetminus X)\,,$$

and from Corollary II.3 we conclude that $card\,(A) = card\,(A \smallsetminus X)$. Let Y be a proper subset of $A \smallsetminus X$ such that

$$card\,(Y) = card\,(X)\,.$$

From (0.2) we know that $card\,(X)$ is not finite and

$$card\,(X \times Y) = card\,(X) \cdot card\,(Y) = card\,(X) \cdot card\,(X) = card\,(X)\,.$$

Moreover,

$$card\,(Y \times Y) = card\,(Y) \cdot card\,(Y) = card\,(X) \cdot card\,(X) = card\,(X) = card\,(Y)\,.$$

Since $X \times Y$, $Y \times X$ and $Y \times Y$ are pairwise disjoint and have the same cardinality, from Theorem II.2, we have

$$card\,(D) = card\,(Y)\,,$$

with

$$D := (X \times Y) \cup (Y \times X) \cup (Y \times Y)\,.$$

Let $M : D \to Y$ be a bijection and define

$$R : \quad (X \times X) \cup D \quad \longrightarrow \quad X \cup Y$$
$$x \quad \longmapsto \quad \begin{cases} M(x) & \text{if } \quad x \in D \\ h(x) & \text{if } \quad x \in X \times X. \end{cases}$$

Since

$$D \cup (X \times X) = (X \cup Y) \times (X \cup Y)\,,$$

it follows that $R \in \Upsilon$ and $h < R$, a contradiction. We thus conclude that

$$card\,(A) = card\,(X)$$

and the proof is completed. □

Corollary II.5. *If β is an infinite cardinal, then*

$$\aleph_0 \cdot \beta = \beta.$$

Proposition II.6. *The following equality holds:* $2^{\aleph_0} = \mathfrak{c}$.

Proof. From the Cantor-Bernstein-Schroeder theorem I.3, it suffices to show that $2^{\aleph_0} \leq \mathfrak{c}$ and $\mathfrak{c} \leq 2^{\aleph_0}$. The map $h : \mathbb{R} \longrightarrow 2^{\mathbb{Q}}$ defined by $h(x) = h_x$ with $h_x : \mathbb{Q} \to \{0,1\}$ given by

$$h_x(y) = \begin{cases} 0 & \text{if} \quad y < x \\ 1 & \text{if} \quad x \leq y \end{cases}$$

is injective and, thus,

$$\mathfrak{c} \leq 2^{\aleph_0}.$$

On the other hand, the map

$$\begin{array}{rcl} g: \ \{0,1\}^{\mathbb{N}} & \longrightarrow & \mathbb{R} \\ (a_i) & \longmapsto & \sum_{i \in \mathbb{N}} a_i . 10^{-i} \end{array}$$

is injective and therefore $2^{\aleph_0} \leq \mathfrak{c}$. $\square$

Lemma II.7. *If α is an infinite cardinal and β is a cardinal number such that $2 \leq \beta \leq 2^{\alpha}$, then*

$$\beta^{\alpha} = 2^{\alpha}.$$

Proof. Notice that, since $\alpha \cdot \alpha = \alpha$, we have

$$2^{\alpha} \leq \beta^{\alpha} \leq (2^{\alpha})^{\alpha} = 2^{\alpha \cdot \alpha} = 2^{\alpha}$$

and thus, again from the Cantor-Bernstein-Schroeder theorem we conclude that $\beta^{\alpha} = 2^{\alpha}$. $\square$

Proposition II.8. $\mathfrak{c}^{\aleph_0} = \mathfrak{c}$.

Proof. From the previous lemma, since $2 < \mathfrak{c} = 2^{\aleph_0}$ we have

$$\mathfrak{c}^{\aleph_0} = 2^{\aleph_0} = \mathfrak{c},$$

and we are done. $\square$

The famous (and also simple) pigeonhole principle states that if r items are put into s containers, with $r > s$, then at least one container contains more than one item. This is sometimes called, by obvious reasons, *the drawer principle*. We shall choose its statement with pigeons. We are mainly interested in one of its generalizations to infinite cardinal numbers; see Theorem II.9 below. This result is known as the Infinite Pigeonhole Principle and will be used later in Theorem III.4.

Theorem II.9. *If S, T are sets such that S is infinite, $card(S) > card(T)$ and $f : S \to T$ is any map, then there is $t \in T$ such that $f^{-1}(t)$ is infinite.*

Proof. It suffices to show that if S is infinite and $f : S \to T$ is such that $f^{-1}(t)$ is finite for all $t \in T$, then $card(S) \leq card(T)$.

So, let us suppose that $f : S \to T$ is such that $f^{-1}(t)$ is finite for all $t \in T$. It is plain that in this case T is infinite. For each t in $f(S)$, let

$$f^{-1}(t) = \{s_{t,1}, \ldots, s_{t,n_t}\}$$

and

$$C_{t,n} = \{s_{t,n}\}, \text{ if } n \leq n_t$$
$$C_{t,n} = \phi, \text{ if } n > n_t.$$

Then

$$S = \bigcup_{t \in T} f^{-1}(t) = \bigcup_{(t,n) \in T \times \mathbb{N}} C_{t,n}.$$

Since each $C_{t,n}$ has no more than one element and the $C_{t,n}$ are pairwise disjoint, the map

$$\Psi : \bigcup_{(t,n) \in T \times \mathbb{N}} C_{t,n} \longrightarrow T \times \mathbb{N}$$

defined by $\Psi(s_{t,n}) = (t, n)$ is an injection. Thus, from Corollary II.5,

$$card(S) = card \left(\bigcup_{(t,n) \in T \times \mathbb{N}} C_{t,n} \right) \leq card(T \times \mathbb{N}) = card(T) \cdot \aleph_0 = card(T),$$

which concludes the proof. □

As a matter of fact, a more subtle version of the Infinite Pigeonhole Principle is valid. It states that if α, β are infinite cardinal numbers with $\alpha > \beta$ and α pigeons are "stuffed" in β pigeonholes, there will exist at least one pigeonhole having α pigeons "stuffed" in it.

III Basic concepts and results of abstract and linear algebra

In this section we introduce some concepts from Abstract Algebra that will be needed to simplify the forthcoming definition of algebras and also to help the reader to have clear and unambiguous concepts. We prefer this approach because, at the same time, we recall the notions from Abstract Algebra and simplify the statement of our definitions. We also recall some basic results (some of them not easy to find in the literature) from Linear Algebra. We begin with the notions of group, ring and field.

Definition III.1. A nonempty set G is a *group* if there is an operation $\bullet : G \times G \to G$, such that:

(i) Given $a, b, c \in G$, then $a \bullet (b \bullet c) = (a \bullet b) \bullet c$.

(ii) There is an element $e \in G$ such that $a \bullet e = e \bullet a = a$ for all $a \in G$.

(iii) For every $a \in G$ there is an element $b \in G$ such that $a \bullet b = b \bullet a = e$.

If, in addition, $a \bullet b = b \bullet a$ for all $a, b \in G$, then G is an *Abelian group*.

The set of all integers with the addition is, of course, a group. For the notion of *ring* two operations are needed:

Definition III.2. A nonempty set R is a *ring* if there are two operations defined in R, say $+ : R \times R \to R$ and $\bullet : R \times R \to R$, with:

(i) $(R, +)$ is a group.

(ii) If $a, b, c \in R$, then $a \bullet (b \bullet c) = (a \bullet b) \bullet c$.

(iii) If $a, b, c \in R$, then $a \bullet (b + c) = (a \bullet b) + (a \bullet c)$.

(iv) $(b + c) \bullet a = (b \bullet a) + (c \bullet a)$.

If, in addition, there exists an element $1 \in R$ with $a \bullet 1 = 1 \bullet a = a$ for all $a \in R$ we say that R is a ring with unity (or a unitary ring, or ring with identity). If the operation $\bullet$ is commutative, R is called a *commutative ring*.

Finally, a commutative ring F with unity and such that, for every nonzero element $a \in F$, there is an element $a^{-1} \in F$, such that $a \bullet a^{-1} = 1$, is called a *field*. The elements in a field are usually called *scalars*.

The set of all polynomials with the usual operations is a classical example of ring and $\mathbb{Q}$, $\mathbb{R}$ and $\mathbb{C}$ are standard examples of fields. The notions of *vector space* (or *linear space*, or *linear manifold*) over a field $\mathbb{K}$ as well as the concept of *vector subspace* (or *linear subspace* or *linear submanifold*) or simply *subspace* or *submanifold* are very well known in Linear Algebra and we omit them.

Let V be a vector space over a field $\mathbb{K}$ and S be a subset of V. A vector $x \in V$ is a *linear combination* of the elements of S if there are $x_1, \ldots, x_r \in S$ and scalars $\lambda_1, \ldots, \lambda_r \in \mathbb{K}$ such that

$$x = \lambda_1 x_1 + \cdots + \lambda_r x_r.$$

We recall that a set S contained in V is called *linearly independent* if for all positive integers n and all pairwise distinct elements $x_1, \ldots, x_n$ in S we have $\lambda_1 = \cdots = \lambda_n = 0$ whenever $\lambda_1, \ldots, \lambda_n$ are scalars such that $\lambda_1 x_1 + \cdots + \lambda_n x_n = 0$. Sometimes we just say that $x_1, \ldots, x_n$ are linearly independent instead of saying that the set $\{x_1, \ldots, x_n\}$ is linearly independent.

The subspace of V consisting of all linear combinations of elements of a given subset S of V is denoted by $span\,(S)$ and called the *linear span of S*. We say that S generates V if

$$span\,(S) = V.$$

We recall that a *Hamel basis* of a vector space V is a set $\mathcal{B}$ such that $span\,(\mathcal{B}) = V$ and $\mathcal{B}$ is linearly independent. Any vector space V has a Hamel basis and, moreover, every linearly independent subset of a vector space can be complemented to form a Hamel basis of V. These facts are simple and folkloric consequences of the Zorn's lemma. We say that V is finite dimensional when there is a Hamel basis $\mathcal{B}$ in V, with $card\,(\mathcal{B}) < \infty$. In this case it is well known, by any undergraduate Linear Algebra student, that if $\mathcal{B}_1$ and $\mathcal{B}_2$ are Hamel bases of V, then $card\,(\mathcal{B}_1) = card\,(\mathcal{B}_2)$. So, in this case ($V$ is finite dimensional), we can define the dimension of V, denoted by $\dim V$, as

$$\dim V := card\,(\mathcal{B}), \tag{0.3}$$

where $\mathcal{B}$ is any Hamel basis of V. If V is not finite dimensional we say that V is infinite dimensional.

This result–that is, the invariance of the cardinal number of a Hamel basis of a given finite dimensional linear space–is, in addition, still valid in a more general context: For any vector space V, if $\mathcal{B}_1$ and $\mathcal{B}_2$ are Hamel bases of V, then $card\,(\mathcal{B}_1) = card\,(\mathcal{B}_2)$. Thus, we can use the definition (0.3) in general.

Next, we can finally prove the above result: any two Hamel bases of a vector space have the same cardinality. This assertion is nontrivial and is usually omitted in Linear Algebra texts. We shall need a preliminary lemma. For any set S, let us define

$$F(S) = \{A \subset S : card\,(A) < \infty\}.$$

Lemma III.3. *If S is not finite, then*

$$card\,(F(S)) = card\,(S).$$

Proof. It is plain that there is an injective map from S to $F(S)$. Using the Cantor-Bernstein-Schroeder theorem (Theorem I.3), it suffices to prove that $card\,(F\,(S)) \leq card\,(S)$.

For $n \geq 1$, let $S^n = S \times \cdots \times S$ (n times). For all n, define

$$f_n : S^n \to F(S)$$

by

$$f_n(s_1, \ldots, s_n) = \{s_1, \ldots, s_n\},$$

and the surjection

$$f : \bigcup_{n=1}^{\infty} S^n \to F(S)$$

given by $f(s_1, \ldots, s_n) = f_n(s_1, \ldots, s_n)$. Note that from Definition II.1 and Theorem II.4, we have

$$card(S^n) = card(S)^n = card(S).$$

One can also verify that

$$card\left(\bigcup_{n=1}^{\infty} S^n\right) = card\left(\mathbb{N} \times S\right).$$

Thus, from Corollary II.5, we have

$$card\left(\bigcup_{n=1}^{\infty} S^n\right) = card\left(\mathbb{N} \times S\right) = \aleph_0 \cdot card(S) = card(S)$$

and therefore

$$card(\mathrm{Im}(f)) \leq card\left(\bigcup_{n=1}^{\infty} S^n\right) = card(S).$$

Since

$$\{\phi\} \cup \mathrm{Im}(f) = F(S)$$

and $F(S)$ is infinite, we have

$$card\left(F(S)\right) = card(\mathrm{Im}(f)) \leq card(S),$$

and we are done. $\qquad\square$

Theorem III.4. *Any two Hamel bases of a vector space V have the same cardinality.*

Proof. The case of vector spaces of finite dimension is well known and, thus, we shall suppose that V is infinite dimensional. Using the Cantor-Bernstein-Schroeder theorem, it suffices to prove that if $A, B \subset V$ and $span\,(B) = V$ and A is linearly independent, then $card(A) \leq card(B)$.

Since $span\,(B) = V$, it is obvious that B is infinite. For each x in A, there is a unique finite set $Z \subset B$ such that

$$x = \sum_{z \in Z} a_z z,$$

with $a_z \neq 0$ for all z, and we define

$$f : A \to F(B) : x \mapsto Z,$$

where $F(B)$ is the set of all finite subsets of B. Since B is infinite, if we had $card(A) > card(B)$, by Lemma III.3, we would have $card(A) > card(F(B))$

and thus, by the Infinite Pigeonhole Principle (Theorem II.9), it would exist $Z \in F(B)$ such that $f^{-1}(Z)$ is infinite. But

$$span \left(f^{-1}(Z) \right) \subset span \left(Z \right),$$

and $span \left(Z \right)$ is finite dimensional, a contradiction. We thus conclude that $card(A) \leq card(B)$.   □

The above result allows us to define the *dimension* of a vector space V as $card \left(\mathcal{B} \right)$, where $\mathcal{B}$ is any Hamel basis of V.

We recall that a norm in a vector space V is a function $\|\cdot\|$ from V to $\mathbb{R}$ satisfying

(N1) $\|x\| \geq 0$ for all $x \in V$, and $\|x\| = 0$ if and only if $x = 0$.

(N2) $\|ax\| = |a| \cdot \|x\|$ for all scalars a and all $x \in V$.

(N3) $\|x + y\| \leq \|x\| + \|y\|$ for all $x, y \in V$.

Sometimes, and to avoid confusion, the norm of a vector space V is denoted by $\|\cdot\|_V$. A vector space V with a norm is called *normed vector space*. It becomes a metric space (see Section IV) when we consider the natural metric generated by the norm, that is, the metric or distance d given by

$$d(x, y) = \|x - y\| \, .$$

When this metric space is complete then the normed space $(V, \| \cdot \|)$ is called a *Banach space*. It is well known that the dimension of infinite dimensional Banach spaces is at least $\mathfrak{c}$. The proof does not depend on the Continuum Hypothesis (CH), but for the sake of simplicity the proof presented here uses the CH.

Proposition III.5. *If V is an infinite dimensional Banach space, then* $\dim V \geq \mathfrak{c}$.

Proof. Let us suppose that there is a countable Hamel basis $\mathcal{B} = \{v_j : j \in \mathbb{N}\}$ for an infinite dimensional Banach space V. Thus $V = \bigcup_{n=1}^{\infty} F_n$, where each F_n is the subspace generated by $\{v_1, \ldots, v_n\}$. It is not difficult to see that, since each F_n is finite dimensional, it is closed. Thus by the Baire category theorem (see Section IV) there exists a positive integer n_0 such that F_{n_0} has non empty interior, and this is a contradiction (proper subspaces always have empty interior). Thus, if we assume the Continuum Hypothesis we conclude that $\dim V \geq \mathfrak{c}$.   □

An *algebra* over a field $\mathbb{K}$ (sometimes called a linear algebra) is a set A with operations

$$+ : A \times A \to A$$
$$\circ : A \times A \to A$$
$$\cdot : \mathbb{K} \times A \to A$$

such that $(A, +)$ is an abelian group, $(A, +, \circ)$ is a ring, $(A, +, \cdot)$ is a linear space over $\mathbb{K}$ and

$$\lambda \cdot (a \circ b) = (\lambda \cdot a) \circ b = a \circ (\lambda \cdot b).$$

The definition of algebra can "vary" depending on the context and, more generally, algebras can be considered over a ring and not over a field. From now on when we refer to algebras we are considering the definition above and $\mathbb{K} = \mathbb{R}$ or $\mathbb{C}$ will be fixed. If the ring $(A, +, \cdot)$ has a unity, then A is called *unital algebra*.

A subset $A_1 \subset A$ is a *subalgebra* if A_1 is an algebra with the operations inherited from A. If A is unital we ask that the subalgebras have unity, and the same unity as that of A.

If C is a subset of a (unital) algebra A we define the algebra generated by C (denoted by $\mathcal{A}(C)$) as the intersection of all (unital) subalgebras of A containing C. We say that $S = \{z_\alpha \ \alpha \in \Gamma\}$ is a minimal set of generators of an (unital) algebra B if $B = \mathcal{A}(S)$ is the algebra generated by S, and for every $\alpha_0 \in \Gamma$, we have $z_{\alpha_0} \notin \mathcal{A}(S \setminus \{z_{\alpha_0}\})$.

An *homomorphism of algebras* $f : A \to B$ is a map such that

$$f(a_1 + a_2) = f(a_1) + f(a_2),$$
$$f(a_1 \circ a_2) = f(a_1) \circ f(a_2),$$
$$f(\lambda \cdot a) = \lambda \cdot f(a).$$

Above, for the sake of simplicity, we are using the same symbols for the operations in A and B.

When A and B are unital algebras an homomorphism $f : A \to B$ such that $f(1_A) = 1_B$ is called unital homomorphism. When a (unital) homomorphism is bijective, it is called (unital) isomorphism, or simply isomorphism.

Given a cardinal κ, we say that an (unital) algebra A over $\mathbb{K}$ is a κ-generated free algebra if there exists a subset $Z \subset A$, with $card(Z) = \kappa$, such that any function f from Z to some (unital) algebra A_0 over $\mathbb{K}$ can be uniquely extended to an (unital) homomorphism from A into A_0. Then Z is called a *set of free generators* of the algebra A.

Example III.6. *The algebra of polynomials*

$$\mathbb{K}[X] = \{p : \ p \text{ is a polynomial in } X \text{ with scalars in } \mathbb{K}\}$$

is a 1-generated free algebra. In fact, let $Z = \{X\}$ *and let* $f : Z \to A_0$ *be a map from* Z *to a unital algebra* A_0 *over* $\mathbb{K}$*. Then the unique unital homomorphism* g *from* $\mathbb{K}[X]$ *to* A_0 *which extends* f *is given by*

$$g(a_0 + a_1 X + \cdots + a_n X^n) = a_0 1_{A_0} + a_1 f(X) + \cdots + a_n f(X)^n.$$

Example III.7. *Any 1-generated free unital algebra A is isomorphic to $\mathbb{K}[X]$. In fact, let $Z = \{a\}$ and let $f : Z \to \mathbb{K}[X]$ be such that $f(a) = X$. Then the unique unital homomorphism g from A to $\mathbb{K}[X]$ which extends f is such that*

$$g(a^n) = f(a^n) = f(a)^n = X^n.$$

We easily conclude that g is a bijection. In particular, since $\{X, X^2, \dots\}$ is linearly independent, we conclude that $\{a, a^2, \dots\}$ is linearly independent.

Example III.8. *The space ℓ_∞ of bounded sequences in $\mathbb{K}$ with the coordinatewise operations is a unital algebra. The set*

$$c_{00} := \left\{ (x_n) \in \mathbb{K}^{\mathbb{N}} : \ card\,(j : x_j \neq 0)\} < \infty \right\}$$

is a subalgebra of ℓ_∞ but it is not a 1-generated free algebra. In fact, if $a \in c_{00}$, then $\{a, a^2, \dots\}$ is not linearly independent.

IV   Residual subsets

In this short section, a number of elementary concepts and results from General Topology are recalled. The main result will be the Baire category theorem, which is an important tool to discover existence of, at first glance, strange objects, mainly in Functional Analysis. Although its content may be found in any reasonable book on Topology (see, e.g., [319]), we shall provide, for the sake of convenience, a short proof of it.

Assume that X is a nonempty set. A mapping $d : X \times X \to [0, \infty)$ is said to be a *metric* or a *distance* on X provided that it satisfies the following properties for all $x, y, z \in X$: $d(x, y) = 0$ implies $x = y$; $d(x, y) = d(y, x)$; and $d(x, y) \leq d(x, z) + d(y, z)$. If X is endowed with a distance d then X (or more properly, the pair (X, d)) is called a *metric space*. A distance d defined on a set X generates a topology in it, namely, the topology whose open sets are the unions of open balls $B(a, r) := \{x \in X : d(x, a) < r\}$ ($a \in X$, $r > 0$). The corresponding closed balls $\{x \in X : d(x, a) \leq r\}$ are denoted by $\overline{B}(a, r)$. A topological space (X, τ) is said to be *metrizable* whenever there is a distance d on X that generates the topology τ.

A sequence (x_n) in a metric space (X, d) is called a *Cauchy sequence* provided that, given $\varepsilon > 0$, there exists $n_0 = n_0(\varepsilon) \in \mathbb{N}$ such that $d(x_m, x_n) < \varepsilon$ for all $m, n \in \mathbb{N}$ with $n \geq n_0$. And a metric space (X, d) is said to be *complete* whenever every Cauchy sequence $(x_n) \subset X$ converges to some element $x_0 \in X$. That (x_n) converges to x_0 means that, given $\varepsilon > 0$, there exists

$n_0 = n_0(\varepsilon) \in \mathbb{N}$ such that $d(x_n, x_0) < \varepsilon$ for all $n \in \mathbb{N}$ with $n \geq n_0$. A topological space (X, τ) is said to be *completely metrizable* provided that there is a distance d generating τ such that the metric space (X, d) is complete.

Now, we state and prove the *Baire Category Theorem*, which asserts that countable intersections of large open sets inside a complete metric space are still large.

Theorem IV.1. *In a complete metric space X, any countable intersection of dense open subsets is dense in X.*

Proof. Let U be a nonempty open subset of X. Assume that $\{G_1, G_2, \dots\}$ is a sequence of open dense subsets of X. Then $B \cap G_1$ is open and nonempty; therefore $B \cap G_1$ contains the closure $\overline{B_1}$ of an open ball B_1. In the same way, $B_1 \cap G_2$ contains the closure $\overline{B_2}$ of an open ball B_2, and so on. Since X is complete, the decreasing sequence of closed balls $\overline{B_1}, \overline{B_2}, \dots$ contains some point, x_0: indeed, taking any sequence (x_n) with $x_n \in \overline{B_n}$ for all $n \in \mathbb{N}$, it is easy to see that (x_n) is a Cauchy sequence, so it converges to some point $x_0 \in X$. Observe that x_0 belongs to $U, G_1, G_2, \dots$. Therefore the intersection $\bigcap_{n=1}^{\infty} G_n$ intersects U. Since U is arbitrary, this intersection is dense in X, which concludes the proof. $\square$

Notice that, due to the purely topological nature of the properties "open" and "dense" of the last result, its conclusion holds in fact for any completely metrizable topological space. A topological space is said to be a *Baire space* if any countable intersection of dense open subsets is dense in X. Hence completely metrizable spaces are examples of Baire spaces. Also locally compact topological spaces are Baire, yet this fact will not be used in the sequel.

Baire's theorem provides us with an instrument that allows to assert the *topological genericity* of a given property or, in other words, allows to discover sets that are large in a topological sense. This is a usual technique in Functional Analysis, and in this book this topological genericity (when it takes place) will be compared to *algebraic genericity*. We will find examples where both kinds of genericity live together, but also examples in which both genericities are not compatible at all.

Recall that a subset of a topological space is called a G_δ set (F_σ set, resp.) if it is a countable intersection of open sets (a countable union of closed sets, resp.). The following concepts are useful to measure the "topological size" of a subset of a topological space.

Definition IV.2. Let X be a topological space. A subset $A \subset X$ is called

- *rare* or *nowhere dense* if $(\overline{A})^0 = \varnothing$.

- *meager* or *of first category* whenever A is a countable union of nowhere dense sets.

- *of second category* if it is not of first category.

- *residual* or *comeager* if $X \setminus A$ is of first category.

For instance, if $X = \mathbb{R}$, then we have that $\mathbb{N}$ is nowhere dense, $\mathbb{Q}$ is of first category, the interval $(0, 1)$ is of second category but not residual, while $\mathbb{R} \setminus \mathbb{Q}$ is residual and of second category. If X is an arbitrary topological space, then every subset of a meager subset is also meager, and a countable union of meager sets is also meager.

Remark IV.3. *The category of a set gives us an idea of its size. The sets of first category may be thought of as "small,", while their complements–that is, the residual sets–may be thought of as "large." In order that this makes sense, it is necessary for a set not to be small and large simultaneously. Baire's theorem provides a kind of space in which this cannot happen.*

Proposition IV.4. *If X is a Baire space space then X is of second category in itself. In particular, any completely metrizable topological space is of second category in itself.*

Proof. Assume, by way of contradiction, that X is meager. There there exist rare subsets R_n such that $X = \bigcup_{n=1}^{\infty} R_n$. Then, setting $F_n := \overline{R_n}$ $(n \geq 1)$, it follows that $X = \bigcup_{n=1}^{\infty} F_n$, where each F_n is a closed subset satisfying $F_n^0 = \varnothing$. Let $A_n := F_n^c$. Then each A_n is open and dense and, consequently, $\varnothing = \bigcap_{n=1}^{\infty} A_n$ is dense, which is absurd. □

The following battery of properties is easy to check and help in understanding the idea of topological size. Their proofs are left as an (easy) exercise:

(a) A topological space is Baire if and only if every nonempty open subset is of second category.

(b) A topological space is Baire if and only every countable union of closed sets with empty interior has empty interior.

(c) If X is a normed space, then X is Baire if and only if X is of second category in itself.

(d) In a Baire space, a subset is residual if and only if it contains some dense G_δ set; and a subset is meager if and only if it is contained in some F_σ set with empty interior.

(e) In a Baire space, every residual subset is of second category.

(f) In a Baire space, a countable intersection of residual subsets is still residual.

V Lineability, spaceability, algebrability and their variants

We recall that a *topological vector space* is a vector space with a topology so that the operations are continuous. The "linear size" of a subset in such a space can be described with the notions of lineability and spaceability, given in the next definition.

Definition V.1. Let X be a topological vector space, M be a subset of X and μ be a cardinal number.

(1) M is said to be μ-*lineable* if $M \cup \{0\}$ contains a vector space of dimension μ.

(2) M is said to be μ-*spaceable* if $M \cup \{0\}$ contains a closed vector space of dimension μ.

(3) M is said to be μ-*dense-lineable* if $M \cup \{0\}$ contains a dense vector space of dimension μ.

It will be usual to simply refer to the set M as *lineable*, or *spaceable* or *dense-lineable*, if the respective existing subspace is infinite dimensional, that is, if $\mu \geq \aleph_0$. Note that if we are in the framework of vector spaces (with no topology involved) then the notion of lineability is still valid.

The maximum of the cardinal numbers μ such that M is μ-lineable might not exist. The following simple example may be illuminating.

Example V.2. *Let* $j_1 \leq k_1 < j_2 \leq \cdots \leq k_m < j_{m+1} \leq \ldots$ *be integers with* $k_n - j_n \to \infty$ $(n \to \infty)$, *and let* $M := \bigcup_m \{\sum_{i=j_m}^{k_m} a_i X^i : a_i \in \mathbb{K}\}$. *The sets* $\{\sum_{i=j_m}^{k_m} a_i X^i : a_i \in \mathbb{K}\}$ $(m \in \mathbb{N})$ *are obviously pairwise disjoint, and thus* M *is n-lineable for all n but is not* $\aleph_0$-*lineable in* $\mathbb{K}[X]$.

A set is *maximal lineable* or *maximal spaceable* if the dimension of the existing linear space equals $\dim(X)$.

The notion of *algebrability* follows the same idea of that of lineability, replacing linear spaces by algebras.

Definition V.3. Given a Banach algebra $\mathcal{A}$, a subset $\mathcal{B} \subset \mathcal{A}$ and two cardinal numbers α and β, we say that:

(1) $\mathcal{B}$ is *algebrable* if there is a subalgebra $\mathcal{C}$ of $\mathcal{A}$ so that $\mathcal{C} \subset \mathcal{B} \cup \{0\}$ and the cardinality of any system of generators of $\mathcal{C}$ is infinite.

(2) $\mathcal{B}$ is *dense-algebrable* if, in addition, $\mathcal{C}$ can be taken dense in $\mathcal{A}$.

(3) $\mathcal{B}$ is (α, β)-*algebrable* if there is an algebra $\mathcal{B}_1$ so that $\mathcal{B}_1 \subset \mathcal{B} \cup \{0\}$, $\dim(\mathcal{B}_1) = \alpha$, $\operatorname{card}(S) = \beta$ and S is a minimal system of generators of $\mathcal{B}_1$.

(4) At times we shall say that $\mathcal{B}$ is, simply, κ-*algebrable* if there exists a κ-generated subalgebra $\mathcal{C}$ of $\mathcal{A}$ with $\mathcal{C} \subset \mathcal{B} \cup \{0\}$. If, in addition, $\mathcal{C}$ is dense in $\mathcal{A}$, then $\mathcal{B}$ is said to be κ-*dense-algebrable*.

(5) If $\kappa = \dim(\mathcal{A})$ in (4), then we shall say, respectively, that $\mathcal{B}$ is *maximal algebrable* and *maximal dense-algebrable*.

Definition V.4. Let α be a cardinal number. Let $\mathcal{L}$ be a commutative algebra and $S \subset \mathcal{L}$. We say that:

(1) S is *strongly α-algebrable* if $S \cup \{0\}$ contains an α-generated algebra which is isomorphic to a free algebra. We say simply that S is *strongly algebrable* if the last condition is satisfied with $\alpha = \aleph_0$.

(2) S is *densely* (*closely,* resp.) *strongly α-algebrable* if a topology is given on $\mathcal{L}$ and $S \cup \{0\}$ contains a dense (closed, resp.) α-generated algebra which is isomorphic to a free algebra. We say simply that S is *densely* (*closely,* resp.) *strongly algebrable* if the last condition is satisfied with $\alpha = \aleph_0$.

From Example III.8 one can verify that c_{00} is algebrable in ℓ_∞ but it is not strongly 1-algebrable.

This book is entirely devoted to the study of lineability, dense-lineability, spaceability, algebrability and strong algebrability in different fields of Mathematics. Thus, plenty of examples of these concepts will be given along the text.

VI   Notes and remarks

Section I. The notion of cardinalities appeared around 1880, and is due to Georg Cantor. Cantor is also the creator of Set Theory. The Cantor-Bernstein-Schroeder theorem was stated without proof by Cantor in 1883; the first proofs are due to E. Schroeder and F. Bernstein in 1896 and 1897, respectively.

The Axiom of Choice was first conceived by E. Zermelo in 1904. Although at a first glance it may seem almost tautological, it is absolutely not. It has many surprising consequences. For instance, take the Banach–Tarski Paradox (although it is not a paradox): it states that a 3-dimensional ball can be decomposed into a finite number of disjoint subsets, which can be put back together to construct two identical copies of the original ball.

The Continuum Hypothesis (CH) is due to Cantor (1878). The validity or not of the CH is the first problem of the famous list of 23 problems posed by D. Hilbert in 1900. The answer is a combination of results of K. Gödel (1940) and P. Cohen (1963): it is independent in the standard axiomatic set theory (ZFC model). This means that it can not be proved or refuted within the standard axiomatic set theory of ZFC. Thus, the CH or its negation can be added as an axiom to the ZFC model without causing any logical problem.

Section II. The Axiom of Choice also implies the existence of a smallest cardinal number, denoted by $\aleph_1$, that is greater than $\aleph_0$; so the Continuum Hypothesis is equivalent to assert that $2^{\aleph_0} = \aleph_1$.

It is not true that in general $\alpha^{\aleph_0} = \alpha$ for all $\alpha > \mathfrak{c}$. For some cardinal numbers this is valid and the question of whether this is valid or not for a certain $\alpha > \mathfrak{c}$ is quite technical and subtle and relies on the notion of *cofinality*.

Section III. The term Hamel Basis is named after Georg Hamel. He obtained (via Axiom of Choice) a basis for the vector space of the real numbers over the rationals. It is not difficult to see that this is an infinite dimensional vector space (for instance, using that π is a transcendental number it is simple to observe that the set $\{\pi, \pi^2, \dots\}$ is linearly independent). Moreover with a little bit more of effort one can verify that its dimension is $\mathfrak{c}$.

Section V. The concepts of lineability and spaceability are due to R. Aron, V.I. Gurariy and J.B. Seoane, although the essence of these notions was treated before, with other terminology (see [28, 241, 360]). The notion of maximal lineability was introduced by L. Bernal in [83]. Algebrability has its roots in the papers [33, 35, 360] and the notion of strongly algebrability was, later, introduced by A. Bartoszewicz and S. Głąb (see [47]).

Chapter 1

Real Analysis

In this chapter we deal with lineability in the context of real functions. This is one of the most fruitful environments to investigate lineability issues. The range of material is vast, from continuous functions with special properties to wildly noncontinuous functions.

1.1 What one needs to know

This is a very homogeneous chapter: everything is related to real functions. This is also the biggest chapter of the book and it would be not an easy task to make a compendium of everything that one needs to know to read and understand all the content of all of its sections. The prerequisites are essentially Real Analysis and some rudiments of the theory of Metric Spaces. In Section 1.7 we will invoke a famous topological result, the Borsuk-Ulam Theorem.

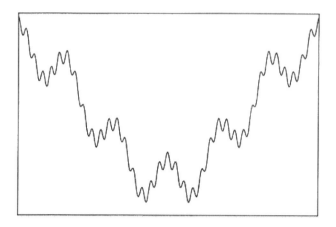

FIGURE 1.1: A sketch of Weierstrass' monster .

1.2   Weierstrass' monsters

It came as a general shock when in 1872, and during a presentation before the Berlin Academy, K. Weierstrass provided the classical example of a function that was continuous everywhere but differentiable nowhere (see Figure 1.1). The particular example was defined as

$$f(x) = \sum_{k=0}^{\infty} a^k \cos(b^k \pi x),$$

where $0 < a < 1$, b is any odd integer and $ab > 1 + 3\pi/2$.

The apparent *shock* was a consequence of the general thought that most mathematicians shared: that a continuous function must have derivatives at a significant set of points (even A.M. Ampère attempted to give a theoretical justification for this). Although the first published example is certainly due to Weierstrass, already in 1830 the Czech mathematician B. Bolzano exhibited a continuous nowhere differentiable function. Let us make a brief overview of the appearance throughout history of "Weierstrass' monsters" (see, e.g., [372] for a thorough study of the citations below):

Discoverer	Year
B. Bolzano	$\approx$1830
M.Ch. Cellérier	$\approx$1830
B. Riemann	1861
H. Hankel	1870
K. Weierstrass	1872

After 1872 many other mathematicians also constructed similar functions. Just to cite a partial list of these, we have: H.A. Schwarz (1873), M.G. Darboux (1874), U. Dini (1877), K. Hertz (1879), G. Peano (1890), D. Hilbert (1891), T. Takagi (1903), H. von Koch (1904), W. Sierpiński (1912), G.H. Hardy (1916), A.S. Besicovitch (1924), B. van der Waerden (1930), S. Mazurkiewicz (1931), S. Banach (1931), S. Saks (1932) and W. Orlicz (1947) (we would also like to refer the interested reader to [10, 258] for some recent results on this class of continuous nowhere differentiable functions).

As a nice application of the Baire category theorem, Banach obtained in 1931 that *most* continuous functions are nowhere differentiable; see, e.g., [330]. Specifically, the set of all continuous but nowhere differentiable functions on $\mathbb{R}$ is residual in $\mathcal{C}(\mathbb{R})$, when endowed with the topology of uniform convergence in compacta.

Later, at the end of the twentieth century and nowadays, there are also authors who have, as well, constructed Weierstrass' monsters with even additional "pathologies." The lineability of this type of functions has been thoroughly studied in the last years. The very first result in this direction was due to V.I. Gurariy in 1966 ([239, 240]), who showed that the set of continuous nowhere differentiable functions on $[0, 1]$ is lineable. The lineability of this class of functions has been studied in depth, as we summarize next. V. Fonf, V. Gurariy and V. Kadeč [203], in 1999, showed that the set of *nowhere* differentiable functions on $[0, 1]$ is spaceable. But much more is true: L. Rodríguez-Piazza showed that the X in [203] can be chosen to be isometrically isomorphic to any separable Banach space [345]. Several authors have invested plenty of time on the study of this special set of functions since the ending of the twentieth century. For instance, S. Hencl [245] showed in 2000 that any separable Banach space is isometrically isomorphic to a subspace of $\mathcal{C}[0, 1]$ whose nonzero elements are nowhere approximately differentiable (recall that if $I \subset \mathbb{R}$ is an interval and $x_0 \in \mathbb{R}$ then a function $f : I \to \mathbb{R}$ is called approximately differentiable at x_0 provided that there is $\alpha \in \mathbb{R}$ such that, for each $\varepsilon > 0$, the set $\{x \in E : |\frac{f(x)-f(x_0)}{x-x_0} - \alpha| < \varepsilon\}$ has x_0 as a density point) and nowhere Hölder. And Bayart and Quarta [63] produced the following result, that is related to the algebraic structure of this special set.

Theorem 1.2.1. *The set of continuous nowhere Hölder functions on $[0, 1]$ contains (except for the null function) an infinitely generated algebra. Moreover, this algebra can be chosen to be dense in $\mathcal{C}[0, 1]$. In other words, the set of continuous nowhere Hölder functions on $[0, 1]$ is $\aleph_0$-dense-algebrable.*

From this last assertion one can infer that the set of continuous nowhere differentiable functions on $[0, 1]$ is dense-lineable and, in particular, $\aleph_0$-lineable. Just recently, the authors of [258] provided the first constructive proof of the $\mathfrak{c}$-lineability of this set.

Finally, and in the vein of Gurariy's results on the class of Weierstrass' monsters, let us mention that if $\mathcal{C}^\infty(I)$ denotes the space of infinitely differentiable real functions on an interval $I \subset \mathbb{R}$ then, obviously, $\mathcal{C}^\infty([0,1])$ is not spaceable in $\mathcal{C}[0,1]$. In spite of this fact, the class $\mathcal{C}^\infty((0,1))$ is spaceable in $\mathcal{C}((0,1))$. For this, see [82], where, in addition, the use of Müntz sequences allows to prove that the family of continuous functions on $[0,1]$ which are analytic in $(0,1)$ is spaceable in $\mathcal{C}[0,1]$. Hence $\mathcal{C}[0,1] \cap \mathcal{C}^\infty((0,1))$ is spaceable as well.

Clearly, the set of everywhere differentiable functions on $\mathbb{R}$ is linear and, thus, $\mathfrak{c}$-lineable since it is itself a vector space. V. I. Gurariy showed in [239] that this cannot be improved: the set of everywhere differentiable functions on $[0,1]$ is not spaceable. In this section we shall discuss some special subsets of differentiable functions.

1.3   Differentiable nowhere monotone functions

The existence of differentiable functions on $\mathbb{R}$ which are nowhere monotone (denoted $\mathcal{DNM}(\mathbb{R})$ from now on) is a well known fact since the appearance of the example by Katznelson and Stromberg in 1974, [270]. Several more examples and constructions have followed since. One of the most recent constructions, if not the most, of such a function can be found in [28], where the authors make use of several technical lemmas in order to achieve two main goals: firstly, the construction of one such function and, secondly, the following result (see [28]).

Theorem 1.3.1. *The set $\mathcal{DNM}(\mathbb{R})$ of differentiable functions on $\mathbb{R}$ which are nowhere monotone is $\aleph_0$-lineable in $\mathcal{C}(\mathbb{R})$.*

The previous result was improved in [26], where the authors proved that the set of differentiable nowhere monotone function on any compact interval of $\mathbb{R}$ is actually dense-lineable, so showing that the vector space in [28] can be chosen to be dense.

Recall that $\dim(\mathcal{C}[0,1])$ and $\dim(\mathcal{C}(\mathbb{R}))$ are both equal to $\mathfrak{c}$. Thus we could also wonder whether the set $\mathcal{DNM}(\mathbb{R})$ is $\mathfrak{c}$-lineable. The answer is yes, and it was obtained in [209], where the authors use approximately continuous functions and the properties of the density topology[1] to obtain the above statement by means of the following statement. Recall that if $I \subset \mathbb{R}$ is an interval then a function $f : I \to \mathbb{R}$ is said to be approximately continuous whenever, for every open set $U \subset \mathbb{R}$, the set $f^{-1}(U)$ is Lebesgue-measurable and has Lebesgue density one at each of its points. The following result, extracted from [209], settles the lineability problem for this previous class of functions.

---

[1]The density topology can be defined as the initial topology for the approximately continuous functions.

Theorem 1.3.2. *The set of bounded approximately continuous functions on $\mathbb{R}$ that are positive in a dense subset of $\mathbb{R}$ and negative in another dense subset of $\mathbb{R}$ is $\mathfrak{c}$-lineable.*

Proof. Let (by Lemma 3.4 in [209]) $f_0 : \mathbb{R} \to [0,1]$ be an approximately continuous mapping $f_0 \colon \mathbb{R} \to [0,1]$ satisfying the following properties:

1. Z_{f_0} (its zero set) is a G_δ, dense set with Lebesgue measure zero.

2. f_0 is discontinuous at every $x \in \mathbb{R} \setminus Z_{f_0}$.

Now consider the set $\{f_\alpha : \alpha \in (0,\infty)\}$, where $f_\alpha(x) = e^{-\alpha|x|} f_0(x)$, for all $x \in \mathbb{R}$. Then $\{f_\alpha : \alpha \in (0,\infty)\}$ is clearly a linearly independent set of bounded mappings since $\mathbb{R} \ni x \mapsto e^{-\alpha|x|}$, with $\alpha \in (0,\infty)$, are linearly independent and bounded, and f_0 is bounded. Moreover, if

$$f(x) := \sum_{k=1}^{n} \lambda_k f_{\alpha_k}(x) = \left(\sum_{k=1}^{n} \lambda_k e^{-\alpha_k|x|} \right) f_0(x),$$

for all $x \in \mathbb{R}$, with $\lambda_k \in \mathbb{R}$ for all $k = 1, \ldots, m$, then f is bounded. Since the mapping defined by $g(x) := \sum_{k=1}^{n} \lambda_k e^{-\alpha_k|x|}$ for all $x \in \mathbb{R}$ is continuous, then the set of continuity points of f is contained in the union of the set of continuity points of f_0 and Z_g. This concludes the proof since Z_g is a finite set, and therefore its measure is zero. $\qquad\square$

The last statement from the previous proof comes from the following simple fact. If $\alpha_1, \ldots, \alpha_n > 0$ and $\lambda_1, \ldots, \lambda_n \in \mathbb{R}$, then the mapping defined by $p(x) := \sum_{k=1}^{n} \lambda_k x^{\alpha_k}$ for every $x \geq 0$ has infinite zeros only if $\lambda_1 = \cdots = \lambda_n = 0$. Notice that the latter is equivalent to the fact that $\{x^\alpha : \alpha > 0\}$ is a linearly independent set of continuous functions on $[0,\infty)$.

Next, since each of the bounded approximately continuous and bounded mappings is the derivative of a differentiable function (see [146, Theorem 5.5(a), p. 21]), they are an easy consequence of Theorem 1.3.2 the following assertion [209].

Theorem 1.3.3. *The set $\mathcal{DNM}(\mathbb{R})$ of differentiable functions on $\mathbb{R}$ that are nowhere monotone is $\mathfrak{c}$-lineable.*

Due to the result by Gurariy previously discussed, the set $\mathcal{DNM}[0,1]$ cannot be spaceable.

1.4 Nowhere analytic functions and annulling functions

Let $J \subset \mathbb{R}$ be a non-degenerated interval. Recall that $\mathcal{C}^\infty(J)$ denotes the vector space of all infinitely differentiable (or "smooth") functions $J \to \mathbb{K}$,

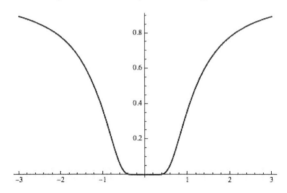

FIGURE 1.2: The function φ, belonging to $\mathcal{C}^\infty(\mathbb{R})$ but not analytic at 0.

where $\mathbb{K} = \mathbb{R}$ or $\mathbb{C}$, and that $\mathcal{C}^\infty(J)$ becomes an F-space –that is, a topological vector space whose topology can be defined by a complete translation-invariant distance– if it is endowed with the distance

$$d(f,g) = \sum_{n=1}^\infty \frac{1}{2^n} \frac{\max_{0 \le k \le n} \max_{x \in K_n} |f^{(k)}(x) - g^{(k)}(x)|}{1 + \max_{0 \le k \le n} \max_{x \in K_n} |f^{(k)}(x) - g^{(k)}(x)|},$$

where $\{K_n\}_{n \ge 1}$ is any fixed increasing sequence of compact sets of $\mathbb{R}$ such that $J = \bigcup_{n \ge 1} K_n$ (of course, one can take $K_n = J$ for all n if J is compact itself). Note that d defines the topology of uniform convergence in compacta of functions and their derivatives of all orders. Under this topology, $\mathcal{C}^\infty(J)$ is separable, and polynomials form a dense subset of it. For the sake of simplicity, most results in this section will be established for $J = I$ or $J = \mathbb{R}$, but it is easy to prove them for any interval J. Here $I := [0,1]$, the closed unit interval. The topology $\mathcal{C}^\infty(I)$ is generated by the seminorms $p_n(f) := \max_{x \in I} |f^{(n)}(x)|$ $(n = 0, 1, 2, \dots)$.

As usual, a function $f : J \to \mathbb{R}$ with $f \in \mathcal{C}^\infty(J)$ is said to be *analytic* at a point $x_0 \in J$ provides that f agrees with its Taylor series

$$\sum_{n \ge 0} \frac{f^n(x_0)}{n!}(x - x_0)^n$$

in a neighborhood of x_0. And f is called *analytic* in J if it is analytic at each point of J. There exist $\mathcal{C}^\infty$-functions that are not analytic, as the following well known function shows (see Figure 1.2):

$$\varphi(x) = \begin{cases} e^{-1/x^2} & \text{if } x \ne 0, \\ 0 & \text{if } x = 0. \end{cases}$$

Since $\varphi^{(n)}(0) = 0$ $(n \ge 0)$, the above function only agrees with its Taylor series expansion at $x = 0$. Hence φ belongs to $\mathcal{C}^\infty(\mathbb{R})$ but it is not analytic

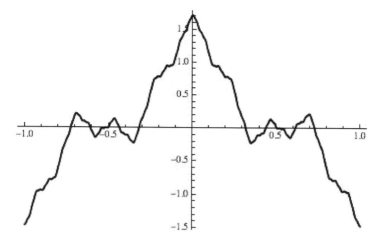

FIGURE 1.3: A sketch of f.

at 0. But more is true: in 1876, du Bois Reymond constructed a function belonging to $\mathcal{S}(\mathbb{R})$, the family of all everywhere *singular* functions, that is, the class of $\mathcal{C}^\infty$-functions on $\mathbb{R}$ that are analytic at *no* point of $\mathbb{R}$. A nice example is the following one due to Lerch [286] (see Figure 1.3):

$$f(x) = \sum_{n=1}^{\infty} \frac{\cos\left(a^n x\right)}{n!},$$

where $a > 1$ is an odd positive integer. In general, $\mathcal{S}(J)$ will stand for the set of $\mathcal{C}^\infty$-functions on J that are nowhere analytic in the interval J. Again, this phenomenon is topologically generic: in 1954, Morgenstern [314] established the residuality of the family $\mathcal{S}(I)$ of everywhere singular functions on I in the space $\mathcal{C}^\infty(I)$.

As a matter of fact, smooth functions may be non-analytic in a stronger sense. To be more precise, if x_0 is a point of an interval J, then a function $f \in \mathcal{C}^\infty(J)$ is said to have a *Pringsheim singularity* at x_0 whenever the radius of convergence $\rho(f, x_0)$ of the Taylor series of f at x_0 is zero. Recall that this radius is given by $\rho(f, x_0) = 1/\limsup_{n\to\infty} |\frac{f^{(n)}(x_0)}{n!}|^{1/n}$. For instance, the function φ above is not analytic at $x_0 = 0$, but this point is not a Pringsheim singularity for φ. Let us denote by $\mathcal{PS}(J)$ the family of smooth functions on J having a Pringsheim singularity at every point of J. Obviously, $\mathcal{PS}(J) \subset \mathcal{S}(J)$; the inclusion being strict. Explicit examples of functions in $\mathcal{PS}(J)$ are known, see for instance [351, Chapter 19]. Moreover, a classical result due to Zahorski [385] asserts that if $f \in \mathcal{S}(J)$ then the set $\{x \in J : f$ has a Pringsheim-singularity at $x\}$ is never empty.

In 1955, Salzmann and Zeller [357] established the residuality of $\mathcal{PS}(I)$ in the space $\mathcal{C}^\infty(I)$, so solving a question posed by Steinhaus and Marczewski. Of course, this implies the residuality of the bigger set $\mathcal{S}(I)$. In the next theorem, we establish the topological genericity of $\mathcal{PS}(I)$ in a more general way given in [82]. Roughly speaking, it will be proved that most functions have arbitrarily large derivatives of high orders at every point. Apart from its intrinsic interest, this will allow us to analyze lineability properties of $\mathcal{PS}(I)$.

Theorem 1.4.1.   (a) *Let $(c_n) \subset (0, \infty)$ be a sequence of positive real numbers, and M be an infinite subset of $\mathbb{N}_0$. Then the set*

$$\mathcal{M}((c_n), M) := \{f \in C^\infty(I) : \text{there are infinitely many } n \in M \text{ such that}$$
$$\max\{|f^{(n)}(x)|, |f^{(n+1)}(x)|\} > c_n \text{ for all } x \in I\}$$

is residual in $C^\infty(I)$.

(b) *The set $\mathcal{PS}(I)$ is residual in $C^\infty(I)$.*

Proof. (a) Let (c_n), M be as in the hypothesis. The assertion of the theorem is equivalent to say that the set $\mathcal{A} := C^\infty(I) \setminus \mathcal{M}((c_n), M)$ is of first category. To prove this, observe that $\mathcal{A} = \bigcup_{N \in M} A_N$, where $A_N := \bigcap_{\substack{k > N \\ k \in M}} B_k$ and

$$B_k := \{f \in C^\infty(I) : \text{there exists } x = x(f) \in I \text{ such that}$$
$$\max\{|f^{(k)}(x)|, |f^{(k+1)}(x)|\} \leq c_k\}.$$

Now, each map

$$\Phi_k : (f, x) \in C^\infty(I) \times I \mapsto \max\{|f^{(k)}(x)|, |f^{(k+1)}(x)|\} \in [0, \infty) \quad (k \in \mathbb{N}_0)$$

is continuous. Therefore the set $\Phi_k^{-1}([0, c_k])$ is closed in $C^\infty(I) \times I$. Consequently, its projection on $C^\infty(I)$ is closed, because it is a projection that is parallel to I, which is compact. But such a projection is precisely B_k, so B_k is closed. Since A_N is an intersection of certain sets B_k, we obtain that each A_N is also closed. It follows that $\mathcal{A}$ is an F_σ set.

It is enough to show that each A_N has empty interior. By way of contradiction, let us assume that $A_N^0 \neq \emptyset$. Then there would exist a basic neighborhood $U(g, \alpha, m) := \{h \in C^\infty(I) : p_j(h - g) < \alpha \text{ for all } j = 0, 1, \ldots, m\}$ such that $U(g, \alpha, m) \subset A_N$, for certain $g \in C^\infty(I)$, $\alpha > 0$ and $m \in \mathbb{N}$. By the density of the set of polynomials in $C^\infty(I)$, there are a polynomial P, a number $\varepsilon \in (0, 1)$ and a positive integer n with $U(P, \varepsilon, n) \subset U(g, \alpha, m)$, so

$$U(P, \varepsilon, n) \subset A_N. \tag{1.1}$$

Let us choose $k \in M$ with $k > \max\{n, N, \text{degree}(P)\}$, and let

$$b := (1 + c_k)(4/\varepsilon).$$

So $b > 1$. Now, we define the function

$$f(x) := P(x) + \frac{\varepsilon \sin(b\,x)}{2b^n}.$$

Note that the absolute value of the mth-derivative of the function $\varphi(x) := \sin(b\,x)$ is $b^m |\sin(b\,x)|$ (if m is even) or $b^m |\cos(bx)|$ (if m is odd). Then we have, for every $j \in \{0, 1, \ldots, n\}$, that

$$p_j(f - P) = \sup_{x \in I} \left| \frac{\varepsilon \varphi^{(j)}(x)}{2\, b^n} \right| \le \frac{\varepsilon}{2} b^{j-n} \le \frac{\varepsilon}{2} < \varepsilon.$$

Thus, $f \in U(P, \varepsilon, n)$. On the other hand, we have for any $x \in I$ that either $|\sin(b\,x)| \ge 1/2$ or $|\cos(bx)| \ge 1/2$, because of the basic law $\sin^2 t + \cos^2 t = 1$. Fix $x \in I$. Then we can select for each $j \in \mathbb{N}_0$ one number $m(j) \in \{j, j+1\}$ such that $|\varphi^{(m(j))}(x)| \ge b^{m(j)}/2$. Consequently,

$$\max\{|f^{(k)}(x)|, |f^{(k+1)}(x)|\} \ge |f^{(m(k))}(x)| = \left| P^{(m(k))}(x) + \frac{\varepsilon}{2b^n} \varphi^{(m(k))}(x) \right|$$

$$= \left| \frac{\varepsilon}{2b^n} \varphi^{(m(k))}(x) \right| \ge \frac{\varepsilon}{4} b^{k-n} \ge \frac{\varepsilon}{4} b = 1 + c_k > c_k.$$

To summarize, we have found $k \in M$ with $k > N$ such that $\max\{|f^{(k)}|, |f^{(k+1)}|\} > c_k$ on I, a contradiction with (1.1).

(b) The residuality of $\mathcal{PS}(I)$ is a simple consequence of (a) by choosing $M = \mathbb{N}$ and $c_n = (n+1)!(n+1)^{n+1}$. Take into account that $\rho(f, x) = 1/\limsup_{n \to \infty} \left| \frac{f^{(n)}(x)}{n!} \right|^{1/n}$. $\quad\square$

Concerning lineability, García, Palmberg, and Seoane [216] demonstrated that there actually exists an uncountably infinitely generated algebra every nonzero element of which is in $\mathcal{C}^\infty(\mathbb{R})$ and nonanalytic at a prescribed point x_0. Cater [155] showed in 1984 that, although the set of nowhere analytic functions on $[0, 1]$ is clearly not a linear space, there exists a vector space in $\mathcal{S}([0,1]) \cup \{0\}$ of dimension $\mathfrak{c}$. Recently Bernal [82] proved that there is a dense linear subspace in $\mathcal{C}^\infty(I)$ every nonzero element of which is nowhere analytic, and even Pringsheim-singular. In fact, we have the following set of results.

Theorem 1.4.2.   (a) *Let (c_n) be a sequence in $(0, \infty)$. Then the set*

$$\mathcal{A}((c_n)) := \left\{ f \in \mathcal{C}^\infty(I) : \limsup_{n \to \infty} \frac{|f^{(n)}(x)|}{c_n} = \infty \text{ for all } x \in I \right\}$$

is dense-lineable in $\mathcal{C}^\infty(I)$.

(b) *The set $\mathcal{PS}(I)$ is dense-lineable in $\mathcal{C}^\infty(I)$.*

(c) *The set $\mathcal{S}(I)$ is maximal dense-lineable in $\mathcal{C}^\infty(I)$.*

(d) *The set $\mathcal{PS}(I)$ is maximal lineable if the arrival space $\mathbb{K}$ is $\mathbb{C}$.*

Proof. (a) Since the set of all polynomials is dense in $\mathcal{C}^\infty(I)$, this metric space is separable, so second-countable. It follows that one can find a countable open basis $\{V_n : n \in \mathbb{N}\}$ for its topology. Let $M_0 := \mathbb{N}$ and $d_n := \max\{c_n, c_{n+1}\}$

$(n \geq 0)$. According to Theorem 1.4.1, the set $\mathcal{M}((n(1+d_n)), M_0)$ is residual, hence dense. This allows us to choose a function

$$f_1 \in \mathcal{M}((n(1+d_n)), M_0) \cap V_1.$$

Then there is an infinite subset $M_1 \subset M_0$ such that

$$\max\{|f_1^{(n)}(x)|, |f_1^{(n+1)}(x)|\} > n(1+d_n)$$

for all $n \in M_1$ and all $x \in I$. For a function $g : I \to \mathbb{K}$, denote $\|g\|_I :=$ $\sup_{x \in I} |g(x)|$. Again by Theorem 1.4.1, the set $\mathcal{M}((n(1+d_n)(1+\|f_1^{(n)}\|_I + \|f_1^{(n+1)}\|_I)), M_1)$ is dense, so we can pick a function

$$f_2 \in \mathcal{M}((n(1+d_n)(1+\|f_1^{(n)}\|_I + \|f_1^{(n+1)}\|_I)), M_1) \cap V_2.$$

An induction procedure leads us to the construction of a sequence of functions $\{f_k : k \in \mathbb{N}\} \subset \mathcal{C}^\infty(I)$ and of a nested sequence of infinite sets $M_0 \supset M_1 \supset M_2 \supset M_3 \supset \cdots$ satisfying, for all $k \in \mathbb{N}$, $x \in I$, $n \in M_k$, that

$$f_k \in V_k \tag{1.2}$$

and

$$\max\{|f_k^{(n)}(x)|, |f_k^{(n+1)}(x)|\} > n(1+d_n)\left(1 + \sum_{j=1}^{k-1}(\|f_j^{(n)}\|_I + \|f_j^{(n+1)}\|_I)\right) \tag{1.3}$$

where the last sum is defined as 0 if $k = 1$.

Next, let us define

$$\mathcal{D} := \mathrm{span}(\{f_k : k \in \mathbb{N}\}).$$

It is derived from (1.2) that $\{f_k : k \in \mathbb{N}\}$ is dense, so $\mathcal{D}$ is dense linear subspace of $\mathcal{C}^\infty(I)$. It remains to prove that every function $f \in \mathcal{D}\setminus\{0\}$ belongs to $\mathcal{A}((c_n))$. To prove it, note that for such a function there exist $N \in \mathbb{N}$ and real constants a_k $(k = 1, \ldots, N)$ such that $a_N \neq 0$ and $f = a_1 f_1 + \cdots + a_N f_N$. Since $a_N \neq 0$, we can find $n_0 \in \mathbb{N}$ such that $n|a_N| \geq a_k$ for all $n \geq n_0$ and all $k = 1, \ldots, N-1$. If $x \in I$ is fixed, by (1.3) we can select for each $n \in M_N$ one value $m(n) \in \{n, n+1\}$ such that $|f_N^{(m(n))}(x)| > n(1+d_n)(1+\sum_{j=1}^{N-1}(\|f_j^{(n)}\|_I +$

$\|f_j^{(n+1)}\|_I)$). It follows, for all $x \in I$ and all $n \in M_N$ with $n \geq n_0$, that

$$\frac{|f^{(m(n))}(x)|}{d_n} \geq \frac{1}{d_n}|a_N|\|f_N^{(m(n))}(x)| - \frac{1}{d_n}\sum_{k=1}^{N-1}|a_k|\|f_k^{(m(n))}(x)|$$

$$\geq \frac{1}{d_n}\Big[n|a_N|(1+d_n)\Big(1+\sum_{k=1}^{N-1}(\|f_k^{(n)}\|_I + \|f_k^{(n+1)}\|_I)\Big)$$

$$-\sum_{k=1}^{N-1}|a_k|\|f_k^{(m(n))}\|_I\Big]$$

$$\geq \frac{1}{d_n}\Big[nd_n|a_N| + \sum_{k=1}^{N-1}(n|a_N| - |a_k|)\|f_k^{(m(n))}\|_I\Big] \geq n|a_N|.$$

Hence $\lim_{\substack{n \to \infty \\ n \in M_N}} \frac{|f^{(m(n))}(x)|}{c_{m(n)}} = \infty$ for each $x \in I$, because $c_{m(n)} \leq d_n$ ($n \geq 1$).
Consequently, $\limsup_{n \to \infty} \frac{|f^{(n)}(x)|}{c_n} = \infty$ ($x \in I$), so $f \in \mathcal{A}((c_n))$.

(b) Simply take $c_n = n!\,n^n$ in (a).

(c) Let us fix a translation-invariant distance d defining the topology of $\mathcal{C}^\infty(I)$.
Fix also a function $\varphi \in \mathcal{S}(I)$. Let $\{P_n\}_{n \geq 1}$ be an enumeration of the polynomials with coefficients in $\mathbb{Q}$ (if $\mathbb{K} = \mathbb{R}$) or in $\mathbb{Q} + i\mathbb{Q}$ (if $\mathbb{K} = \mathbb{C}$). Then $\{P_n\}_{n \geq 1}$ is a dense subset of $\mathcal{C}^\infty(I)$. Consider, for each $\alpha \in \mathbb{R}$, the function $e_\alpha(x) := \exp(\alpha x)$. The continuity of the scalar multiplication in the topological vector space $\mathcal{C}^\infty(I)$ allows to assign to each $\alpha > 0$ a number $\varepsilon_\alpha > 0$ such that $d(0, \varepsilon_\alpha e_\alpha \varphi) < 1/\alpha$. Denote $\varphi_\alpha := \varepsilon_\alpha e_\alpha \varphi$ and $f_{n,\alpha} := P_n + \varphi_\alpha$ ($\alpha > 0$, $n \in \mathbb{N}$). It follows that

$$d(P_n, f_{n,\alpha}) < \frac{1}{\alpha} \quad (\alpha > 0, \; n \in \mathbb{N}).$$

Now, let us define

$$\mathcal{D} := \mathrm{span}(\{f_{n,\alpha} : \alpha \in [n, n+1), \; n \in \mathbb{N}\}).$$

It is clear that $\mathcal{D}$ is a linear subspace of $\mathcal{C}^\infty(I)$. Observe that $\mathcal{D} \supset \{f_{n,n}\}_{n \geq 1}$ and that the set $\{f_{n,n}\}_{n \geq 1}$ is dense because $\{P_n\}_{n \geq 1}$ is and $d(P_n, f_{n,n}) < 1/n \to 0$ ($n \to \infty$). Therefore $\mathcal{D}$ is also dense.

Now, we need to show that, for each nonempty subset $A \subset \mathbb{R}$, the functions e_α ($\alpha \in A$) are linearly independent. Suppose that this is not the case. Then there would exist a number $N \in \mathbb{N}$, scalars $c_1, \ldots, c_N$ with $c_N \neq 0$ and values $\alpha_1 < \cdots < \alpha_N$ in A such that $c_1 e_{\alpha_1} + \cdots + c_N e_{\alpha_N} = 0$ on I. From the Analytic Continuation Principle, we obtain that the last equality holds on the whole line $\mathbb{R}$. We can suppose that $N \geq 2$. This implies

$$0 \neq c_N = -(c_1 e_{\alpha_1 - \alpha_N}(x) + \cdots + c_N e_{\alpha_{N-1} - \alpha_N}(x)) \longrightarrow 0 \quad \text{as } x \to +\infty,$$

which is absurd. This shows the claimed linear independence.

In order to demonstrate that $\dim(\mathcal{D}) = \mathfrak{c}$, it is enough to show that, for each polynomial P and any nonempty subset $A \subset \mathbb{R}$, the functions $P + \varphi_\alpha$ $(\alpha \in A)$ are linearly independent. Indeed, since $\mathcal{D} \supset \{P_1 + \varphi_\alpha : \alpha \in [0, 1)\}$, we would have $\dim(\mathcal{D}) \geq \operatorname{card}([0, 1)) = \mathfrak{c}$, from which the equality $\dim(\mathcal{D}) = \mathfrak{c}$ follows. So, fix P and A as above. Assume, by way of contradiction, that the functions $P + \varphi_\alpha$ $(\alpha \in A)$ are not linearly independent. Then there would exist $N \in \mathbb{N}$, $c_1, \ldots, c_N$ with $c_N \neq 0$ and $\alpha_1 < \cdots < \alpha_N$ in A such that $c_1(P + \varphi_{\alpha_1}) + \cdots + c_N(P + \varphi_{\alpha_N}) = 0$ on I. Let $\psi := c_1 \varepsilon_{\alpha_1} e_{\alpha_1} + \cdots + c_N \varepsilon_{\alpha_N} e_{\alpha_N}$. Due to the linear independence of the functions e_α and to the continuity of ψ, there is an open interval $J \subset I$ such that $\psi(x) \neq 0$ for all $x \in J$. Therefore

$$\varphi(x) = -\frac{(\sum_{j=1}^{N} c_j) P(x)}{\psi(x)} \qquad (x \in J).$$

Hence φ would be analytic on J. This is the desired contradiction.

Finally, fix a function $f \in \mathcal{D} \setminus \{0\}$. Suppose, again by way of contradiction, that $f \notin \mathcal{S}(I)$. Then $S(f) \neq I$, where $S(f)$ denotes the subset of singular points for f, which is clearly a closed subset of J. Moreover, there exist numbers $N \in \mathbb{N}$, $m_1, \ldots, m_N \in \mathbb{N}$, scalars $c_{j,k}$ and values $\alpha(j, k) \in [j, j+1)$ $(k = 1, \ldots, m_j; \ j = 1, \ldots, N)$ satisfying $\alpha(j, 1) < \cdots < \alpha(j, m_j)$ for all $j = 1, \ldots, N$, $c_{N, m_N} \neq 0$ and $f = \sum_{j=1}^{N} \sum_{k=1}^{m_j} c_{j,k}(P_j + \varphi_{\alpha(j,k)})$. The key point is that the values $\alpha(j, k)$ are pairwise distinct. Let us set

$$h := \sum_{j=1}^{N} \sum_{k=1}^{m_j} c_{j,k} \varepsilon_{\alpha(j,k)} e_{\alpha(j,k)}.$$

By the claim proved above, this function is not identically zero on I. Also, thanks to the Analytic Continuation Principle, the set Z of zeros of h in the compact interval I cannot be infinite. Then $I \setminus (S(f) \cup Z)$ is a nonempty relatively open subset of I. Consequently, there is an interval $J \subset I$ where f is analytic and h vanishes at no point. Moreover, we have $f = Q + h\varphi$, where Q is the polynomial $Q = \sum_{j=1}^{N} (\sum_{k=1}^{m_j} c_{j,k}) P_j$. It is derived that

$$\varphi(x) = \frac{f(x) - Q(x)}{h(x)}$$

on J. But this would force the analyticity of φ on such an interval, a contradiction. Then $\mathcal{D} \setminus \{0\} \subset \mathcal{S}(I)$, which proves the maximal dense-lineability of $\mathcal{S}(I)$.

(d) We follow the notation of part (c). Fix a function $f \in \mathcal{PS}(I)$ and consider

$$\mathcal{D} := \operatorname{span}(\{f e_\alpha : \alpha \in I\}).$$

Obviously, $\mathcal{D}$ is a linear submanifold of $\mathcal{C}^\infty(I)$. Let us show that $\dim(\mathcal{D}) = \mathfrak{c}$.

For this, it is enough to prove the linear independence of the functions fe_α ($\alpha \in I$). This follows from the following facts: the functions e_α are linearly independent, a finite linear combination of these functions is analytic, the set of zeros in I of an analytic function on I is finite and, finally, a Pringsheim singular function cannot vanish identically on an interval.

Therefore, our task is to select $f \in PS(I)$ such that $\mathcal{D} \setminus \{0\} \subset PS(I)$. Let us define inductively a pair of suitable sequences $(c_n), (b_n) \subset (0, \infty)$. Firstly, set $c_1 := 4$, $b_1 := 2 + c_1$. Assume now that, for some integer $n \geq 2$, the numbers $c_1, \ldots, c_{n-1}, b_1, \ldots, b_{n-1}$ have already been determined. Then we define

$$c_n := 4 + 2 \sum_{k=1}^{n-1} b_k^{n+1-k} + (2n)!(2n)^{2n} + (2n)! \, n \sum_{k=1}^{n-1} c_k$$

$$\text{and} \quad b_n := 2 + c_n + \sum_{k=1}^{n-1} b_k^{n+1-k}.$$

Note that $b_n > 2$ for all $n \in \mathbb{N}$. Secondly, we define f as

$$f(x) := \sum_{k=1}^{\infty} b_k^{1-k} \exp(ib_k x).$$

By using the Weierstrass M-test for the series of the derivatives of any order, as well as elementary results of derivation of series and the triangle inequality, one concludes that $f \in C^\infty(I)$ and $|f^{(n)}(x)| > c_n$ ($n \in \mathbb{N}$, $x \in I$). Moreover, since $c_n > n! \, n^n$, we get $\rho(f, x) = 0$ ($x \in I$), so $f \in PS(I)$. On the other hand, we have for $n \in \mathbb{N}$ and $x \in I$ that

$$|f^{(n)}(x)| \leq |b_n^{n+1-n} i^n \exp(ib_n x)| + \sum_{k \neq n} |b_k^{n+1-k} i^k \exp(ib_k x)|$$

$$= b_n + \sum_{k=1}^{n-1} b_k^{n+1-k} + (1 + b_{n+2}^{-1} + b_{n+3}^{-2} + \cdots)$$

$$< b_n + \sum_{k=1}^{n-1} b_k^{n+1-k} + 2 = 4 + c_n + 2 \sum_{k=1}^{n-1} b_k^{n+1-k} \leq 2c_n.$$

Finally, fix $g \in \mathcal{D} \setminus \{0\}$. Then there are $N \in \mathbb{N}$, complex constants $a_1, \ldots, a_N$ and numbers $\alpha_1 < \alpha_2 < \cdots < \alpha_N$ in I with $h := \sum_{j=1}^{N} a_j e_{\alpha_j} \not\equiv 0$ and $g = fh$. Observe that

$$|h^{(n)}(x)| \leq e \sum_{j=1}^{N} |a_j| =: \beta \quad (n \in \mathbb{N}_0, \, x \in I).$$

Let $x_0 \in I$, and let $p \in \mathbb{N}_0$ be the smallest integer such that $h^{(p)}(x_0) \neq 0$ (p exists because h is analytic). Denote $\gamma := |h^{(p)}(x_0)|$, $c_0 := \|f\|_I$, and fix

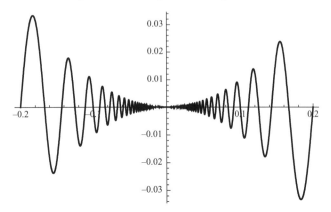

FIGURE 1.4: The function g.

$n > 2p + 2\beta\gamma^{-1}$ (so $n - p > 2\beta\gamma^{-1}$, $2n - 2p > n$ and $(2n - 2p)! > n! \geq \binom{n}{k}$) for all $k \in \{0, 1, \ldots n\}$). From Leibniz's formula, we arrive to

$$|g^{(n)}(x_0)| = \left| \binom{n}{p} f^{(n-p)}(x_0) h^{(p)}(x_0) + \sum_{k=0}^{n-p-1} \binom{n}{k} f^{(k)}(x_0) h^{(n-k)}(x_0) \right|$$

$$\geq \gamma c_{n-p} - 2\beta \sum_{k=0}^{n-p-1} \binom{n}{k} c_k$$

$$\geq \gamma[c_{n-p} - (2n-2p)!(n-p) \sum_{k=1}^{n-p-1} c_k - (n-p)c_0]$$

$$\geq \gamma[(2n-2p)!(2n-2p)^{2n-2p} - (n-p)c_0] \geq \gamma[n! \, n^n - nc_0].$$

Then $\rho(g, x_0) = 0$. Consequently, $g \in \mathcal{PS}(I)$, as desired. $\qquad\square$

If an entire function has infinitely many zeros with an accumulation point, then by the Identity theorem (also called Analytic Continuation Principle, that has been already used in the proof of the preceding theorem), it must be the zero function. For real differentiable functions this holds no longer. For instance, the differentiable function $g : \mathbb{R} \to \mathbb{R}$ (Figure 1.4) given by

$$g(x) = \begin{cases} x^2 \sin(\pi/x) & \text{if } x \neq 0, \\ 0 & \text{if } x = 0, \end{cases}$$

has the infinite set $Z = \{\frac{1}{n} : n \in \mathbb{N}\} \cup \{0\}$ as its set of zeros Z has an accumulation point (namely, 0) but, obviously, $g \neq 0$.

This raises the natural question of the existence of smooth functions with infinitely many zeros which, in addition, are nowhere analytic. In fact, as Conejero *et al.* [174] have recently proved constructively, there is a plethora of such functions, and this family enjoys a nice algebraic structure. This is the content of our next theorem, with which we conclude this section. Continuous functions with infinitely many zeros in a closed bounded interval–or equivalently, with some accumulation point of zeros–are known as *annulling functions*; see [196, Definition 2.1].

Theorem 1.4.3. *There is an algebra $\mathcal{A}$ of functions $\mathbb{R} \to \mathbb{R}$ enjoying, simultaneously, each of the following properties:*

(a) $\mathcal{A} \subset \mathcal{C}^{\infty}(\mathbb{R})$.

(b) $\mathcal{A}$ *is uncountably infinitely generated, that is, the cardinality of a minimal system of generators of $\mathcal{A}$ is uncountable.*

(c) *Each nonzero member of $\mathcal{A}$ is nowhere analytic. Hence $\mathcal{S}(\mathbb{R})$ is $\mathfrak{c}$-algebrable.*

(d) *Every element of $\mathcal{A}$ is annulling.*

(e) *For every $f \in \mathcal{A} \setminus \{0\}$ and $n \in \mathbb{N}$, the nth derivative $f^{(n)}$ is also $\mathcal{C}^{\infty}$, nowhere analytic and annulling.*

Proof. An abridged proof will be furnished, the details being left as an exercise to the interested reader. First at all, consider the function $\phi : \mathbb{R} \to \mathbb{R}$ given by

$$\phi(x) = \begin{cases} e^{\frac{-1}{x^2}} \cdot e^{\frac{-1}{(x-1)^2}} & \text{if } 0 < x < 1, \\ 0 & \text{elsewhere.} \end{cases}$$

This is a popular function, well known from an elementary course of calculus, which is positive in $(0, 1)$, zero outside $(0, 1)$, infinitely differentiable in $\mathbb{R}$ and analytic except at $x = 0$ and $x = 1$. Moreover, ϕ is *flat* at the points 0 and 1, that is, $\phi^{(k)}(0) = 0 = \phi^{(k)}(1)$ for all $k \in \mathbb{N}_0$.

Denote by $[x]$, as usual, the integer part of each $x \in \mathbb{R}$. Let $\mathcal{H}$ be a basis of the real numbers $\mathbb{R}$, considered as a $\mathbb{Q}$-vector space. Without loss of generality, we can assume that $\mathcal{H}$ consists only of positive real numbers. Let us now consider the smallest algebra $\mathcal{A}$ of $\mathcal{C}(\mathbb{R})$ that contains the family of functions $\{\rho_\alpha\}_{\alpha \in \mathcal{H}}$ with $\rho_\alpha : \mathbb{R} \to \mathbb{R}$ defined as follows:

$$\rho_\alpha(x) = \sum_{j=1}^{\infty} \frac{\lambda_j(x)}{\mu_j} \phi\left(2^j x - [2^j x]\right) \alpha^j,$$

where, for every $j \in \mathbb{N}$, we have denoted

$$\lambda_j(x) = \begin{cases} 1 & \text{if } |x| \geq \frac{1}{2^j}, \\ 0 & \text{otherwise,} \end{cases}$$

and $\mu_j = s_k!$ if $s_{k-1} < j \leq s_k$. Here s_k is the sum of the first k positive

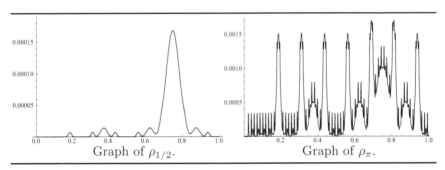

Graph of $\rho_{1/2}$. Graph of ρ_π.

FIGURE 1.5: Graphs of ρ_α for some choices of α.

integers, that is, $s_k = \frac{k(k+1)}{2}$. A sketch of what these ρ_α's look like can be seen in Figure 1.5.

If we replace x by $2^j x - [2^j x]$, then the behavior of $\phi(x)$ over the interval $[0,1]$ is replicated by $\phi(2^j x - [2^j x])$ on any dyadic interval of the form $\left[\frac{m-1}{2^j}, \frac{m}{2^j}\right]$ for all $m \in \mathbb{Z}$. In particular, the last function equals 0 together with all its derivatives at the points $\frac{1}{2^k}$ ($0 \le k \le j$) and 0. Let $\phi_j(x) := \frac{\lambda_j(x)}{\mu_j}\phi(2^j x - [2^j x])\alpha^j$. Notice that, due to the flatness of ϕ at 0 and 1, ϕ_j is smooth everywhere (and analytic everywhere but at $x = \frac{m}{2^l}$ for all $m \in \mathbb{Z}$ and $0 \le l \le j$). Now, $\sum_{j=1}^\infty \phi_j^{(k)}$ is uniformly bounded for all $k \in \mathbb{N}$. Recall that if $\{f_j(x)\}_{j=1}^\infty$ is a sequence of continuously differentiable functions on $\mathbb{R}$ such that $\sum_{j=1}^\infty f_j(x)$ converges pointwise to $f(x)$ and $\sum_{j=1}^\infty f_j'(x)$ converges uniformly on $\mathbb{R}$, then $f(x)$ is differentiable and $f'(x) = \sum_{j=1}^\infty f_j'(x)$. From this and the Weierstrass M-test, it follows that ρ_α is smooth for all α's considered. Since $\lambda_j(x) = 0$ if $|x| < \frac{1}{2^j}$, we get that each term $\frac{\lambda_j(x)}{\mu_j}\phi\left(2^j x - [2^j x]\right)\alpha^j$ is flat at all points of $\{0\} \cup \left\{\frac{1}{2^j} : j \in \mathbb{N}\right\}$. Hence each ρ_α is flat at the same points. From the rules of differentiation, we get (a) and (d). Moreover, if f is smooth then every derivative $f^{(n)}$ is also smooth. Consequently, for each $f \in \mathcal{A}$ and every $n \in \mathbb{N}$, we have that $f^{(n)}$ is also $\mathcal{C}^\infty$ and annulling, which is part of (e).

In order to prove that ρ_α is nowhere analytic, let us assume, by the way of contradiction, that ρ_α is analytic at a point, so it is also analytic on an interval. Since the dyadics form a dense subset in $\mathbb{R}$, the function ρ_α would be analytic at some $x_0 = \frac{m}{2^n}$, with m odd. If $1 \le j \le n-1$, then ϕ_j is analytic at x_0 and hence

$$\widehat{\rho}_\alpha(x) := \sum_{j=n}^\infty \phi_j(x)$$

is also analytic at x_0. However, $\widehat{\rho}_\alpha^{(k)}(x_0) = 0$ for all integers $k \ge 0$. This contradicts the fact that $\widehat{\rho}_\alpha(x)$ is positive in some punctured neighborhood of x_0. Then ρ_α is nowhere analytic and hence every derivative $\rho_\alpha^{(k)}$ is nowhere analytic. From this one can derive that, for every nonzero polynomial P of n real variables without constant term and every set $\{a_1, \ldots, a_n\}$ of distinct points of $\mathcal{H}$, the function $P(\rho_{a_1}, \ldots, \rho_{a_n})$ is either 0 or nowhere analytic. Moreover, the same holds for $(P(\rho_{a_1}, \ldots, \rho_{a_n}))^{(k)}$ if $k \in \mathbb{N}$; indeed, an elementary integration argument shows that, for a smooth function f and a point x_0, it holds that f is analytic at x_0 if and only if some –or any– derivative $f^{(k)}$ is analytic at x_0. This shows (c) and completes (e).

Concerning (b), the aim is proving that if a function $h := P(\rho_{a_1}, \ldots, \rho_{a_n})$ as before vanishes identically then the coefficients of P are all 0. We give an idea of the proof by showing that if h has the special form $h(x) := \sum_{k=1}^{N} b_k \rho_{a_k}^{m_k}(x)$ (with the a_k's as above and each $m_k \in \mathbb{N}$) and $h \equiv 0$, then all the coefficients b_k are 0. For this, we evaluate $h(x)$ at the points $x_j = \frac{3}{2^{j+1}}$, for $j = s_{n-1}+1, \ldots, s_n$. Evaluating the function $\rho_{a_k}^{m_k}$ at the points x_j, the infinite sum is reduced to

$$\rho_{a_k}^{m_k}\left(\frac{3}{2^{j+1}}\right) = \left(\frac{\phi(1/2)a_k^j}{s_n!}\right)^{m_k} = \frac{e^{-8m_k}a_k^{jm_k}}{(s_n!)^{m_k}}.$$

Therefore, if we consider the system of equations obtained from the conditions $h\left(\frac{3}{2^{j+1}}\right) = 0$ for $j = s_{n-1}+1, \ldots, s_n$, we obtain the following:

$$\begin{pmatrix} \frac{e^{-8m_1}a_1^{m_1(s_{n-1}+1)}}{(s_n!)^{m_1}} & \cdots & \frac{e^{-8m_n}a_n^{m_n(s_{n-1}+1)}}{(s_n!)^{m_n}} \\ \vdots & \ddots & \vdots \\ \frac{e^{-8m_1}a_1^{m_1 s_n}}{(s_n!)^{m_1}} & \cdots & \frac{e^{-8m_n}a_n^{m_n s_n}}{(s_n!)^{m_n}} \end{pmatrix} \begin{pmatrix} b_1 \\ \vdots \\ b_n \end{pmatrix} = \begin{pmatrix} 0 \\ \vdots \\ 0 \end{pmatrix}.$$

If (for all $j = 1, \ldots, n$) we multiply the jth-column of the above matrix by $\frac{(s_n!)^{m_j}e^{8m_j}}{a_j^{m_j(s_{n-1}+1)}}$, we have that the former system is equivalent to a system with the $n \times n$ Vandermonde-type matrix V whose jth row is $a_1^{j-1}, \ldots, a_n^{j-1}$. This matrix is non-singular since the a_k's are pairwise different. Therefore, $b_k = 0$ for $k = 1, \ldots, n$, as required. $\qquad\square$

1.5 Surjections, Darboux functions and related properties

The class of Darboux functions is, probably, the one that has produced the most lineability results within the framework of real analysis. This class is quite large and one of the first lineability findings concerned this class (more precisely, the calls of everywhere surjective functions, see [28], Example 1.5.2, and Definition 1.5.1). This section does not pretend to cover all results within this vast class, since there are (at the moment) ongoing research within it. However, we refer to [167, 323] for some recent advances inside this class. We would like to emphasize that this section simply pretends to give a general overview of some lineability results concerning the mentioned class. However, in Section 7.6, we shall get back to this class.

Of course, this class is also related to many other *pathological* ones. Let us recall some of them that, although well known in Real Analysis, can be found in, e.g., [204, 261].

Definition 1.5.1. Let $f \in \mathbb{R}^{\mathbb{R}}$. We say that:

(1) $f \in ES(\mathbb{R})$ (f is *everywhere surjective*) if $f(I) = \mathbb{R}$ for every nontrivial interval I.

(2) $f \in SES(\mathbb{R})$ (f is *strongly everywhere surjective*) if f takes all values $\mathfrak{c}$ times on any interval.

(3) $f \in PES(\mathbb{R})$ (f is *perfectly everywhere surjective*) if for every perfect set P, $f(P) = \mathbb{R}$.

(4) $f \in AC(\mathbb{R})$ (f is *almost continuous,* in the sense of J. Stallings [369]) if every open set containing the graph of f contains also the graph of some continuous function.

(5) If $h \colon X \to \mathbb{R}$, where X is a topological space, $h \in \mathrm{Conn}(X)$ (h is a *connectivity function*) if the graph of $h|_C$ is connected for every connected set $C \subset X$. (If $h \in \mathbb{R}^{\mathbb{R}}$, it is equivalent to say that its graph is connected.)

(6) $f \in Ext(\mathbb{R})$ (f is *extendable*) if there is a connectivity function $g \colon \mathbb{R}^2 \to [0,1]$ such that $f(x) = g(x,0)$ for every $x \in \mathbb{R}$.

(7) $f \in PR(\mathbb{R})$ (f is a *perfect road function*) if for every $x \in \mathbb{R}$ there is a perfect set $P \subset R$ such that x is a bilateral limit point of P and $f|_P$ is continuous at x.

(8) $f \in PC(\mathbb{R})$ (f is *peripherally continuous*) if for every $x \in \mathbb{R}$ and pair of open sets $U, V \subset \mathbb{R}$ such that $x \in U$ and $f(x) \in V$ there is an open neighborhood W of x with $\overline{W} \subset U$ and $f(\partial W) \subset V$.

(9) $f \in J(\mathbb{R})$ (f is *Jones function*) if its graph intersects every closed subset of $\mathbb{R}^2$ with uncountable projection on the x-axis (see [260]).

(10) $f \in \mathcal{Q}(\mathbb{R})$ if f is a $\mathbb{Q}$-linear function on $\mathbb{R}$.

(11) $f \in Gr(\mathbb{R})$ (f is a *dense-graph function*) if its graph $\{(x, f(x)) : x \in \mathbb{R}\}$ is dense in $\mathbb{R}^2$.

(12) $f \in \mathcal{D}(\mathbb{R})$ (f is a *Darboux function*) if it has the "intermediate value property," that is, for any two values a and b in the domain of f, and any y between $f(a)$ and $f(b)$, there is some c between a and b with $f(c) = y$.

In order to make all the above definitions clearer to the reader, we can picture some of them in the following diagram, which links most of the above classes (the proofs of the below implications are either trivial, or can be found in [204, 209, 210, 261]). In what follows, $A \to B$ means that the class A is a subset of B.[2]

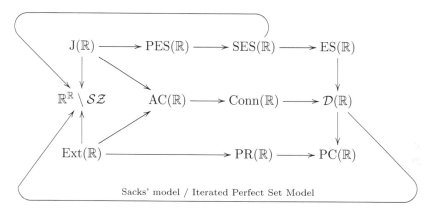

Lebesgue [221, 222, 281] was probably the first to show a somewhat surprising example of a function in ES($\mathbb{R}$). In [28] the authors proved that the set of such *everywhere surjective* functions is $2^{\mathfrak{c}}$-lineable, which is the *best* possible result in terms of dimension. One could think that, in terms of surjectivity, these everywhere surjective functions are in some sense the most exotic. As the above diagram shows, this is very far from being true.

For the sake of completeness, and for the beauty of the construction, we provide here a proof of the existence of a function $f \in$ ES($\mathbb{R}$).

Example 1.5.2 (An everywhere surjective function, [209, Example 2.2]). *Let $(I_n)_{n \in \mathbb{N}}$ be the collection of all open intervals with rational endpoints. The interval I_1 contains a Cantor type set, call it C_1. Now, $I_2 \setminus C_1$ also contains a Cantor type set, call it C_2. Next, $I_3 \setminus (C_1 \cup C_2)$ contains, as well, a Cantor type set, C_3. Inductively, we construct a family of pairwise disjoint Cantor type sets, $(C_n)_{n \in \mathbb{N}}$, such that for every $n \in \mathbb{N}$, $I_n \setminus (\bigcup_{k=1}^{n-1} C_k) \supset C_n$. Now, for every $n \in \mathbb{N}$, take any bijection $\phi_n : C_n \to \mathbb{R}$, and define $f : \mathbb{R} \to \mathbb{R}$ as*

$$f(x) = \begin{cases} \phi_n(x) & \text{if } x \in C_n, \\ 0 & \text{otherwise.} \end{cases}$$

---

[2]Due to the highly technical set theoretical background that it would require, we refer the interested reader to [172] for a complete and modern study of the so-called *Iterated Perfect Set Model*.

*Then f is clearly everywhere surjective (and also zero almost everywhere!).
Indeed, let I be any interval in $\mathbb{R}$. There exists $k \in \mathbb{N}$ such that $I_k \subset I$. Thus
$f(I) \supset f(I_k) \supset f(C_k) = \phi_k(C_k) = \mathbb{R}$.*

Let us now provide a brief account on some of the known lineability and algebrability results of several of the above classes (under ZFC). In the table below, the letter λ stands for the maximal (known) dimension of lineability of the given class.

CLASS	λ	Ref.	CLASS	λ	Ref.
$AC(\mathbb{R})$	$2^{\mathfrak{c}}$	[210]	$PC(\mathbb{R})$	$2^{\mathfrak{c}}$	[28]
$Conn(\mathbb{R})$	$2^{\mathfrak{c}}$	[210]	$\mathcal{DNM}(\mathbb{R})$	$\mathfrak{c}$	[28, 209]
$Ext(\mathbb{R})$	$\geq \mathfrak{c}^+$	[210]	$PES(\mathbb{R}) \setminus J(\mathbb{R})$	$2^{\mathfrak{c}}$	[209]
$AC(\mathbb{R}) \setminus Ext(\mathbb{R})$	$2^{\mathfrak{c}}$	[210]	$SES(\mathbb{R}) \setminus PES(\mathbb{R})$	$2^{\mathfrak{c}}$	[209]
$PR(\mathbb{R})$	$\geq \mathfrak{c}^+$	[210]	$ES(\mathbb{R}) \setminus SES(\mathbb{R})$	$2^{\mathfrak{c}}$	[209]
$\mathcal{SZ}$	$d_{\mathfrak{c}}$ ("$2^{\mathfrak{c}}$")	[210, 212]	$\mathcal{D}(\mathbb{R}) \setminus ES(\mathbb{R})$	$2^{\mathfrak{c}}$	[209]
$J(\mathbb{R})$	$2^{\mathfrak{c}}$	[208]	$PC(\mathbb{R}) \setminus \mathcal{D}(\mathbb{R})$	$2^{\mathfrak{c}}$	[209]
$PES(\mathbb{R})$	$2^{\mathfrak{c}}$	[209]	$ES(\mathbb{R}) \cap \mathcal{Q}(\mathbb{R})$	$2^{\mathfrak{c}}$	[219]
$SES(\mathbb{R})$	$2^{\mathfrak{c}}$	[209]	$Gr(\mathbb{R}) \cap \mathcal{Q}(\mathbb{R}) \setminus ES(\mathbb{R})$	$2^{\mathfrak{c}}$	[219]
$ES(\mathbb{R})$	$2^{\mathfrak{c}}$	[28]	$ES(\mathbb{R}) \setminus \mathcal{Q}(\mathbb{R})$	$2^{\mathfrak{c}}$	[219]
$\mathcal{D}(\mathbb{R})$	$2^{\mathfrak{c}}$	[28]	$Gr(\mathbb{R}) \setminus (ES(\mathbb{R}) \cup \mathcal{Q}(\mathbb{R}))$	$2^{\mathfrak{c}}$	[219]

The quotation marks above ("$2^{\mathfrak{c}}$", next to the class $\mathcal{SZ}$) refer to Theorem 6.6.5. This class $\mathcal{SZ}$ will be studied in detail in Section 6.6. Also, and as usual, $\mathfrak{a}^+$ denotes the successor cardinal of $\mathfrak{a}$.

Some of the results above follow from a result involving the concept of *additivity* [210, 261], that will be presented in detail in Chapter 7 of this monograph.

Let us show one of the earliest results in lineability, namely, let us provide the proof of the $2^{\mathfrak{c}}$-lineability of $ES(\mathbb{R})$ (found in [28]). The proof makes use of the following set theoretical observation, of independent interest:

Lemma 1.5.3. *Let $C_1, C_2, \ldots, C_m$ be distinct non-empty sets. Then there exists an integer $k \in \{1, 2, \ldots, m\}$ such that $C_k \setminus C_i \neq \emptyset$ for all $i \neq k$.*

Proof. Suppose that for every $k \in \{1, 2, \ldots, m\}$ there exists $i \neq k$ such that $C_k \setminus C_i = \emptyset$. By relabelling the sets, we would then have $C_1 \subset C_2 \subset C_3 \subset \cdots \subset C_{m-1} \subset C_m$, which is a contradiction. $\square$

Also, the following proposition shall be needed.

Proposition 1.5.4. *There exists a vector space Λ of functions $\mathbb{R} \to \mathbb{R}$ with the following two properties:*

(i) *Every non-zero element of Λ is an onto function, and*

(ii) $dim(\Lambda) = 2^{\mathfrak{c}}$.

Proof.   Let $r \in \mathbb{R}, r \neq 0$, be fixed. For a non-empty subset $C \subset \mathbb{R}$, define $H_C : \mathbb{R}^{\mathbb{N}} \longrightarrow \mathbb{R}$ by

$$H_C(y, x_1, x_2, x_3, \dots) = \varphi_r(y) \cdot \prod_{i=1}^{\infty} I_C(x_i).$$

Here, I_C is the indicator function and $\varphi_r(y) = e^{ry} - e^{-ry}$, the most important property of this function being the fact that for any $r \neq 0$, $\varphi_r : \mathbb{R} \to \mathbb{R}$ is onto.

First of all, we show that the family $\{H_C : \emptyset \neq C \subset \mathbb{R}\}$ is linearly independent. Indeed, let us take m distinct subsets $C_1, C_2, \cdots, C_m$ of $\mathbb{R}$. Suppose that for some choice of scalars, the function

$$\sum_{j=1}^{m} \lambda_j H_{C_j}$$

is identically 0. Clearly, we may assume (by Lemma 1.5.3) that no λ_j is 0, and since all the C_j's are different we may further assume that for every $j < m$, there exists a point $x_j \in C_m \setminus C_j$. Evaluating at the point

$$\bar{x} = (1, x_1, x_2, \dots, x_{m-2}, x_{m-1}, x_{m-1}, x_{m-1}, \dots),$$

we obtain

$$0 = \sum_{j=1}^{m} \lambda_j H_{C_j}(\bar{x}) = \varphi_r(1) \cdot \sum_{j=1}^{m} \left[\lambda_j \cdot \prod_{i=1}^{\infty} H_{C_j}(x_i) \right] = \varphi_r(1) \cdot \lambda_m.$$

Therefore, we obtain that $\lambda_m = 0$, which is a contradiction. Hence the family $\{H_A : A \subset \mathbb{R}, A \neq \emptyset\}$ is linearly independent.

We now show that every $h \in \Gamma := \text{span}\{H_A : A \subset \mathbb{R}, A \neq \emptyset\}, h \neq 0$, is onto. First, given $s \in \mathbb{R}$ choose $b \in \mathbb{R}$ such that $\varphi_r(b) = s$. Then $H_A(b, a, a, a, \dots) = s$, where $a \in A$ is arbitrary, which shows that each H_A is onto. If $h \in \Gamma \setminus \{0\}$ then for some non-zero real numbers $\lambda_1, \lambda_2, \dots, \lambda_m$ and some distinct non-empty subsets of $\mathbb{R}$, $C_1, C_2, \dots, C_m$, we have

$$h = \sum_{j=1}^{m} \lambda_j H_{C_j}.$$

Arguing as before, we first find some C_j (which, without loss, is C_m) such that for each $j = 1, \dots, m - 1$, there is a point $x_j \in C_m \setminus C_j$. Given $s \in \mathbb{R}$, let $\bar{x} = (a, x_1, x_2, \dots, x_{m-2}, x_{m-1}, x_{m-1}, \dots)$, where $\varphi_r(a) = \frac{s}{\lambda_m}$. It is straightforward to verify that $h(\bar{x}) = s$.

Thus, every $h \in \Gamma \setminus \{0\}$ is onto. It is clear that $\dim(\Gamma) = 2^{\mathfrak{c}}$ since $\mathrm{card}(\{A : A \subset \mathbb{R}, A \neq \emptyset\}) = 2^{\mathfrak{c}}$. Since there is a bijection between $\mathbb{R}$ and $\mathbb{R}^{\mathbb{N}}$, we can also construct a vector space Λ of onto functions $f : \mathbb{R} \to \mathbb{R}$ having dimension $2^{\mathfrak{c}}$. $\qquad\square$

Next, as a consequence of the previous proposition, we can obtain the result. Indeed, choose Λ as in the preceding proposition, and fix any everywhere surjective function f. We claim that the vector space

$$\Delta := \{H \circ f : H \in \Lambda\}$$

satisfies the required conditions. First, $\dim(\Delta) = 2^{\mathfrak{c}}$ since $\dim(\Lambda) = 2^{\mathfrak{c}}$. To see this it is enough to show that given m linear independent functions $H_j \in \Lambda \setminus \{0\}$, $j = 1, \ldots, m$, then the family $\{H_j \circ f\}_{j=1}^{m}$ is also linear independent. Take m nonzero real numbers $\{\lambda_j\}_{j=1}^{m}$, and suppose that the function $h = \sum_{j=1}^{m} \lambda_j \cdot (H_j \circ f)$ is identically zero. By construction h can be written as $h = G \circ f$, where G is onto. Take any $0 \neq s \in \mathbb{R}$. There exists $d \in \mathbb{R}$ with $G(d) = s$, and there also exists $a \in \mathbb{R}$ so that $f(a) = d$. Thus $h(a) = s \neq 0$, which is a contradiction. Next, let us take $g \in \Delta \setminus \{0\}$, $s \in \mathbb{R}$, and any interval $(a, b) \subset \mathbb{R}$. We need to find $\ell \in (a, b)$ such that $g(\ell) = s$. We can express g as $g = G \circ f$, with G being an onto function. Thus, for $s \in \mathbb{R}$ there exists $d \in \mathbb{R}$ such that $G(d) = s$. Now, we can find $\ell \in (a, b)$ with $f(\ell) = d$ (since f is our previously fixed everywhere surjective function). The proof is completed.

Notice that some of the above classes are themselves subsets of the set of surjective functions from $\mathbb{R}$ to $\mathbb{R}$ and all of these subsets are $2^{\mathfrak{c}}$-lineable. One could think that a similar result would hold for one-to-one functions. In [360] (see, also, [8, 209]) a negative answer was given to this question.

Theorem 1.5.5. *The set of injective functions is not lineable. Moreover, if V is a vector space, every nonzero element of which is an injective function on $\mathbb{R}$, then $\dim(V) = 1$.*

The algebrability of the above classes (or modifications of them) has also been considered by many authors. As we have already mentioned, to obtain algebrability is highly more complex than to obtain lineability. For instance, it is out of place to consider algebras of functions in $\mathrm{ES}(\mathbb{R})$, since given any $f \in \mathbb{R}^{\mathbb{R}}$ we have that $f^2 \notin \mathrm{ES}(\mathbb{R})$. The same happens for the classes $\mathrm{PES}(\mathbb{R})$, $\mathrm{SES}(\mathbb{R})$, $\mathrm{J}(\mathbb{R})$ and many others. But sometimes a dual of the previous classes can be certainly constructed in $\mathbb{C}^{\mathbb{C}}$. For instance, and just to cite very recent results on algebrability, let us look at the following table:

CLASS	ALGEBRABILITY DIMENSION	Ref.
$\mathrm{ES}(\mathbb{C})$, $\mathrm{PES}(\mathbb{C})$	$2^{\mathfrak{c}}$	[24, 35], [50]
$\mathrm{SES}(\mathbb{C}) \setminus \mathrm{PES}(\mathbb{C})$	$2^{\mathfrak{c}}$	[46]
$\mathcal{EDD}(\mathbb{R})$	$2^{\mathfrak{c}}$	[46, 50]
$\mathcal{EDF}(\mathbb{R})$	$2^{\mathfrak{c}}$	[46]
$\mathcal{EDC}(\mathbb{R})$	$2^{\mathfrak{c}}$	[46]

In the table above the classes $\mathcal{EDD}(\mathbb{R})$, $\mathcal{EDF}(\mathbb{R})$ and $\mathcal{EDC}(\mathbb{R})$ denote, respectively, the set of everywhere discontinuous Darboux function, the set of nowhere continuous functions having finitely many values, and the set of nowhere continuous functions mapping compact sets to compact sets. These last two classes were also thoroughly studied in [211] (see Theorem 1.6.1 from Section 1.6 in this monograph).

Remark 1.5.6. *Turning to surjectivity, but this time in the setting of families of functions defined on complex domains, a simple Baire-category argument shows that if X is an infinite dimensional Banach space and $f : \mathbb{D} \to X$ is continuous* ($\mathbb{D} := \{z \in \mathbb{C} : |z| < 1\}$, *the open unit disc) then f cannot be surjective. Let $H(\mathbb{D}, X) := \{f : \mathbb{D} \to X : f$ is holomorphic on $\mathbb{D}\}$. In 1976, Glovebnik [225] and independently Rudin [350] proved that the set $\mathcal{D} := \{f \in H(\mathbb{D}, X) : f(\mathbb{D})$ is dense in $X\}$ is not empty. Very recently, López-Salazar [293] has been able to demonstrate the lineability of $\mathcal{D}$.*

1.6 Other properties related to the lack of continuity

As any undergrad math major knows, continuous functions transform compact or connected sets into compact or connected sets, respectively. It is interesting to ask whether this characterizes continuity or not. In other words, if a function $f : \mathbb{R} \to \mathbb{R}$ transforms compact sets into compact sets and connected sets into connected sets, can we expect f to be continuous? This question was positively answered in the late 1960's by Hamlett [242] and White [380] (see also Velleman [377, Theorem 2]). Thus, the fact that f satisfies only one of the following two conditions:

1. f transforms compact sets into compact sets, or

2. f transforms connected sets into connected sets

might be very far away from making f continuous, which is what Gámez, Muñoz and Seoane proved in [211]. Let us see this result in more depth.

Theorem 1.6.1. *There exist $2^{\mathfrak{c}}$-dimensional linear spaces U and V of $\mathbb{R}^{\mathbb{R}}$ such that:*

(a) *Every nonzero element of U is nowhere continuous and transforms connected sets into connected sets.*

(b) *Every nonzero element of V is nowhere continuous and transforms any set into a compact set.*

It is shown in [28, Proposition 4.2] that there exists a 2^c-dimensional linear space of functions from $\mathbb{R}$ onto $\mathbb{R}$ mapping every nontrivial interval onto $\mathbb{R}$ (that is, everywhere surjections). It is easy to see that any function satisfying that property is nowhere continuous and transforms connected sets into connected sets. Hence we already have the linear space U from the statement of the theorem. To see part (b), let us first recall the following remark.

Remark 1.6.2. *It is a well known fact (at an undergraduate level) that $\mathbb{R}$ and $\mathbb{R}^{\mathbb{N}}$ have the same cardinality, namely $\mathfrak{c}$, which (as we will see) implies that $\mathbb{R}$ and $\mathbb{R}^{\mathbb{N}}$ are isomorphic when seen as $\mathbb{Q}$-linear spaces. Actually, in a more general frame, it can be shown that if V is an infinite-dimensional vector space over $\mathbb{Q}$, then $\mathrm{card}(V) = \dim(V)$. Indeed, if H is a Hamel basis of V over $\mathbb{Q}$ then every element v of V can be written as $v = \lambda_1 e_1 + \cdots + \lambda_n e_n$ where $\lambda_i \in \mathbb{Q}$ and $e_i \in H$ for $i = 1, \ldots, n$ and for some $n \in \mathbb{N}$. Hence v belongs to the finite-dimensional $\mathbb{Q}$-linear space generated by $\{e_1, \ldots, e_n\}$. This shows that*

$$V = \bigcup_{A \in \mathcal{P}_f(H)} [A],$$

where $\mathcal{P}_f(H)$ denotes the set of finite subsets of H and $[A]$ stands for the $\mathbb{Q}$-linear span of the set A. Finally, since $[A]$ is countable,

$$\mathrm{card}(V) \leq \mathrm{card}(\mathcal{P}_f(H)) = \mathrm{card}(H) = \dim(V) \leq \mathrm{card}(V).$$

Thus, applying this argument to $\mathbb{R}$ and $\mathbb{R}^{\mathbb{N}}$, we obtain

$$\dim(\mathbb{R}) = \mathrm{card}(\mathbb{R}) = \mathfrak{c} = \mathrm{card}(\mathbb{R}^{\mathbb{N}}) = \dim(\mathbb{R}^{\mathbb{N}}),$$

proving that $\mathbb{R}$ and $\mathbb{R}^{\mathbb{N}}$ are $\mathbb{Q}$-linearly isomorphic, as we wished.

Next, let H be a Hamel basis of $\mathbb{R}$ over $\mathbb{Q}$ and suppose that $\varphi : \mathbb{R} \to \mathbb{R}^{\mathbb{N}}$ is a $\mathbb{Q}$-linear isomorphism, which exists due to Remark 1.6.2. Observe that if $\alpha \neq 0$ then $\alpha \cdot \mathbb{Q}$ is a dense set in $\mathbb{R}$. For each subset A of H we define $f_A : \mathbb{R} \to \mathbb{R}$ by

$$f_A(x) := \chi_{([A] \setminus \{0\})^{\mathbb{N}}}(\varphi(x)),$$

for all $x \in \mathbb{R}$. Next, choose $h_0 \in H$ and consider the set

$$F = \{f_A : \varnothing \neq A \in \mathcal{P}(H), \, h_0 \notin A\}.$$

We have that the cardinality of F is 2^c. Also, by Exercise 1.6, F is linearly independent. Now let $g = \sum_{i=1}^{n} \lambda_i f_{A_i}$ be a linear combination of distinct elements of F with $\lambda_i \neq 0$ for $i = 1, \ldots, n$. The linear independence of the elements of F ensures that $g \neq 0$. Also, since $g(\mathbb{R})$ is clearly finite, g maps any set into a compact set.

To conclude the proof we just need to show that g is nowhere continuous. Since $g(0) = 0$ and $g \neq 0$, there must exist $x_1 \in \mathbb{R} \setminus \{0\}$ such that $g(x_1) \neq 0$. Now, notice that if $x_2 = \varphi^{-1}(h_0, h_0, h_0, \ldots)$ then $x_2 \neq 0$ and $g(x_2) = 0 \neq$

$g(x_1)$. Also, observe that for all $x \in \mathbb{R}$ and $q \in \mathbb{Q} \setminus \{0\}$ we have $g(qx) = g(x)$. Therefore $\mathbb{Q} \cdot x_1$ and $\mathbb{Q} \cdot x_2$ are dense (and disjoint) subsets of $\mathbb{R}$ satisfying

$$g(\mathbb{Q} \cdot x_1) = \{g(x_1)\} \neq \{0\} = \{g(x_2)\} = g(\mathbb{Q} \cdot x_2),$$

which implies that g is nowhere continuous, as desired.

Next, one could also study noncontinuous functions focusing on the structure of their sets of discontinuities. To start with, the set of points at which a function is continuous is always a G_δ set, so the set of discontinuities is an F_σ set. Moreover, the set of discontinuities of a monotonic function is at most countable (Froda's theorem). In this direction, several results have been obtained in the last years. Recall that a point $x_0 \in \mathbb{R}$ is called a removable discontinuity of a function $f : \mathbb{R} \to \mathbb{R}$ whenever there exists $\lim_{x \to x_0} f(x)$ and is finite. The following result is due to García-Pacheco, Palmberg and Seoane [216].

Theorem 1.6.3. *The set M of functions on $\mathbb{R}$ with a dense set of points of removable discontinuity is $(\mathfrak{c}, \mathfrak{c})$-algebrable and, thus, $\mathfrak{c}$-lineable.*

Let us give the proof of the previous result, which shall be useful to illustrate a usual technique that one can use in order to obtain algebrability:

Proof. For every $\beta > 0$ we define the function f_β (closely related to the Dirichlet function, Figure 1.6) by

$$f_\beta(x) = \begin{cases} n^{-\beta} & \text{if } x \in \mathbb{Q}, \\ 0 & \text{if } x \in \mathbb{R} \setminus \mathbb{Q}, \end{cases}$$

where $n \in \mathbb{N}$ is defined to be the smallest positive integer such that $x = t/n$ for some $t \in \mathbb{Z}$ (see Figure 1). Clearly, it is discontinuous at every point $x \in \mathbb{Q}$ and, hence, it has a dense set of points of discontinuity. In order to see that all discontinuities are removable, we simply have to notice that for any $x_0 \in \mathbb{Q}$,

$$\lim_{x \to x_0^-} f_\beta(x) = \lim_{x \to x_0^+} f_\beta(x) = 0.$$

Thus, we can redefine the function as the function identically equal to zero, which is, of course, continuous.

Let $\mathcal{H}$ be a positive Hamel basis of $\mathbb{R}$ as a $\mathbb{Q}$-vector space. Also, let A be the algebra generated by

$$\{f_\gamma : \gamma \in \mathcal{H}\}.$$

We shall begin by showing that the family $\{f_\gamma : \gamma \in \mathcal{H}\}$ is algebraically independent. Thus, take any polynomial

$$P(f_{\gamma_1}, \ldots, f_{\gamma_k}) := \sum_{i=1}^{m} \alpha_i \prod_{j=1}^{k} f_{\gamma_j}^{n_{j,i}},$$

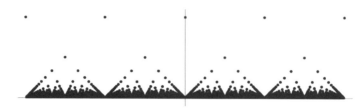

FIGURE 1.6: The graph of f_1 (the Dirichlet function) on $[-2, 2]$.

where $\{\alpha_i\}_{i=1}^m \subset \mathbb{R} \setminus \{0\}$, m, k and all the $n_{j,i}$'s are natural numbers and where $n_{j,a} = n_{j,b}$ for all $j \in \{1, \ldots, k\}$ if and only if $a = b$. A straightforward calculation shows that

$$P(f_{\gamma_1}, \ldots, f_{\gamma_k})(x) = \sum_{i=1}^m \alpha_i n^{-(n_{1,i}\gamma_1 + \ldots + n_{k,i}\gamma_k)} =: \sum_{i=1}^m \alpha_i n^{-\beta_i}.$$

Since $\{\gamma_i\}_{i=1}^k \subset \mathcal{H}$, we get that all the β_i's are different positive real numbers. Thus,

$$P(f_{\gamma_1}, \ldots, f_{\gamma_k})(x) = \sum_{i=1}^m \alpha_i f_{\beta_i}(x) =: F(x),$$

where we, without loss of generality, may assume that $0 < \beta_1 < \ldots < \beta_m$.

Assume that $F \equiv 0$. Then, by evaluating F at

$$1, \frac{1}{2}, \ldots, \frac{1}{2^{m-1}}$$

we obtain the following linear system of equations:

$$
\begin{pmatrix}
1 & 1 & 1 & \cdots & 1 \\
2^{-\beta_1} & 2^{-\beta_2} & 2^{-\beta_3} & \cdots & 2^{-\beta_m} \\
\left(2^{-\beta_1}\right)^2 & \left(2^{-\beta_2}\right)^2 & \left(2^{-\beta_3}\right)^2 & \cdots & \left(2^{-\beta_m}\right)^2 \\
\vdots & \vdots & \vdots & \ddots & \vdots \\
\left(2^{-\beta_1}\right)^{m-1} & \left(2^{-\beta_2}\right)^{m-1} & \left(2^{-\beta_3}\right)^{m-1} & \cdots & \left(2^{-\beta_m}\right)^{m-1}
\end{pmatrix}
\cdot
\begin{pmatrix}
\alpha_1 \\
\alpha_2 \\
\alpha_3 \\
\vdots \\
\alpha_m
\end{pmatrix}
=
\begin{pmatrix}
0 \\
0 \\
0 \\
\vdots \\
0
\end{pmatrix}.
$$

Since $2^{-\beta_i} = 2^{-\beta_j}$ if and only if $i = j$ and since the matrix is a Vandermonde-type matrix we obtain that it is non-singular. Therefore, $\alpha_1 = \ldots = \alpha_m = 0$, which is a contradiction. Finally, we have that F has a dense set of points of removable discontinuity. $\qquad\square$

In this same direction, the following battery of results is also related to the set of discontinuities of functions in $\mathbb{R}$ (see [3, 216]).

Theorem 1.6.4. (1) *Given a closed set $F \subset \mathbb{R}$, the set H of all functions $\mathbb{R} \to \mathbb{R}$ whose set of points of discontinuity is F is lineable with $\lambda(H) \geq \mathfrak{c}$. Moreover, if the interior $F^\circ \neq \varnothing$ then $\lambda(H) = 2^{\mathfrak{c}}$.*

(2) *Given any nonclosed F_σ set F, the set of functions whose set of points of discontinuity is F is coneable.*

(3) *Given a closed set F of measure zero contained in an interval $[a, b]$, the set of all Riemann integrable functions whose set of points of discontinuity is F is lineable.*

(4) *Given any nonclosed F_σ set F of measure zero contained in an interval $[a, b]$, the set of all Riemann integrable functions whose set of points of discontinuity is exactly F is coneable.*

(5) *Let I be any nontrivial interval and consider a point $a \in I$. Let K denote the set of all functions from I to $\mathbb{R}$ having a removable discontinuity at a. Then, $\lambda(K) = 1$. If L denotes the set of all functions from I to $\mathbb{R}$ having a jump discontinuity at a, then $\lambda(L) = 1$. Also, if H denotes the set of all functions from I to $\mathbb{R}$ having either a removable or jump discontinuity at a, then $\lambda(H) = 2$.*

1.7 Continuous functions that attain their maximum at only one point

From now on $\widehat{C}(X)$ represents the set of continuous functions $f : X \to \mathbb{R}$ which attain its maximum at only one point of their domains. As we shall see in the next sections the lineability of $\widehat{C}(X)$ has a straight dependence on X. Later on, we shall see that this dependence has an intriguing topological component, that becomes more visible in higher dimensions.

Continuous functions on $[a, b)$ or $\mathbb{R}$

The following results deal with the cases $X = [a, b)$ and $X = \mathbb{R}$.

Proposition 1.7.1. *$\widehat{C}[a, b)$ is 2-lineable.*

Proof. Let us consider the trigonometric maps sin and cos defined in $[0, 2\pi)$. It is obvious that sin and cos are linearly independent and attain their maximum at exactly one point in $[0, 2\pi)$. We shall show that any nontrivial linear combination of sin and cos also attains its maximum at exactly only one point. In fact, for any $(\alpha, \beta) \in \mathbb{R}^2 \setminus \{(0, 0)\}$, there is $\theta \in [0, 2\pi)$ such that

$$(\alpha, \beta) = (\sqrt{\alpha^2 + \beta^2} \cos \theta, \sqrt{\alpha^2 + \beta^2} \sin \theta).$$

Hence

$$\alpha \cos x + \beta \sin x = \sqrt{\alpha^2 + \beta^2} \cos \theta \cos x + \sqrt{\alpha^2 + \beta^2} \sin \theta \sin x$$
$$= \sqrt{\alpha^2 + \beta^2} \cos (x - \theta).$$

But the function cos has only one maximum in $[-\theta, 2\pi - \theta)$ and thus $\alpha \cos x + \beta \sin x$ has only one maximum in $[0, 2\pi)$. The extension of this result from $[0, 2\pi)$ to $[a, b)$ is immediate; it suffices to compose each function $f \in \widehat{C}[0, 2\pi)$ with

$$g : [a, b) \to [0, 2\pi), \ g(x) = \frac{2\pi(x - a)}{b - a}.$$

$\square$

When $X = \mathbb{R}$ we also have 2-lineability; however, the reasoning is quite different.

Proposition 1.7.2. $\widehat{C}(\mathbb{R})$ *is 2-lineable.*

Proof. Consider $x(t)$, $y(t)$ (see Figure 1.7) defined by

$$x(t) := \mu(t) \cos(4 \arctan |t|)$$

and

$$y(t) := \mu(t) \sin(4 \arctan |t|),$$

where tan is defined in $(-\frac{\pi}{2}, \frac{\pi}{2})$ and μ is a continuous function given by

$$\mu(t) = \begin{cases} e^t, & \text{if} \quad t \leq 0, \\ 1, & \text{if} \quad t \geq 0. \end{cases}$$

Note that $x(t)$ and $y(t)$ are linearly independent. In fact, if $\alpha, \beta \in \mathbb{R}$ and

$$\alpha\mu(t) \cos(4 \arctan |t|) + \beta\mu(t) \sin(4 \arctan |t|) = 0,$$

since $\mu(t) \neq 0$ for all $t \in \mathbb{R}$, we have

$$\alpha \cos(4 \arctan |t|) + \beta \sin(4 \arctan |t|) = 0.$$

Choosing $t = 0$ we obtain $\alpha = 0$. Also, with a suitable choice of t we obtain $\beta = 0$. Since $\cos(4 \arctan |t|) \leq 1$, $\sin(4 \arctan |t|) \leq 1$ and $\mu(t) \geq 0$, for all $t \in \mathbb{R}$, it follows that $x(t) \leq \mu(t)$ and $y(t) \leq \mu(t)$. Since $\mu(t) \leq 1$ for all $t \in \mathbb{R}$, the functions $x(t)$ and $y(t)$ are bounded (by above) by 1. Let us show that $x(t)$ and $y(t)$ attain the maximum 1 at only one point. For $t = 0$,

$$x(0) = 1 \cos(4 \arctan 0) = \cos 0 = 1.$$

For $t \neq 0$, we have $4 \arctan |t| \in (0, 2\pi)$ and thus $\cos(4 \arctan |t|) < 1$. Since $\mu(t) \leq 1$ it follows that $x(t) < 1$, for all $t \neq 0$. So, we conclude that $x(t) \in \widehat{C}(\mathbb{R})$.

Regarding $y(t)$, a similar argument shows that it has only one point of maximum.

Now we shall show that any nontrivial linear combination of x and y has only one point of maximum. Let $\alpha, \beta \in \mathbb{R}$ and $\alpha x(t) + \beta y(t)$ be a nontrivial

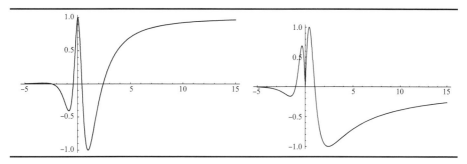

FIGURE 1.7: Plots of $x(t)$ and $y(t)$, respectively.

linear combination. We already know that for all $(\alpha, \beta) \in \mathbb{R}^2 \setminus \{(0,0)\}$, there is $\theta \in [0, 2\pi)$ such that

$$(\alpha, \beta) = (\sqrt{\alpha^2 + \beta^2} \cos \theta, \sqrt{\alpha^2 + \beta^2} \sin \theta).$$

Therefore

$$
\begin{aligned}
\alpha x\,(t) + \beta y(t) &= \sqrt{\alpha^2 + \beta^2} \cos \theta\, x\,(t) + \sqrt{\alpha^2 + \beta^2} \sin \theta\, y\,(t) \\
&= \sqrt{\alpha^2 + \beta^2} \cos \theta \cdot \mu(t) \cos(4 \arctan |t|) \\
&\quad + \sqrt{\alpha^2 + \beta^2} \sin \theta \cdot \mu(t) \sin(4 \arctan |t|) \\
&= \sqrt{\alpha^2 + \beta^2}\, \mu(t) \cdot \big(\cos \theta \cdot \cos(4 \arctan |t|) \\
&\quad + \sin \theta \cdot \sin(4 \arctan |t|)\big) \\
&= \sqrt{\alpha^2 + \beta^2}\, \mu(t) \cdot \cos\left(4 \arctan |t| - \theta\right).
\end{aligned}
$$

Note that for $t \geq 0$, $\mu(t) = 1$ and $4 \arctan t \in [0, 2\pi)$. Since $\cos x$ admits only one maximum in $[-\theta, 2\pi - \theta)$, it follows that

$$\alpha x(t) + \beta y(t) = \sqrt{\alpha^2 + \beta^2} \cos(4 \arctan t - \theta)$$

attains only one maximum when $t \geq 0$. For $t < 0$, $\mu(t) = e^t < 1$, and thus

$$\alpha x(t) + \beta y\,(t) < \alpha x\,(-t) + \beta y\,(-t),$$

i.e., the maximum is not attained. We thus conclude that any nontrivial linear combination $\alpha x(t) + \beta y(t)$ belongs to $\widehat{C}(\mathbb{R})$. $\qquad \square$

Continuous functions on $[a, b]$

The following notations are useful. Given a topological space X and $f \in C(X)$, we set

- $M(f) := \sup_{t \in X} f(t)$;

- $m(f) := \inf_{t \in X} f(t)$;

- $M_f := \{t \in X : f(t) = M(f)\}$;

- $m_f := \{t \in X : f(t) = m(f)\}$.

In all this section, for each $f \in C(X)$ we consider

$$\|f\| := \sup_{t \in X} |f(t)|.$$

Note that

$$\|f\| = M(|f|).$$

Definition 1.7.3. Let $f_1, ..., f_n \in C[0,1]$.

(a) A point $t \in [0,1]$ is *ignorable for* $f_1, ..., f_n$ if, for any $\alpha_1, ..., \alpha_n \in \mathbb{R}_+^*$, we have $t \notin M_{\sum_{i=1}^n \alpha_i f_i}$.

(b) A point $t \in [0,1]$ is said to be *a fence between* t_1 *and* t_2 in $[0,1]$ for $f_1, ..., f_n$ if t belongs to the open interval of extreme points t_1 and t_2, and t is ignorable for $f_1, ..., f_n$.

Definition 1.7.4. A pair of functions $\{f, g\}$ in $C[0,1]$ is said to be *canonical* if there are $t_f \in M_f, t_g \in M_g$ and $\bar{t} \in]t_f, t_g[$ or $\bar{t} \in]t_g, t_f[$ such that $m_f = \{\bar{t}\}$ or $m_g = \{\bar{t}\}$.

Lemma 1.7.5. *In the above conditions* $\bar{t}$ *is a fence between* t_f *and* t_g *for* f, g.

Proof. Let us suppose that $m_f = \{\bar{t}\}$. Then $f(\bar{t}) < f(t)$ for all $t \in [0,1] \setminus \{\bar{t}\}$; in particular $f(\bar{t}) < f(t_g)$. Since $t_g \in M_g$, it follows that

$$g(t_g) = M(g) = \sup_{t \in [0,1]} g(t).$$

Hence

$$g(\bar{t}) \leq g(t_g)$$

and thus

$$\alpha f(\bar{t}) + \beta g(\bar{t}) < \alpha f(t_g) + \beta g(t_g)$$

for all $\alpha, \beta \in \mathbb{R}_+^*$. Therefore $\bar{t} \notin M_{\alpha f + \beta g}$. If $m_g = \{\bar{t}\}$ the argument is similar. $\square$

A canonical pair of functions is not a basis of a 2-dimensional subspace V such that $V \setminus \{0\} \subseteq \hat{C}[0,1]$. This fact shall be proved in Proposition 1.7.7 and needs the use of following lemma.

From now on, for any map $\phi : [0,1] \to C[0,1]$ we will denote $\phi(x)$ by ϕ_x.

Lemma 1.7.6. *If ϕ is a continuous map from $[0,1]$ to $C[0,1]$ such that for all $x \in [0,1]$, M_{ϕ_x} is a singleton $\{t_x\}$, then the function*

$$\mu : \quad [0,1] \quad \to \quad [0,1]$$
$$x \quad \mapsto \quad \mu(x) = t_x$$

is continuous.

Proof. Let us suppose that there exists $(x_n)_{n=1}^{\infty} \in [0,1]$ such that $x_n \to x_0$, but $\mu(x_n) = t_{x_n}$ does not converge to $t_{x_0} = \mu(x_0)$. Since $[0,1]$ is compact, up to a subsequence, if necessary, $(t_{x_n})_{n=1}^{\infty}$ converges to some $\bar{t} \in [0,1]$. Note that

$$
\begin{aligned}
|\phi_{x_n}(t_{x_n}) - \phi_{x_0}(\bar{t})| &= |\phi_{x_n}(t_{x_n}) - \phi_{x_0}(t_{x_n}) + \phi_{x_0}(t_{x_n}) - \phi_{x_0}(\bar{t})| \\
&\leq |\phi_{x_n}(t_{x_n}) - \phi_{x_0}(t_{x_n})| + |\phi_{x_0}(t_{x_n}) - \phi_{x_0}(\bar{t})| \\
&\leq \sup_{y \in [0,1]} |\phi_{x_n}(y) - \phi_{x_0}(y)| + |\phi_{x_0}(t_{x_n}) - \phi_{x_0}(\bar{t})| \\
&= \|\phi_{x_n} - \phi_{x_0}\| + |\phi_{x_0}(t_{x_n}) - \phi_{x_0}(\bar{t})|
\end{aligned}
$$

and thus

$$\phi_{x_n}(t_{x_n}) \longrightarrow \phi_{x_0}(\bar{t})$$

when $x_n \longrightarrow x_0$, since ϕ and ϕ_{x_0} are continuous. However, we also have

$$\phi_{x_n}(t_{x_n}) \longrightarrow \phi_{x_0}(t_{x_0}).$$

Since the limit is unique, we have

$$\phi_{x_0}(\bar{t}) = \phi_{x_0}(t_{x_0}) = M(\phi_{x_0}),$$

and since $M_{\phi_{x_0}} = \{t_{x_0}\}$, we have $\bar{t} = t_{x_0}$ and thus μ is continuous. $\qquad\square$

Proposition 1.7.7. *For any canonical pair of functions $\{f, g\}$ in $C[0,1]$ there exist two positive real numbers α and β such that the function $\alpha f + \beta g$ has at least two absolute maxima.*

Proof. Suppose that there is a pair of canonical functions $\{f, g\}$ such that for all $(\alpha, \beta) \in \mathbb{R}_+^2 \setminus \{(0,0)\}$, $M_{\alpha f + \beta g}$ is a singleton. Consider the maps

$$\phi : \quad [0,1] \quad \to \quad C[0,1]$$
$$x \quad \mapsto \quad \phi_x = (1-x)f + xg$$

and

$$\mu : \quad [0,1] \quad \to \quad [0,1]$$
$$x \quad \mapsto \quad \mu(x) = t_x,$$

where $\{t_x\} = M_{\phi_x}$. By the previous lemma, μ is continuous and thus it assumes all the values between $\mu(0) = t_0$ and $\mu(1) = t_1$, where $\{t_0\} = M_f$ and $\{t_1\} = M_g$. However, it contradicts Lemma 1.7.5, that asserts that there is a fence between t_0 and t_1, i.e., a point $\tilde{t} \in]t_0, t_1[$ or $]t_1, t_0[$ such that $\tilde{t} \notin M_{\alpha f + \beta g}$ for all $(\alpha, \beta) \in \mathbb{R}_+^2 \setminus \{(0,0)\}$. $\qquad\square$

Theorem 1.7.8. *The set $\widehat{C}[0,1]$ is strongly non lineable, i.e., $\widehat{C}[0,1]$ is 1-lineable but not 2-lineable.*

Proof. We shall show that to any two pairs of linearly independent functions $\{f,g\} \subset C[0,1]$, there is a pair $(\alpha, \beta) \in \mathbb{R}^2 \setminus \{(0,0)\}$ such that $\alpha f + \beta g$ admits at least two points of maxima. If we suppose this is not the case, let $f, g \in C[0,1]$ be such that for all $(\alpha, \beta) \in \mathbb{R}^2 \setminus \{(0,0)\}$, $M_{\alpha f + \beta g}$ is a singleton. Since $[0,1]$ is compact and the function $\alpha f + \beta g$ is continuous, then both $M_{\alpha f + \beta g}$ and $m_{\alpha f + \beta g}$ are singletons: the case of $M_{\alpha f + \beta g}$ is immediate and $m_{\alpha f + \beta g}$ is a singleton because it is precisely $M_{-(\alpha f + \beta g)}$, that is itself a singleton. Let us define

$$\epsilon(f,g) := M_f \cup M_g \cup m_f \cup m_g.$$

It is simple to verify that $card\,(\epsilon(f,g)) \leq 4$, because $card\,(M_i) = card\,(m_j) = 1$, whenever $\{i,j\} = \{f,g\}$. Let us consider two cases:

(1) If $card\,(\epsilon(f,g)) \geq 3$, note that at least one of the four pairs $\{f,g\}, \{f,-g\}, \{-f,g\}, \{-f,-g\}$ is canonical. This is a contradiction, by Proposition 1.7.7.

(2) If $card\,(\epsilon(f,g)) = 2$, we shall consider two cases:

$$M_f = m_g \text{ and } m_f = M_g$$

or

$$M_f = M_g \text{ and } m_f = m_g.$$

Note that we can not have $M_f = m_f$, because in this case f would not belong to $\widehat{C}[0,1]$. If $M_f = m_g$ and $m_f = M_g$, considering the same functions ϕ and μ used in the Proposition 1.7.7, we would find $\alpha \in]0,1[$ such that $M_{(1-\alpha)f + \alpha g}$ is different from M_f and $m_f = M_g$. Therefore

$$card\,(\epsilon(f, (1-\alpha)f + \alpha g)) \geq 3$$

and by the case (1) we get a contradiction. If we suppose $M_f = M_g$ and $m_f = m_g$, instead of M_g and m_g, we work with m_{-g} and M_{-g} and proceeding in a similar way we again get a contradiction.

So, we just need to consider $card\,(\epsilon(f,g)) = 1$. In this case $M_f = m_f = m_g = M_g$, and f, g are constant maps and thus linearly dependent, a contradiction. $\square$

Corollary 1.7.9. *If $a < b$, $\widehat{C}[a,b]$ is 1-lineable.*

Higher dimensions: a topological flavor

The results of the previous sections are summarized as:

(A) $\widehat{C}[a,b)$ contains, up to the origin, a 2-dimensional linear subspace of $C[a,b)$.

(B) $\widehat{C}(\mathbb{R})$ contains, up to the origin, a 2-dimensional linear subspace of $C(\mathbb{R})$.

(C) There is no 2-dimensional linear subspace of $C[a,b]$ contained in $\widehat{C}[a,b] \cup \{0\}$.

In the next sections we shall see how lineability behaves when we replace $[a,b), \mathbb{R}$ and $[a,b]$ by higher dimensional prototypes. While analytic techniques were used to prove (A), (B) and (C) we shall see that the nature of this kind of problem in higher dimensions has a strong topological flavor; for example, the Borsuk-Ulam theorem is invoked.

Functions defined on preimages of Euclidean spheres

In this section we show that in (A) the interval $[a,b)$ can be replaced by preimages D of Euclidean spheres.

By $\langle \cdot, \cdot \rangle$ we denote the usual inner product in the Euclidean spaces, by $\| \cdot \|_2$ we mean the Euclidean norm on $\mathbb{R}^n$ and by B we denote the closed unit ball of $\mathbb{R}^n$ with the Euclidean norm. The next theorem is a generalization of the result (A) above.

Theorem 1.7.10. *Let $n \geq 2$ be a positive integer and D a topological space such that there is a continuous bijection from D to S^{n-1}. Then $\widehat{C}(D)$ is n-lineable.*

Proof. Let $\pi_i : S^{n-1} \to \mathbb{R}$ be a projection over the ith coordinate, $i = 1, ..., n$, and $G : D \to S^{n-1}$ be a continuous bijection. Let us show that the functions $\pi_i, i = 1, ..., n$, are linearly independent. In fact, consider any linear combination

$$\sum_{i=1}^{n} a_i \pi_i.$$

If $a_i \neq 0$ for some i, we have

$$\sum_{i=1}^{n} a_i^2 \neq 0,$$

and thus

$$z = \frac{(a_1, ..., a_n)}{\left(\sum_{i=1}^{n} a_i^2\right)^{\frac{1}{2}}} \in S^{n-1}.$$

Therefore

$$\sum_{i=1}^{n} a_i \pi_i(z) = \frac{\sum_{i=1}^{n} a_i^2}{\left(\sum_{i=1}^{n} a_i^2\right)^{\frac{1}{2}}} \neq 0.$$

Now let us show that any nontrivial linear combination of the functions $\pi_i, i = 1, ..., n$, has a unique maximum. Consider $\sum_{i=1}^{n} a_i \pi_i$, with $a_i \neq 0$ for some $i = 1, ..., n$, and $y \in S^{n-1}$. Thus

$$\sum_{i=1}^{n} a_i \pi_i(y) = \langle a, y \rangle,$$

with $a = (a_1, ..., a_n)$. Now we consider the map

$$f : \quad \mathbb{R}^n \quad \to \quad \mathbb{R}$$
$$y \quad \mapsto \quad \langle a, y \rangle.$$

Note that $f|_{S^{n-1}}$ attains its maximum at $z \in S^{n-1}$ if, and only if,

$$z = \frac{(a_1, ..., a_n)}{\left(\sum_{i=1}^{n} a_i^2\right)^{\frac{1}{2}}}.$$

In fact, note that $f(S^{n-1})$ attains its maximum because its domain is a compact. Also, S^{n-1} can be written as the set of points x such that $\varphi(x) = 1$, where

$$\varphi : \quad \mathbb{R}^n \quad \to \quad \mathbb{R}$$
$$x \quad \mapsto \quad \|x\|_2^2.$$

Using the Lagrange Multipliers theorem $p \in S^{n-1}$ is a critical point of $f|_{S^{n-1}}$ if, and only if, there is a $\mu \in \mathbb{R}$ such that the gradients of f and φ behave as

$$\operatorname{grad} f(p) = \mu \cdot \operatorname{grad} \varphi(p).$$

Since $\operatorname{grad} f = a$ and $\operatorname{grad} \varphi = 2x$, making $\lambda = 2\mu$, it follows that

$$a = \lambda x.$$

Since $\|x\|_2^2 = 1$ and

$$\|a\|_2^2 = \|\lambda x\|^2$$

we conclude that

$$\lambda = \pm \|a\|_2.$$

We thus conclude that the critical points of $f|_{S^{n-1}}$ are $x = \pm \frac{a}{\|a\|_2}$ and it is not difficult to see that $z = \frac{a}{\|a\|_2}$ is a point of maximum of $f|_{S^{n-1}}$.

Now, consider the compositions

$$\pi_i \circ G : D \to \mathbb{R}$$

for $i = 1, ..., n$. It is simple to see that the set

$$\{\pi_i \circ G : i = 1, ..., n\}$$

is linearly independent. Let $h = \sum_{i=1}^{n} b_i(\pi_i \circ G)$ be a nontrivial linear combination of the functions $\pi_i \circ G$. Since $\sum_{i=1}^{n} b_i \pi_i$ attains its maximum in one unique point $x_0 \in S^{n-1}$ and since G is a bijection, it follows that h attains its maximum in one unique point, $G^{-1}(x_0)$. Hence, the subspace span$\{\pi_i \circ G\}, i = 1, ..., n$, provides the n-lineability of $\widehat{C}(D)$. $\qquad\square$

Functions on spaces including $\mathbb{R}$

The next theorem has a somewhat complicated statement but it can be easily seen as an extension of (B) above for higher dimensions (see Remark 1.7.12).

Theorem 1.7.11. *Let $n \geq 2$ be a positive integer and D be a topological space containing a closed set Y such that there is a continuous bijection $F : Y \to S^{n-1}$ and a continuous extension of F, denoted by $G : D \to \mathbb{R}^n$, such that $\|G(x)\|_2 < 1$ for all $x \notin Y$. Then $\widehat{C}(D)$ is n-lineable.*

Proof. Let π_i be the projection over the i-th coordinate of $\mathbb{R}^n$. Note that $S^{n-1} \subset G(D)$, since

$$S^{n-1} = F(Y) = G(Y) \subset G(D).$$

Besides, $G(D) \subset \{x \in \mathbb{R}^n : \|x\|_2 \leq 1\}$. Let us see that any nontrivial combination

$$f := \sum_{i=1}^{n} b_i \pi_i : G(D) \to \mathbb{R}$$

attains its maximum in only one point $x_0 \in G(D)$ and that this point belongs to S^{n-1}. Restricting $\sum_{i=1}^{n} b_i \pi_i$ to $S^{n-1} \subset G(D)$, the same argument used in the previous proof shows that there is a unique $x_0 \in S^{n-1}$ such that $f|_{S^{n-1}}$ attains a maximum, i.e.,

$$\sum_{i=1}^{n} b_i \pi_i(x) \leq \sum_{i=1}^{n} b_i \pi_i(x_0), \qquad (1.1)$$

for all $x \in S^{n-1}$. Since $x \in S^{n-1}$, we have $-x \in S^{n-1}$, and thus

$$\sum_{i=1}^{n} b_i \pi_i(-x) \leq \sum_{i=1}^{n} b_i \pi_i(x_0).$$

Hence

$$-\sum_{i=1}^{n} b_i \pi_i(x) \leq \sum_{i=1}^{n} b_i \pi_i(x_0). \qquad (1.2)$$

From (1.1) and (1.2) we observe that

$$\left|\sum_{i=1}^{n} b_i \pi_i(x)\right| \le \sum_{i=1}^{n} b_i \pi_i(x_0) \tag{1.3}$$

for all $x \in S^{n-1}$. Now, we just need to prove that if $y \in G(D) \setminus S^{n-1}$, then

$$\sum_{i=1}^{n} b_i \pi_i(y) < \sum_{i=1}^{n} b_i \pi_i(x_0).$$

The case $y = 0$ is trivial, and thus we consider $y \ne 0$. Since $\|y\|_2 < 1$, using (1.3), we have

$$\sum_{i=1}^{n} b_i \pi_i(y) \le \left|\sum_{i=1}^{n} b_i \pi_i(y)\right|$$

$$= \|y\|_2 \left|\sum_{i=1}^{n} b_i \pi_i\left(\frac{y}{\|y\|_2}\right)\right|$$

$$< \left|\sum_{i=1}^{n} b_i \pi_i\left(\frac{y}{\|y\|_2}\right)\right|$$

$$\le \sum_{i=1}^{n} b_i \pi_i(x_0).$$

Up to now we have proved that any nontrivial linear combination

$$f = \sum_{i=1}^{n} b_i \pi_i : G(D) \to \mathbb{R}$$

attains its maximum in one unique $x_0 \in G(D)$ and that this point belongs to S^{n-1}. Now, we claim that any nontrivial linear combination

$$h := \sum_{i=1}^{n} b_i (\pi_i \circ G) : D \to \mathbb{R}$$

attains its maximum only in $F^{-1}(x_0) \in D$. In fact, observe that h attains its maximum in $z \in D$ if, and only if, f attains its maximum in $G(z)$. Then, the only point of maximum of h is $F^{-1}(x_0)$. Hence, the subspace span$\{\pi_i \circ G, ..., \pi_n \circ G\}$ completes the proof. □

Remark 1.7.12. *Let us see how we recover the 2-lineability of $\widehat{C}(\mathbb{R})$ in $C(\mathbb{R})$. For $a, b \in \mathbb{R}^2$, let us denote by $|a, b|$ the open line segment in $\mathbb{R}^2$ from a to b. Let us consider $Y = [0, \infty) \subset \mathbb{R}$, and a continuous bijection from $[0, \infty)$ to*

S^1, *for instance,*

$$F: \quad [0, \infty) \quad \to \quad S^1$$

$$t \quad \mapsto \quad F(t) = \left(\cos\left(2\pi - \frac{1}{t + \frac{1}{2\pi}} \right), \quad \sin\left(2\pi - \frac{1}{t + \frac{1}{2\pi}} \right) \right)$$

and certain homeomorphism

$$g : (-\infty, 0) \to |F(0), (0, 0)| \subset \mathbb{R}^2$$

such that

$$\lim_{x \to 0} g(x) = F(0).$$

Now we consider the following continuous extension of F:

$$G: \quad \mathbb{R} \quad \to \quad \mathbb{R}^2$$

$$x \quad \mapsto \quad G(x) = \begin{cases} F(x) & if \quad x \geq 0, \\ g(x) & if \quad x < 0. \end{cases}$$

Note that $\|G(x)\|_2 < 1$ for all $x \notin Y$. In fact, if $x \notin Y$ then $x \in (-\infty, 0)$, and thus

$$G(x) = g(x) \in |F(0), (0, 0)| \subset \mathbb{R}^2.$$

Since $F(0) = (1, 0)$, we have $\|G(x)\|_2 < 1$ and we conclude that $\widehat{C}(\mathbb{R})$ is 2-lineable.

Functions on compact subsets of Euclidean spaces

From now on we replace the compact $[0, 1]$ by an arbitrary compact K from $\mathbb{R}^m$. The main result of this section generalizes the results from Section 1.7 showing that in general the set $\widehat{C}(K)$ is not $(m + 1)$-lineable.

Lemma 1.7.13. *Let $n \in \mathbb{N}$ and D be a metric space so that there is a vector space V composed by continuous functions from D to $\mathbb{R}$ satisfying*

(1) $\dim(V) = n, \quad n \in \mathbb{N}.$

(2) *Each $0 \neq f \in V$ has only one point of maximum.*

Let $D' \subset D$ be the set of points of maximum of the non zero functions belonging to V. Then the vector space

$$V' := \{f|_{D'} : f \in V\}$$

also satisfies (1) and (2).

Proof. Let us first show that the elements of V' satisfy (2). In fact, note that each non null function $f|_{D'} \in V'$ has only one maximum because $f \in V$ has only one maximum. Now, let us show (1), i.e, $\dim(V') = n$. Let $\{f_1, ..., f_n\}$ be a basis of V. If $a_1, ..., a_n \in \mathbb{R}$ are such that

$$\sum_{i=1}^{n} a_i f_i|_{D'} = 0, \tag{1.4}$$

then

$$-\sum_{i=1}^{n} a_i f_i|_{D'} = 0.$$

Suppose that

$$g := \sum_{i=1}^{n} a_i f_i \neq 0.$$

Since the points of maximum of g and $-g$ belong to D', the image points are 0, because

$$g|_{D'} = -g|_{D'} = 0.$$

Hence $|g| \, |_{D'} = 0$, and we conclude that $g = 0$. Since $\{f_1, ..., f_n\}$ is linearly independent, it follows that $a_1 = \cdots = a_n = 0$ and thus $\{f_1|_{D'}, ..., f_n|_{D'}\}$ is linearly independent. To complete the proof, just note that $\dim(V') \leq \dim(V)$, since

$$\begin{array}{rcl} \varphi: \quad V & \to & V' \\ f & \mapsto & f|_{D'} \end{array}$$

is surjective. $\square$

Lemma 1.7.14. *Keeping the terminology and notation of Lemma 1.7.13 and its proof, consider the continuous function*

$$\begin{array}{rcl} F: \quad D & \to & \mathbb{R}^n \\ y & \mapsto & (f_1(y), ..., f_n(y)). \end{array}$$

Let $X := F(D')$. Then

(1) *For each $v \in S^{n-1}$, the function*

$$\begin{array}{rcl} g_v: \quad X & \to & \mathbb{R} \\ x & \mapsto & \langle x, v \rangle \end{array}$$

has a unique point of maximum.

(2) *For each $x \in X$ there is a $v \in S^{n-1}$ such that x is the unique point of maximum of g_v.*

(3) *If we endow D' with the metric induced by the metric of D and X with the Euclidean metric from $\mathbb{R}^n$, then*

$$F|_{D'} : D' \to X$$

is a continuous bijection.

Proof. (1) Given $v = (a_1, ..., a_n) \in S^{n-1}$, consider the function

$$g_v \circ F : D' \to \mathbb{R}.$$

For each $y \in D'$,

$$g_v \circ F(y) = \langle F(y), v \rangle = \sum_{i=1}^{n} a_i f_i|_{D'}(y).$$

Thus $g_v \circ F \in V'$. From Lemma 1.7.13 we know that $g_v \circ F : D' \to \mathbb{R}$ has only one maximum, say $d \in D'$. Therefore

$$\langle F(y), v \rangle \leq \langle F(d), v \rangle,$$

for all $y \in D'$, and the equality holds only when $y = d$. Then

$$\langle x, v \rangle \leq \langle F(d), v \rangle$$

for all $x \in X$. We conclude that $F(d) \in X$ is a point of maximum of g_v. Let us show that this point is unique. Suppose that g_v has another point of maximum $x' \in X$. So $x' = F(d')$ for some $d' \neq d$ and the function

$$g_v \circ F : D' \to \mathbb{R}$$

has two points of maxima, a contradiction.

(2) For each $x \in X$ there is a $d \in D'$ such that $x = F(d)$. But d is a point of maximum of some non null function

$$h := \sum_{i=1}^{n} a_i f_i|_{D'} : D' \to \mathbb{R}.$$

Without loss of generality, let us suppose that

$$\|(a_1, ..., a_n)\|_2 = 1.$$

In fact, if d is a point of maximum of h, where $(a_1, ..., a_n)$ is non necessarily unitary, then d is also a point of maximum of $\dfrac{h}{\|(a_1, ..., a_n)\|_2}$.

Taking $v := (a_1, ..., a_n) \in S^{n-1}$, we have

$$\sum_{i=1}^{n} a_i f_i|_{D'}(y) = \langle F(y), v \rangle \leq \langle F(d), v \rangle = \langle x, v \rangle$$

for all $y \in D'$. Then $\langle z, v \rangle \leq \langle x, v \rangle$ for all $z \in X$. From (1) it follows that x is the only point of maximum of g_v.

(3) Since each coordinate $f_i|_{D'}$ is continuous and $X = F(D')$, we just need to prove that $F|_{D'}$ is injective. Suppose, by contradiction, that this is not the case. So, if $d'_1, d'_2 \in D'$ and $d'_1 \neq d'_2$ are such that $F(d'_1) = F(d'_2)$, then

$$\begin{aligned} g_v \circ F(d'_1) &= \langle v, F(d'_1) \rangle \\ &= \langle v, F(d'_2) \rangle \\ &= g_v \circ F(d'_2) \end{aligned}$$

for all $v \in \mathbb{R}^n$. We thus have that neither d'_1 nor d'_2 can be a point of maximum of $g_v \circ F : D' \to \mathbb{R}$, for any v, because otherwise we should have $d'_1 = d'_2$. But this contradicts the fact that D' is the set of points of maxima of the functions from V. $\qquad \square$

Definition 1.7.15. Let $n \geq 2$ and X be a subset of $\mathbb{R}^n$ with $card\,(X) > 1$ satisfying the following conditions:

(1) For all $v \in S^{n-1}$, the function

$$\begin{aligned} g_v : \quad X &\to \quad \mathbb{R} \\ x &\mapsto \quad g_v(x) = \langle x, v \rangle \end{aligned}$$

has only one point of maximum, say x_v.

(2) For all $x \in X$ there is a $v_x \in S^{n-1}$ such that x is the only point of maximum of g_{v_x}.

Remark 1.7.16. *Define the function*

$$\begin{aligned} f : \quad S^{n-1} &\to \quad X \\ v &\mapsto \quad f(v) = x_v, \end{aligned}$$

where x_v is as in Definition 1.7.15, (1). From Definition 1.7.15, (2) note that f is surjective. The functions f will be employed in what follows.

Lemma 1.7.17. *Let X and f be as in the above remark. Let $K \subset X$ be a compact subset of $\mathbb{R}^n$. Then $f^{-1}(K) \subset S^{n-1}$ is a compact subset of $\mathbb{R}^n$.*

Proof. As $f^{-1}(K) \subset S^{n-1} \subset \mathbb{R}^n$, it suffices to show that $f^{-1}(K)$ is closed. Let $(v_n)_{n=1}^{\infty}$ be a sequence in $f^{-1}(K)$. As $f^{-1}(K) \subset S^{n-1}$ and S^{n-1} is compact, there is a convergent subsequence

$$v_{n_j} \longrightarrow v \in S^{n-1}.$$

For all j, let $x_j = f(v_{n_j}) \in K$. Since K is compact, there is a subsequence $(x_{j_k})_{k=1}^{\infty}$ such that $x_{j_k} \longrightarrow x \in K$. Now, since x_{j_k} is the only point of maximum of the function $g_{v_{n_{j_k}}}$ in X (because $x_{k_k} = f(v_{n_{j_k}})$), we have

$$\langle f(v), v_{n_{j_k}} \rangle \leq \langle x_{j_k}, v_{n_{j_k}} \rangle.$$

Letting $k \longrightarrow \infty$, we obtain

$$g_v(f(v)) = \langle f(v), v \rangle \leq \langle x, v \rangle = g_v(x).$$

But $f(v) = x_v$ is the only maximum of g_v, and thus $f(v) = x$ and $v \in f^{-1}(K)$. So $f^{-1}(K) \subset S^{n-1}$ is closed and therefore compact. □

Lemma 1.7.18. *The function f from Remark 1.7.16 is continuous if and only if X is compact.*

Proof. Suppose that $f : S^{n-1} \to X$ is continuous. Since f is surjective and S^{n-1} is compact, it follows that $X = f(S^{n-1})$ is compact. Reciprocally, suppose that X is compact. Let B be a closed subset of X. Since X is a metric space, it follows that B is also compact and compact in $\mathbb{R}^n$. From Lemma 1.7.17 we know that $f^{-1}(B)$ is a compact subset of S^{n-1}, in particular closed. Thus f is continuous. □

Finally, let us state and prove the following main result (Theorem 1.7.20). Here we need to recall the famous and deep theorem by Borsuk and Ulam, which can be established as follows.

Theorem 1.7.19. *Let $h : S^n \to \mathbb{R}^n$ be a continuous function. Then there exists $v \in S^n$ such that $h(v) = h(-v)$.*

Theorem 1.7.20. *Let $n \geq 2$ and $m \geq 1$ be positive integers. Then $m < n$ if, and only if, for all compact sets $K \subset \mathbb{R}^m$, $\widehat{C}(K)$ is not n-lineable.*

Proof. Let us suppose that $m < n$ and that there exist a compact $K \subset \mathbb{R}^m$ and a set $V \subset \widehat{C}(K)$, so that $V \cup \{0\}$ is an n-dimensional subspace of $C(K)$, i.e., $\widehat{C}(K)$ is n-lineable. Here, the compact K will be the metric space D of Lemmas 1.7.13 and 1.7.14. Let $\{f_1, ..., f_n\}$ be a basis of $V \cup \{0\}$ and define

$$
\begin{aligned}
F : \quad D &\to \quad \mathbb{R}^n \\
y &\mapsto \quad F(y) = (f_1(y), ..., f_n(y))
\end{aligned}
$$

as in Lemma 1.7.14. As before, let $D' \subset D$ be the set of the points of maximum of the functions belonging to V and let $X := F(D')$. Since $n \geq 2$, it follows that

$$\dim(V \cup \{0\}) \geq 2.$$

Thus V contains a non constant function. If $g \in V$ is a non constant function, it is plain that g and $-g$ possess different points of maximum and thus $\mathrm{card}\,(D') > 1$. From Lemma 1.7.14 (3), the restriction

$$F|_{D'} : D' \to X$$

is bijective, and thus $\mathrm{card}\,(X) > 1$ (this fact will be crucial later on this proof). So, consider the function

$$f : S^{n-1} \to X$$

from Remark 1.7.16. Let us prove that D' is closed in D and, therefore, compact in $\mathbb{R}^m$. Let $(d_k)_{k=1}^\infty$ be a sequence in D' converging to d. From the compactness of D it follows that $d \in D$. From Remark 1.7.16, for all k,

$$F(d_k) = f(v_k)$$

for some $v_k \in S^{n-1}$, and this means that $F(d_k)$ is the only point of maximum in X of the function

$$
\begin{aligned}
g_{v_k} : \quad X &\to \quad \mathbb{R} \\
x &\mapsto \quad g_{v_k}(x) = \langle x, v_k \rangle.
\end{aligned}
$$

From the compactness of S^{n-1}, the sequence $(v_k)_{k=1}^\infty$ possesses a convergent subsequence

$$v_{k_j} \longrightarrow v \in S^{n-1}.$$

Note that

$$\langle F(d_{k_j}), v_{k_j} \rangle \geq \langle F(y), v_{k_j} \rangle$$

for all $y \in D'$. Since the points $y \in D'$ are points of maximum of the functions $g_v \circ F : D \to \mathbb{R}$, the inequality

$$\langle F(d_{k_j}), v_{k_j} \rangle \geq \langle F(z), v_{k_j} \rangle$$

holds for all $z \in D$. Using the continuity of $F : D \to \mathbb{R}^n$ we have

$$
\begin{aligned}
\langle F(d), v \rangle &= \lim_{j \to \infty} \langle F(d_{k_j}), v_{k_j} \rangle \\
&\geq \lim_{j \to \infty} \langle F(z), v_{k_j} \rangle \\
&= \langle F(z), v \rangle
\end{aligned}
$$

for all $z \in D$. Thus d is a point of maximum of the function $g_v \circ F : D \to \mathbb{R}$, and therefore $d \in D'$, and we conclude that D' is closed in D. From Lemma 1.7.14 item (3), we know that the function $F : D' \to X$ is a continuous bijection between the compact D' and X (Hausdoff) and thus F is a homeomorphism (see, for instance, [223, Lemma 5]. As D' is compact, it follows that $X = F(D')$ is compact and from Lemma 1.7.18 the function

$$f : S^{n-1} \to X$$

is continuous. Considering $\mathbb{R}^m$ suitably contained in $\mathbb{R}^{n-1}$, because $m < n$, the function

$$F^{-1} \circ f : S^{n-1} \to D' \subset \mathbb{R}^m \subset \mathbb{R}^{n-1}$$

is continuous. From the Borsuk-Ulam Theorem, there is a pair of antipodal points $v, -v \in S^{n-1}$ such that

$$F^{-1} \circ f(v) = F^{-1} \circ f(-v).$$

From the injectivity of F^{-1} we have $f(v) = f(-v) =: x$. As $f : S^{n-1} \to X$ is defined by $f(v) = x_v$, where x_v is the unique point of maximum of $g_v : X \to \mathbb{R}$ (recall that $g_v(x) = \langle x, v \rangle$), it follows that both g_v and g_{-v} have x as a point of maximum. On the other hand, $g_{-v} = -g_v$ because

$$\begin{aligned} g_{-v}(z) &= \langle z, -v \rangle \\ &= -\langle z, v \rangle \\ &= -g_v(z) \end{aligned}$$

for all $z \in X$. We thus conclude that $-g_v$ and g_v attain the maximum at the same point and therefore g_v is constant, a contradiction, because $card\,(X) > 1$ and g_v attains the maximum at only one point in X.

For the converse, suppose that $m \geq n$ and take the compact $K = S^{n-1} \subset \mathbb{R}^n \subset \mathbb{R}^m$. Choosing, for instance, the identity map from $K = S^{n-1}$ to itself, it follows from Theorem 1.7.11 that $\widehat{C}(K)$ is n-lineable. □

Corollary 1.7.21. *The set $\widehat{C}[a,b]$ is not 2-lineable.*

The following example illustrates what may happen when we replace $\mathbb{R}^m$ by an infinite dimensional space in Theorem 1.7.20.

Below, the symbol ℓ_2 denotes the Banach space of all square summable sequences of real numbers. It is a Banach space when endowed with the norm

$$\left\| (a_j)_{j=1}^\infty \right\|_2 = \left(\sum_{j=1}^\infty |a_j|^2 \right)^{\frac{1}{2}}.$$

More details will be given in Chapter 3, which is devoted to Measure Theory and Integration. For the next example we need to recall the definition of the Hilbert cube. The Hilbert cube is the subset of ℓ_2 defined by $\prod_{n=1}^\infty \left[-\frac{1}{n}, \frac{1}{n} \right]$. It is a well known result of Functional Analysis that the Hilbert cube is a compact subset of ℓ_2.

Example 1.7.22. *Consider the set*

$$D = \left\{ \left(\frac{a_n}{n} \right)_{n=1}^\infty : (a_n)_{n=1}^\infty \in \ell_2 \ and \ \| (a_n)_{n=1}^\infty \|_2 \leq 1 \right\}.$$

Note that D is contained in the Hilbert cube. From the compactness of the Hilbert cube we can infer that D is compact if we prove that it is closed. Let $(v_j)_{j=1}^\infty = \left(\left(\frac{v_n^j}{n} \right)_{n=1}^\infty \right)_{j=1}^\infty$ be a sequence in D converging to $w = (w_n)_{n=1}^\infty \in \ell_2$. We thus have coordinatewise convergence, i.e., $w_n = \lim\limits_j \frac{v_n^j}{n}$, so $n w_n = \lim\limits_j v_n^j$, for every fixed n. Moreover,

$$\sum_{n=1}^k n^2 |w_n|^2 = \sum_{n=1}^k \lim_j |v_n^j|^2 = \lim_j \sum_{n=1}^k |v_n^j|^2 \leq \limsup_j \| (v_n^j)_{n=1}^\infty \|_2^2 \leq 1$$

for every k. So we conclude that $\|(nw_n)_{n=1}^{\infty}\|_2 \leq 1$, and thus $w \in D$. Therefore D is a compact subset of ℓ_2.

Let us prove that $\widehat{C}(D) \cup \{0\}$ contains an infinite dimensional subspace of $\mathcal{C}(D)$. Define the function

$$F \colon D \longrightarrow \ell_2 \ , \ F\left(\left(\frac{a_n}{n}\right)_{n=1}^{\infty}\right) = (a_n)_{n=1}^{\infty}$$

and let $\pi_j \colon \ell_2 \longrightarrow \mathbb{R}$ denote the projection onto the j-th coordinate, $j \in \mathbb{N}$. For each j, the function

$$\pi_j \circ F \colon D \longrightarrow \mathbb{R}$$

is continuous since $\pi_j \circ F = j \cdot \pi_j$. Note that it is plain that the functions $\pi_j \circ F, j \in \mathbb{N}$, are linearly independent. Finally, if we consider the non-trivial linear combination $f := \sum_{j=1}^{k} b_j(\pi_j \circ F)$ and define $b := (b_1, \ldots, b_k, 0, 0, \ldots) \in \ell_2$, then we have

$$f(x) = \langle b, F(x) \rangle$$

for every $x \in D$. Since $b \in \ell_2^k$ and $\|F(x)\|_2 \leq 1$ for every $x \in D$, we have

$$f(x) = \langle b, F(x) \rangle < \left\langle b, \frac{b}{\|b\|_2} \right\rangle$$

whenever $F(x) \neq \frac{b}{\|b\|_2}$ (here we use the same argument of the proof of Theorem 1.7.10). Since F is a bijection onto the closed unit ball of ℓ_2, there is a unique $y \in D$ such that $F(y) = \frac{b}{\|b\|_2}$. This shows that f attains its maximum at y. We can verify that this maximum is unique, by making an adaptation of the argument used in the proof of Theorem 1.7.11.

1.8   Peano maps and space-filling curves

We denote by I the closed unit interval $[0, 1]$. In Section 1.5 we saw how large –in the algebraic sense– the set of surjections $\mathbb{R} \to \mathbb{R}$ can be. In fact, several stronger ways of surjectivity were considered there. Nevertheless, each of these surjections is nowhere continuous. Of course, by using a bijection $\mathbb{R} \to I^2$ or $\mathbb{R} \to \mathbb{R}^2$, surjections $\mathbb{R} \to I^2$ or $\mathbb{R} \to \mathbb{R}^2$ (or even $I \to I^2 := I \times I$) can be constructed, but their continuity is far from being guaranteed. In 1890 G. Peano [331] surprised the mathematical world by constructing a curve that filled in a space, that is, a *surjective continuous mapping* $f : I \to I^2$ (see Figure 1.8, where the construction of a continuous curve filling I^3 is sketched). From this it is not difficult to build a *surjective continuous function* $\mathbb{R} \to \mathbb{R}^2$ as an extension of f: simply divide $\mathbb{R}^2$ into countably many non-overlapping squares $a_n + I^2$ (translations of I^2) and map continuously the interval $[0, 1]$

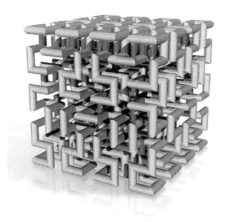

FIGURE 1.8: Sketch of an iteration of a space-filling curve.

on $a_1 + I$, the interval $[2, 3]$ on $a_2 + I^2$, $[4, 5]$ on $a_3 + I^2$ ... and then use the intermediate intervals $[1, 2]$, $[3, 4]$, ... to connect (by segments, for instance) $a_1 + I^2$ to $a_2 + I^2$, $a_2 + I^2$ to $a_3 + I^2$, and so on.

In 2014 Albuquerque [8] undertook the analysis of the lineability properties of this class of functions. In fact, he adopted a more general point of view by studying the class $CS(\mathbb{R}^m, \mathbb{R}^n)$ of continuous surjections $\mathbb{R}^m \to \mathbb{R}^n$, with $m, n \in \mathbb{N}$. In general, we denote by $CS(X, Y)$ the set of all continuous surjections $X \to Y$, where X and Y are topological spaces. Following the approach of [8] we have that, given $m, n \in \mathbb{N}$, if we have a function $f \in CS(\mathbb{R}, \mathbb{R}^n)$ at our disposal then the function $F : (x_1, \dots, x_m) \in \mathbb{R}^m \mapsto f(x_1) \in \mathbb{R}^n$ is, trivially, continuous and surjective. Moreover, if $g \in CS(\mathbb{R}, \mathbb{R}^2)$, so that $g(s) = (g_1(s), g_2(s))$, then the mapping $G_0 : (t, s) \in \mathbb{R}^2 \mapsto (t, g_1(s), g_2(s)) \in \mathbb{R}^3$ is continuous and surjective. Therefore the composite function $G = G_0 \circ f$ belongs to $CS(\mathbb{R}, \mathbb{R}^3)$. Proceeding in an induction manner, we can assure the existence of functions in $CS(\mathbb{R}, \mathbb{R}^n)$ for all $n \geq 1$ (the identity $x \mapsto x$ shows that the case $n = 1$ is trivial). Consequently, we will mainly consider the case $m = 1$, $n = 2$ in our study of the lineability properties of continuous surjections. The extension of these properties to the remaining cases is immediate by using the given constructions or simply by mimicking the proofs.

In [8] it is shown that the set $CS(\mathbb{R}, \mathbb{R}^2)$ is $\mathfrak{c}$-lineable. Observe that, since $\dim(CS(\mathbb{R}, \mathbb{R}^2)) = \mathfrak{c}$, this result is optimal in terms of dimension. The research initiated by Albuquerque has been continued on by Albuquerque, Bernal, Pellegrino and Seoane in [9] and by Bernal and Ordóñez in [104], where the mentioned result of $\mathfrak{c}$-lineability is reinforced and complemented. In Theorems 1.8.1, 1.8.3 and 1.8.6 below, the main assertions of [9] and [104] are collected, respectively. In the vector space $\mathcal{C}(\mathbb{R}, \mathbb{R}^2)$ of continuous functions $\mathbb{R} \to \mathbb{R}^2$, we will consider, as usual, the topology of uniform convergence on compacta, which is generated by the distance $\rho(f,g) = \sum_{n=1}^{\infty} 2^{-n} \frac{\max_{[-n,n]} |f-g|}{1+\max_{[-n,n]} |f-g|}$. Under this topology, $\mathcal{C}(\mathbb{R}, \mathbb{R}^2)$ becomes a complete metrizable separable topological vector space.

Theorem 1.8.1. *The set $CS(\mathbb{R}, \mathbb{R}^2)$ is dense-lineable and spaceable in $\mathcal{C}(\mathbb{R}, \mathbb{R}^2)$.*

Proof. In order to prove the dense-lineability, a good dense subset of $\mathcal{C}(\mathbb{R}, \mathbb{R}^2)$ is needed. It is well known that the set $\mathcal{P}_0$ of functions $P = (P_1, P_2) : \mathbb{R} \to \mathbb{R}^2$ whose components P_1, P_2 are polynomials is dense in $\mathcal{C}(\mathbb{R}, \mathbb{R}^2)$. Since this space is separable, there is a countable subfamily $\mathcal{P}$ of $\mathcal{P}_0$ which is still dense in $\mathcal{C}(\mathbb{R}, \mathbb{R}^2)$. By a direct construction we can easily obtain, for each $P = (P_1, P_2) \in \mathcal{P}$ and each $k \in \mathbb{N}$, a pair of continuous functions $P_{1,k}, P_{2,k} : \mathbb{R} \to \mathbb{R}$ such that $P_{j,k} = P_j$ on $[-k,k]$ and $P_{j,k}(x) = 0$ if $|x| > k+1$ $(j = 1, 2)$. Denote $P_k = (P_{1,k}, P_{2,k})$. Since each compact set $K \subset \mathbb{R}$ is contained in some $[-k,k]$ and the topology of $\mathcal{C}(\mathbb{R}, \mathbb{R}^2)$ is that of uniform converge on compacta, we have that the (countable) set $\{P_k : P \in \mathcal{P}, k \in \mathbb{N}\}$ is dense in $\mathcal{C}(\mathbb{R}, \mathbb{R}^2)$. Let $\{Q_n\}_{n \geq 1}$ be an enumeration of this set.

Now, let us construct a special family of functions in $CS(\mathbb{R}, \mathbb{R}^2)$. For each $a \in \mathbb{R}$ and each $n \in \mathbb{N}$, denote $I_a := [a, a+1]$ and $J_n := [-n,n] \times [-n,n]$. Divide $\mathbb{N}$ into infinitely many strictly increasing sequences $\{m(n,k)\}_{k \geq 1}$ ($n = 1, 2, \dots$). By using translations, dilations, segments in $\mathbb{R}^2$ and Peano's original curve, it is an easy exercise to produce, for every pair $(n,k) \in \mathbb{N} \times \mathbb{N}$, a mapping

$$\varphi_{n,k} \in CS(I_{m(n,k)}, J_k) \quad \text{such that} \quad \varphi_{n,k}(m(n,k)) = (0,0) = \varphi_{n,k}(m(n,k)+1).$$

Then, for each n, the function $\varphi_n : \mathbb{R} \to \mathbb{R}^2$ given by $\varphi_n(x) = \varphi_{n,k}(x)$ if $x \in I_{m(n,k)}$, and $\varphi(x) = (0,0)$ otherwise is well defined and continuous. Moreover, it is surjective because $\varphi_n(\mathbb{R}) \supset \bigcup_{k \geq 1} J_k = \mathbb{R}^2$. In addition, since (J_k) is an increasing exhaustive sequence, the function φ_n satisfies that $\varphi_n(\mathbb{R} \backslash K) = \mathbb{R}^2$ for every compact subset $K \subset \mathbb{R}$. Of course, the function $\alpha \varphi$ also satisfies the same property, for all $\alpha \in \mathbb{R} \backslash \{0\}$. Moreover, the functions φ_n's are linearly independent because they are disjointly supported. By the same reason, any nonzero linear combination $f = c_1 \varphi_1 + \cdots + c_N \varphi_N$ satisfies $f(\mathbb{R} \backslash K) = \mathbb{R}^2$ too, for every compact set $K \subset \mathbb{R}$.

The continuity at the origin of the multiplication by scalars in $CS(\mathbb{R}, \mathbb{R}^2)$ yields, for each $n \in \mathbb{N}$, the existence of an $\varepsilon_n > 0$ such that $\rho(\varepsilon_n, \varphi_n) < 1/n$. But ρ is translation-invariant, so $\rho(Q_n, Q_n + \varepsilon_n \varphi_n) < 1/n \to 0$ as $n \to \infty$. Since (Q_n) is dense in $\mathcal{C}(\mathbb{R}, \mathbb{R}^2)$, the sequence $(Q_n + \varepsilon_n \varphi_n)$ is also dense. Next, we define the sets

$$M_1 := \operatorname{span}\{Q_n + \varepsilon_n \varphi_n : n \geq 1\}$$

and

$$M_2 = \left\{ \sum_{n=1}^{\infty} c_n \varphi_n : c_n \in \mathbb{R} \text{ for all } n \in \mathbb{N} \right\}.$$

Observe that M_2 is well defined because for every $x_0 \in \mathbb{R}$ there is at most one n such that $\varphi_n(x_0) \neq 0$. Evidently, M_1 and M_2 are vector subspaces of $\mathcal{C}(\mathbb{R}, \mathbb{R}^2)$, and M_1 is dense because $M_1 \supset \{Q_n + \varepsilon_n \varphi_n : n \geq 1\}$. Let us prove that M_2 is closed. To this end, assume that a sequence $\{f_j = \sum_{n=1}^{\infty} c_{j,n} \varphi_n\}_{j \geq 1}$ tends to some $f \in \mathcal{C}(\mathbb{R}, \mathbb{R}^2)$. Let $x_0 \in \mathbb{R}$. Since convergence in this space entails pointwise convergence, we have that $f(x_0) = (0,0)$ if $x_0 \in (\mathbb{R} \setminus \bigcup_{n,k} I_{m(n,k)}) \cup \mathbb{N}$. Otherwise, there is a unique pair (m, n) with $x_0 \in I_{m(n,k)}$. Then $f_j(x_0) = c_{j,n} \varphi_n(x_0) \to f(x_0)$ as $j \to \infty$. Taking any $x_0 \in I_{m(n,k)}$ with $\varphi_n(x_0) \neq (0,0)$, we obtain that $\{c_{j,n}\}_{j \geq 1}$ tends to some constant c_n. But $c_{j,n} \varphi_n(x_0) \to c_n \varphi_n(x_0)$ as $j \to \infty$, whence $f = c_n \varphi_n$ on $I_{m(n,k)}$ for every $m(n,k)$. Since all φ_n's vanish at the ends of the non-overlapping intervals $I_{m(n,k)}$, we get $f = \sum_{n=1}^{\infty} c_n \varphi_n \in M_2$, which shows that M_2 is closed.

Our final task is to prove that $M_j \setminus \{0\} \subset CS(\mathbb{R}, \mathbb{R}^2)$ for $j = 1, 2$. Fix $f \in M_1 \setminus \{0\}$. Then there are $N \in \mathbb{N}$ and scalars $c_1, \dots, c_N$ with $c_N \neq 0$ such that $f = \sum_{n=1}^{N} c_n (Q_n + \varepsilon_n \varphi_n)$. By the construction of the Q_n's, we can find a compact set $K \subset \mathbb{R}$ such that $Q_n(x) = 0$ for all $x \notin K$ and all $n = 1, \dots, N$. Hence $f = \sum_{n=1}^{N} c_n \varepsilon_n \varphi_n$ on $\mathbb{R} \setminus K$. As observed before, $f(\mathbb{R} \setminus K) = \mathbb{R}^2$. Thus, $f \in CS(\mathbb{R}, \mathbb{R}^2)$. Now, consider a function $f = \sum_{n=1}^{\infty} c_n \varphi_n \in M_2 \setminus \{0\}$. Then there exists $N \in \mathbb{N}$ with $c_N \neq 0$. Since all φ_n's vanish on all intervals $I_{m(N,k)}$ if $n \neq N$, we get

$$f(\mathbb{R}) \supset f\left(\bigcup_{k \geq 1} I_{m(N,k)}\right) = c_N \varphi_N \left(\bigcup_{k \geq 1} I_{m(N,k)}\right) = c_N \cdot \bigcup_{k \geq 1} J_k = \mathbb{R}^2.$$

This shows the desired surjectivity of f, which concludes the proof. $\qquad\square$

A natural question would be to ask about the algebrability of the set $CS(\mathbb{R}, \mathbb{R}^2)$. Clearly, algebrability cannot be obtained in the real context –that is, if we endow the vector space $\mathbb{R}^2$ with the inner operation of coordinatewise multiplication– since for any $f \in \mathbb{R}^{\mathbb{R}}$, f^2 is always non-negative. However, in the complex frame it is actually possible to obtain *algebrability*.

With this aim, some preparatory results coming from classical complex analysis will be needed. The following definitions and properties can be found, for instance, in Ahlfors' book [2] and Boas' book [124]. Assume that f is an entire function, that is, an analytic function $\mathbb{C} \to \mathbb{C}$. Its *order* $\rho(f)$ is defined as the infimum of all positive real numbers α with the following property: $M(f, r) < e^{r^\alpha}$ for all $r > r(\alpha) > 0$. Note that $\rho(f) \in [0, +\infty]$. Then the order of a constant map is 0 and, if f is non-constant, we have $\rho(f) = \limsup_{r \to +\infty} \frac{\log \log M(f,r)}{\log r}$. If $\sum_{n=1}^{\infty} a_n z^n$ is the MacLaurin series expansion of f then

$$\rho(f) = \limsup_{n \to \infty} \frac{n \log n}{\log (1/|a_n|)}.$$

In particular, given $\alpha > 0$, the function $f_\alpha(z) := \sum_{n=1}^{\infty} \frac{z^n}{n^{n/\alpha}}$ satisfies $\rho(f_\alpha) =$

α. For all entire functions f and g, every $N \in \mathbb{N}$ and every $\alpha \in \mathbb{C} \setminus \{0\}$, one has $\rho(\alpha f^N) = \rho(f)$, $\rho(f \cdot g) \le \max\{\rho(f), \rho(g)\}$ and $\rho(f + g) \le \max\{\rho(f), \rho(g)\}$. Moreover, if f and g have different orders, then $\rho(f + g) = \max\{\rho(f), \rho(g)\} = \rho(f \cdot g)$, where it is assumed $f \not\equiv 0 \not\equiv g$ for the second equality. Finally, as a corollary of Hadamard's theorem, we have that every nonconstant entire function f with $\infty > \rho(f) \notin \mathbb{N}$ is surjective.

As a consequence of the previous properties, we obtain the following lemma concerning the order of a polynomial of several variables evaluated on entire functions with different orders. For a non-constant polynomial in M complex variables $P \in \mathbb{C}[z_1, \dots, z_M]$, let $\mathcal{I}_P \subset \{1, \dots, M\}$ denote the set of indexes k such that the variable z_k explicitly appears in some monomial (with non-zero coefficient) of P; that is, $\mathcal{I}_P = \{n \in \{1, \dots, M\} : \frac{\partial P}{\partial z_n} \not\equiv 0\}$.

Lemma 1.8.2. *Let $f_1, \dots, f_M$ be entire functions such that $\rho(f_i) \ne \rho(f_j)$ whenever $i \ne j$. Then*

$$\rho\left(P\left(f_1, \dots, f_M\right)\right) = \max_{k \in \mathcal{I}_P} \rho\left(f_k\right),$$

for all non-constant polynomials $P \in \mathbb{C}[z_1, \dots, z_M]$. In particular, the set $\{f_1, \dots, f_M\}$ is algebraically independent and generates a free algebra.

Proof. It is clear that we may assume, without loss of generality, that $M > 1$ and the entire functions $f_1, \dots, f_M$ satisfy $\rho(f_1) < \rho(f_2) < \cdots < \rho(f_M)$. Given a non-constant polynomial $P \in \mathbb{C}[z_1, \dots, z_M]$, the properties given before this lemma assure that

$$\rho\left(P\left(f_1, \dots, f_M\right)\right) \le \max_{k \in \mathcal{I}_P} \rho\left(f_k\right).$$

Therefore, we just need to prove that

$$\rho\left(P\left(f_1, \dots, f_M\right)\right) \ge \max_{k \in \mathcal{I}_P} \rho\left(f_k\right).$$

Let $N := \max_{k \in \mathcal{I}_P}$ (so that $\max_{k \in \mathcal{I}_P} \rho\left(f_k\right) = \rho(f_N) > 0$). We can write

$$P(f_1, \dots, f_M) = \sum_{i=0}^{m} P_i(f_1, \dots, f_{N-1}) \cdot f_N^i, \tag{1.5}$$

with some $m > 0$ and $P_m \in \mathbb{C}[z_1, \dots, z_{N-1}] \setminus \{0\}$. Let $\varepsilon > 0$ such that

$$\rho(f_{N-1}) < \rho(f_N) - 2\varepsilon < \rho(f_N) =: \rho_N.$$

Now, the remarks given before this lemma allow us to estimate the order of each one of terms of the sum in (1.5):

$$\rho\left(P_i(f_1, \dots, f_{N-1})\right) \le \rho(f_{N-1}) < \rho_N \quad \text{for all } i = 0, \dots, m \text{ and}$$

$$\rho\left(P_m(f_1, \dots, f_{N-1}) \cdot f_N\right) = \rho_N.$$

By the definition of order, there exist a sequence $(r_n)_{n\geq 1}$ of positive real numbers, going to ∞, and complex numbers z_n, of modulus r_n, such that, for n large enough, the following inequalities hold:

$$|P_m(f_1,\ldots,f_{N-1})(z_n)| \cdot |f_N(z_n)| > e^{r_n^{\rho_N - \varepsilon}} \quad \text{and}$$

$$|P_i(f_1,\ldots,f_{N-1})(z_n)| < e^{r_n^{\rho_N - 2\varepsilon}} \quad \text{for all } i = 0,\ldots,m.$$

In particular, $|f_N(z_n)| > e^{r_n^{\rho_N - \varepsilon} - r_n^{\rho_N - 2\varepsilon}}$ for n large. Thus,

$$|P(f_1,\ldots,f_M)(z_n)| \geq$$

$$\geq |P_m(f_1,\ldots,f_{N-1})(z_n)| \cdot |f_N(z_n)|^m - \sum_{i=0}^{m-1} |P_i(f_1,\ldots,f_{N-1})(z_n)| \cdot |f_N(z_n)|^i$$

$$> e^{r_n^{\rho_N - \varepsilon}} \cdot |f_N(z_n)|^{m-1} - e^{r_n^{\rho_N - 2\varepsilon}} \cdot \sum_{i=0}^{m-1} |f_N(z_n)|^i$$

$$= e^{r_n^{\rho_N - \varepsilon}} \cdot |f_N(z_n)|^{m-1} \cdot \left[1 - e^{r_n^{\rho_N - 2\varepsilon} - r_n^{\rho_N - \varepsilon}} \cdot \sum_{i=0}^{m-1} |f_N(z_n)|^{i-(m-1)} \right].$$

Note that the expression inside the brackets in the last formula tends to 1 as $n \to \infty$: indeed, $e^{r_n^{\rho_N - 2\varepsilon} - r_n^{\rho_N - \varepsilon}} \to 0$ and $|f_N(z_n)|^{-1} < e^{r_n^{\rho_N - 2\varepsilon} - r_n^{\rho_N - \varepsilon}} \to 0$. Thus, it is greater than some constant $C \in (0,1)$ for n large enough. Furthermore, we also have

$$e^{r_n^{\rho_N - \varepsilon}} \cdot |f_N(z_n)|^{m-1} > e^{r_n^{\rho_N - \varepsilon}} \cdot e^{(m-1)r_n^{\rho_N - \varepsilon} - (m-1)r_n^{\rho_N - 2\varepsilon}}$$

$$= e^{m\, r_n^{\rho_N - \varepsilon} - (m-1)r_n^{\rho_N - 2\varepsilon}} > e^{(m/2)\, r_n^{\rho_N - \varepsilon}}$$

for n large enough. Consequently, one has for n large that

$$M\left(P(f_1,\ldots,f_M), r_n\right) \geq C \cdot e^{(m/2)r_n^{\rho_N - \varepsilon}},$$

which implies

$$\rho\left(P(f_1,\ldots,f_M)\right) = \limsup_{r\to+\infty} \frac{\log\log M\left(P(f_1,\ldots,f_M), r\right)}{\log r}$$

$$\geq \limsup_{n\to\infty} \frac{\log\log M\left(P(f_1,\ldots,f_M), r_n\right)}{\log r_n}$$

$$\geq \lim_{n\to\infty} \frac{\log\log(C \cdot e^{(m/2)r_n^{\rho_N - \varepsilon}})}{\log r_n}$$

$$= \rho_N - \varepsilon.$$

Letting $\varepsilon \to 0$, the above inequalities prove $\rho\left(P(f_1,\ldots,f_M)\right) \geq \rho_N = \max_{k\in\mathcal{I}(P)} \rho(f_k)$, as required. $\qquad\square$

Theorem 1.8.3. *The set* $CS(\mathbb{R}, \mathbb{C})$ *is maximal strongly algebrable in* $\mathcal{C}(\mathbb{R}, \mathbb{C})$.

Proof. For each $s > 0$, select an entire function $\varphi_s : \mathbb{C} \to \mathbb{C}$ of order $s > 0$. Let $A := (0, +\infty) \setminus \mathbb{N}$. Lemma 1.8.2 assures that the set $\{\varphi_s\}_{s \in A}$ is a system of cardinality $\mathfrak{c}$ generating a free algebra $\mathcal{A}$.

Next, notice that any element $\varphi \in \mathcal{A} \setminus \{0\}$ may be written as a non-constant polynomial P without constant term evaluated on some $\varphi_{s_1}, \varphi_{s_2}, \ldots, \varphi_{s_N}$:

$$\varphi = P(\varphi_{s_1}, \varphi_{s_2}, \ldots, \varphi_{s_N}) = \sum_{|\alpha| \leq m} c_\alpha \cdot \varphi_{s_1}^{\alpha_1} \cdot \varphi_{s_2}^{\alpha_2} \cdots \varphi_{s_N}^{\alpha_N}$$

for some $m \in \mathbb{N}$, where $|\alpha| := \alpha_1 + \cdots + \alpha_N$ if $\alpha = (\alpha_1, \ldots, \alpha_N)$. By Lemma 1.8.2, there exists $j \in \{1, \ldots, N\}$ such that $\rho(\varphi) = \rho(\varphi_{s_j}) = s_j \notin \mathbb{N}_0$. The fact that s_j is finite and non-integer guarantees that φ is surjective. Finally, take any $F \in CS(\mathbb{R}, \mathbb{C})$ and consider the algebra

$$\mathcal{B} := \{\varphi \circ F\}_{\varphi \in \mathcal{A}}.$$

Then it is plain that $\mathcal{B}$ is freely $\mathfrak{c}$-generated and that $\mathcal{B} \setminus \{0\} \subset CS(\mathbb{R}, \mathbb{C})$, as required. $\qquad\square$

An important topological problem is that of characterizing those topological spaces that are curves, i.e., that are continuous images of the unit interval I. This is solved by the *Hahn–Mazurkiewicz theorem* (see, e.g.,[382, Theorem 31.5] or [251, Chapter 3]): a non-empty Hausdorff topological space is a continuous image of the unit interval if and only if it is a compact, connected, locally connected metrizable topological space. Such topological spaces are called *Peano spaces*. This raises the question of isolating those topological spaces X that are continuous images of the real line, or equivalently: which X's satisfy $CS(\mathbb{R}, X) \neq \varnothing$? With this aim, the notion of σ-Peano space has been introduced in [9], yielding as a product that such spaces generate lineability for the family $CS(\mathbb{R}, X)$ when X is a topological vector space; see Theorem 1.8.6 below.

Definition 1.8.4. A topological space X is a σ-*Peano space* if there exists an increasing sequence of subsets $K_1 \subset K_2 \subset \cdots \subset K_m \subset \cdots \subset X$ such that each K_n is a Peano space (endowed with the topology inherited from X) and its union amounts to the whole space, that is, $\bigcup_{n \in \mathbb{N}} K_n = X$.

Proposition 1.8.5. *Let X be a Hausdorff topological space. The following assertions are equivalent:*

(a) *X is a σ-Peano space.*

(b) *$CS(R, X) \neq \varnothing$.*

Proof. (a) ⇒ (b): Let $K_1 \subset K_2 \subset \cdots$ be an increasing sequence of Peano spaces in X such that its union is the whole X. Fix a point $x_0 \in X$. Without loss of generality, we may suppose that $x_0 \in K_n$, for all $n \geq 1$. Since Peano spaces are arcwise connected [382, Theorem 31.2], for each $n \geq 1$ there is a Peano map $f_n : [n, n+1] \to K_n$, that starts and ends at x_0, i.e. $f_n(n) = x_0 = f_n(n+1)$. Joining all these Peano maps with the constant path $t \in (-\infty, 1] \mapsto x_0 \in K_1$, one obtains a continuous surjective map $F : R \to X$.

(b) ⇒ (a): Let f be a map in $CS(\mathbb{R}, X)$. Therefore

$$X = f(\mathbb{R}) = f\left(\bigcup_{n \in \mathbb{N}} [-n, n]\right) = \bigcup_{n \in \mathbb{N}} f([-n, n]),$$

so X is σ-Peano just by taking $K_n = f([-n, n])$ $(n \geq 1)$. □

Let us give examples of σ-Peano and non-σ-Peano spaces. Trivially, Peano spaces are σ-Peano, and Euclidean spaces $\mathbb{R}^n$ are σ-Peano topological vector spaces. Let X be a separable topological vector space and X^* be its topological dual endowed with the weak*-topology. If X^* is covered by an increasing sequence of weak*-compact subsets, then it is σ-Peano: indeed, when the topological dual is endowed with the weak*-topology, its weak*-compact subsets are metrizable (see, for instance, [352, Theorem 3.16]). Clearly, this holds on the topological dual (endowed with the weak*-topology) of any separable normed space Y because, due to the Banach–Alaoglu theorem (see, e.g., [352]), the unit closed ball B^* of Y^* (hence, every multiple nB^*) is weak*-compact.

On the negative side, observe that σ-Peano spaces are separable, because continuity preserves separability. In particular, the space ℓ_∞ of all bounded sequences –endowed with the supremum norm– is not σ-Peano. Even more, *no infinite dimensional F-space can be written as a countable union of compact spaces* and, therefore, is not σ-Peano. This is a consequence of the Baire category theorem combined with the fact that on infinite dimensional topological vector spaces, compact sets have empty interiors (Riesz's theorem). In particular, no infinite dimensional Banach space is σ-Peano. Moreover there cannot be a continuous surjection $\mathbb{R} \to \mathbb{R}^{\mathbb{N}}$, where $\mathbb{R}^{\mathbb{N}}$ is endowed with its natural product topology. Looking for infinite dimensional smaller subspaces of $\mathbb{R}^{\mathbb{N}}$ that could enjoy the property, we easily find that the space c_{00} of eventually null sequences –with its natural topology induced by the supremum norm– is a σ-Peano space. Indeed, $K_n := [-n, n]^n \times \{0\}^{\mathbb{N}} \subset c_{00}$ $(n \geq 1)$ defines a increasing sequence of Peano spaces in c_{00}, whose union results in the entire space.

Theorem 1.8.6. *If X is a σ-Peano topological vector space with $X \neq \{0\}$ then $CS(R, X)$ is maximal lineable in $\mathcal{C}(\mathbb{R}, X)$.*

Proof. Note first that $\operatorname{card}(\mathcal{C}(\mathbb{R}, X)) = \mathfrak{c}$. Indeed, this is consequence of $\operatorname{card}(X) \leq \mathfrak{c}$ (as an image of the real line), *in tandem* with the fact that the separability of $\mathbb{R}$ implies that each map of $CS(\mathbb{R}, X)$ is uniquely determined by the sequence of its rational images, which defines an injective map $\mathcal{C}(\mathbb{R}, X) \hookrightarrow X^{\mathbb{N}}$ and, therefore, $\mathrm{card}(\mathcal{C}(\mathbb{R}, X)) \le \mathrm{card}\,(X^{\mathbb{N}}) \le \mathfrak{c}$. And, since X has at least two points, we have $\mathrm{card}(\mathcal{C}(\mathbb{R}, X)) \ge \mathrm{card}(2^{\mathbb{N}}) = \mathfrak{c}$.

Therefore, the theorem would be proved as soon as we exhibit a vector space M with $\dim(M) = \mathfrak{c}$ and $M \setminus \{0\} \subset CS(\mathbb{R}, X)$. Take $g : \mathbb{N}_0 \to \mathbb{N} \times \mathbb{N}$ a bijection, and set

$$I_{k,n} := \left[g^{-1}(k, n), g^{-1}(k, n) + 1 \right],$$

for all $k, n \in \mathbb{N}$. Then $\{I_{k,n}\}_{k,n \in \mathbb{N}}$ is a family of compact intervals of $[0, \infty)$ such that $\bigcup_{k,n \in \mathbb{N}} I_{k,n} = [0, \infty)$ and the intervals $I_{k,n}$ have pairwise disjoint interiors. Proceeding as in the construction presented in Proposition 1.8.5, for each n, we can build a continuous surjective map $f_n : \mathbb{R} \to \mathcal{X}$ with the following properties:

- $f_n \left(\bigcup_{k \in \mathbb{N}} I_{k,n} \right) = X$;

- for each $k \in \mathbb{N}$, on the interval $I_{k,n}$, f_n starts and ends at the origin of X and covers the k-th Peano subset of X;

- $f_n \equiv 0$ on $\bigcup_{k \in \mathbb{N}} I_{k,m}$, for all $m \ne n$.

At this point, it is convenient to recall a well known notion from set theory: a family $\{A_\lambda\}_{\lambda \in \Lambda}$ of infinite subsets of $\mathbb{N}$ is called *almost disjoint* (see, also, Section 6.1 and Definition 6.6.3 for a more general approach) if $A_\lambda \cap A_\mu$ is finite whenever $\lambda \ne \mu$. The usual procedure to generate such a family is the following: denote by $\{q_n\}_{n \in \mathbb{N}}$ an enumeration of the rational numbers. For every irrational α, we choose a subsequence $\{q_{n_k}\}_{k \in \mathbb{N}}$ of $\{q_n\}_{n \in \mathbb{N}}$ such that $\lim_{k \to \infty} q_{n_k} = \alpha$ and define $A_\alpha := \{n_k\}_{k \in \mathbb{N}}$. By construction, we obtain that $\{A_\alpha\}_{\alpha \in \mathbb{R} \setminus \mathbb{Q}}$ is an almost disjoint uncountable family of subsets of $\mathbb{N}$.

Now let $\{J_\lambda\}_{\lambda \in \Lambda}$ be an almost disjoint family with cardinality $\mathfrak{c}$ consisting of infinite subsets of $\mathbb{N}$. Define, for each $\lambda \in \Lambda$,

$$F_\lambda := \sum_{n \in J_\lambda} f_n : \mathbb{R} \to X.$$

The pairwise disjointness of the interiors of the intervals $I_{k,n}$ (together with the above properties of f_n) guarantees that F_λ is well defined, as well as continuous. We assert that the set

$$M := \mathrm{span}\,\{F_\lambda\}_{\lambda \in \Lambda}$$

provides the desired maximal lineability. The crucial point is the following argument: let $F_{\lambda_1}, \dots, F_{\lambda_N}$ be distinct and $\alpha_1, \dots, \alpha_N \in \mathbb{R}$, with $\alpha_N \ne 0$. Since $J_{\lambda_N} \setminus \left(\cup_{i=1}^{N-1} J_{\lambda_i} \right)$ is infinite, we may fix $n_0 \in J_{\lambda_N} \setminus \left(\cup_{i=1}^{N-1} J_{\lambda_i} \right)$. Notice that

$$F_{\lambda_1} = \dots = F_{\lambda_{N-1}} \equiv 0 \quad \text{and} \quad F_{\lambda_N} = f_{n_0} \quad \text{on} \quad \bigcup_{k \in \mathbb{N}} I_{k,n_0}.$$

Consequently, $\sum_{k=1}^{N} \alpha_k \cdot F_{\lambda_k} = \alpha_N \cdot f_{n_0}$ on $\bigcup_{k\in\mathbb{N}} I_{k,n_0}$. Then $F := \sum_{k=1}^{N} \alpha_k \cdot F_{\lambda_k}$ is an element of $CS(\mathbb{R}, X)$, because the image of $\mathbb{R}$ under F contains $\alpha_N \cdot f_{n_0}(\bigcup_{k\in\mathbb{N}} I_{k,n_0}) = \alpha_N X = X$. Finally, one may easily prove that the set $\{F_\lambda\}_{\lambda\in\Lambda}$ has $\mathfrak{c}$-many linearly independent elements, and each non-zero element of its linear span M also belongs to $CS(\mathbb{R}, X)$. $\qquad\square$

There are several extensions of the notion of Peano curve on $\mathbb{R}^N$, with $N \geq 2$. Again, since the case $N = 2$ is illuminating enough, we will restrict ourselves to it. For instance, in [354] one can find the notion given in Definition 1.8.7 below. Before this, we need to recall some terminology and properties. Assume that S is a bounded subset of $\mathbb{R}^2$, so that S is included in some rectangle $R = [a, b] \times [c, d]$. Then the inner Jordan content and the outer Jordan content of S are, respectively, given by the following lower and upper Riemann integrals:

$$\underline{c}(S) = \underline{\int}_R \chi_S \, dxdy \quad \text{and} \quad \overline{c}(S) = \overline{\int}_R \chi_S \, dxdy.$$

Here χ_S denotes the characteristic function of S. The set S is said to be *Jordan measurable* provided that $\underline{c}(S) = \overline{c}(S)$, in which case their common value $c(S)$ is called the *Jordan content* of S. This happens if and only if χ_S is Riemann integrable, and if and only if $\lambda(\partial S) = 0$ (λ and ∂S denote, respectively, Lebesgue measure and the boundary of S). Moreover, in this case, S is Lebesgue measurable and $c(S) = \lambda(S)$.

Definition 1.8.7. We say that a continuous function $\varphi : I \to \mathbb{R}^2$ is a *space-filling curve* provided that $\varphi(I)$ is Jordan measurable and $c(\varphi(I)) > 0$.

Note that a bounded set $S \subset \mathbb{R}^2$ is Jordan measurable if and only if χ_S is Riemann integrable, which in turn holds if and only if $\lambda(\partial S) = 0$. In such a case, S is Lebesgue measurable and $\lambda(S) = c(S)$. Then a continuous function $\varphi : I \to \mathbb{R}^2$ is a space-filling curve if and only if the interior $\varphi(I)^0 \neq \varnothing$ and $\lambda(\partial \varphi(I)) = 0$. The mere condition $\lambda(\varphi(I)) > 0$ is not equivalent to the former definition. As a matter of fact, Osgood [329, 354] constructed in 1903 a Jordan arc, that is, a continuous *injective* function $\psi : I \to \mathbb{R}^2$, such that $\lambda(\psi(I)) > 0$; here $\psi(I)$ cannot be Jordan measurable. The symbol $\mathcal{SF}$ will stand for the set of all space-filling curves in the sense of Definition 1.8.7.

The main results about lineability of $\mathcal{SF}$ will be established in Theorem 1.8.10. The following two lemmas are in order to help to prove it. The first of them is elementary, and a proof is suggested in Exercise 1.3.

Lemma 1.8.8. *Assume that $\{f_n\}_{n\geq 1}$ is a sequence in $\mathcal{C}(I, \mathbb{R}^2) \setminus \{0\}$ such that the supports $\{t \in I : f_n(t) \neq 0\}$ $(n = 1, 2, \dots)$ are mutually disjoint. Then $\{f_n\}_{n\geq 1}$ is a basic sequence in $\mathcal{C}(I, \mathbb{R}^2)$, that is, given a member f of its closed linear span, there exists a unique sequence $\{c_n\}_{n=1}^{\infty} \subset \mathbb{R}$ such that $f = \sum_{n=1}^{\infty} c_n f_n$, where the convergence is in the supremum norm.*

Lemma 1.8.9. *Let Y be a Peano space and $[a,b]$ be a closed interval in $\mathbb{R}$. Given $u, v \in Y$, there is a mapping $\Phi \in CS([a,b], Y)$ such that $\Phi(a) = u$ and $\Phi(b) = v$.*

Proof. By the Hahn–Mazurkiewicz theorem, we can select a continuous surjective map $f : I \to Y$. Since Peano spaces are arcwise connected [382, Theorem 31.2], there are continuous mappings $g : [0, 1/3] \to Y$ and $h : [2/3, 1] \to Y$ satisfying $g(0) = u$, $g(1/3) = f(0)$, $h(2/3) = f(1)$ and $h(1) = v$. Define $\varphi : I \to Y$ as

$$\varphi(t) = \begin{cases} g(t) & \text{if } 0 \le t < 1/3 \\ f(3t - 1) & \text{if } 1/3 \le t \le 2/3 \\ h(t) & \text{if } 2/3 \le t \le 1. \end{cases}$$

Then it is evident that the mapping $\Phi : [a,b] \to Y$ given by $\Phi(t) = \varphi\left(\frac{t-a}{b-a}\right)$ does the job. ☐

Theorem 1.8.10. *The family $\mathcal{SF}$ is spaceable in $\mathcal{C}(I, \mathbb{R}^2)$. In particular, it is maximal lineable. Moreover, $\mathcal{SF}$ is strongly algebrable.*

Proof. Denote by $\| \cdot \|$ any fixed norm on $\mathbb{R}^2$, for instance $\|(x, y)\| = |x| + |y|$. Fix any sequence (a_n) with $0 < a_1 < a_2 < \cdots < a_n < \cdots \to 1$. By Lemma 1.8.9, for every $n \in \mathbb{N}$ there is a mapping $f_n \in CS([a_{n-1}, a_n], [-1, 1]^2)$ with $f_n(a_{n-1}) = (0, 0) = f_n(a_n)$, where $a_0 := 0$ (see Figure 1.9). Extend continuously each f_n to I by setting $f_n(t) = (0, 0)$ if $t \in I \setminus [a_{n-1}, a_n]$. Since the supports of these functions are mutually disjoint, Lemma 1.8.8 tells us that $\{f_n\}_{n \ge 1}$ is a basic sequence of $\mathcal{C}(I, \mathbb{R}^2)$. Define

$$M := \overline{\text{span}}\{f_n : n \in \mathbb{N}\}.$$

It is plain that M is a closed vector subspace of $\mathcal{C}(I, \mathbb{R}^2)$. Moreover, it is infinite dimensional because the f_n's, being members of a basic sequence, are linearly independent.

Now, fix a map $f \in M \setminus \{0\}$. Then there is a sequence $(c_n) \subset \mathbb{R}$ with some $c_m \ne 0$ such that $f = \sum_{n=1}^{\infty} c_n f_n$ in $\mathcal{C}(I, \mathbb{R}^2)$. Note that this series converges uniformly on I. Therefore $c_n f_n \to 0$ uniformly on I, that is, $\lim_{n \to \infty} |c_n| \sup_{t \in I} \|f_n(t)\| = 0$. But since $f_n(I) = [-1, 1]^2$, we get $\sup_{t \in I} \|f_n(t)\| = 1$ for all n, hence $c_n \to 0$. Therefore, there exists $p \in \mathbb{N}$ such that $|c_p| = \max\{|c_n| : n \in \mathbb{N}\} > 0$. Consequently,

$$f(I) = \{(0, 0)\} \cup \bigcup_{n \ge 1} (c_n f_n)([a_{n-1}, a_n])$$

$$= \{(0, 0)\} \cup \bigcup_{n \ge 1} |c_n| [-1, 1]^2$$

$$= |c_p| [-1, 1]^2 = [-|c_p|, |c_p|]^2.$$

Then $f(I)$ is, trivially, Jordan measurable and satisfies $f(I)^0 \ne \varnothing$. Thus, $f \in \mathcal{SF}$, which shows the spaceability of this set. The maximal lineability

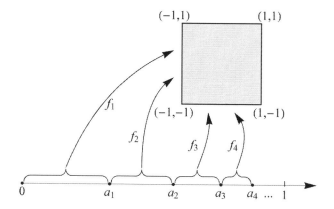

FIGURE 1.9: A step in the construction of the space M.

of $\mathcal{SF}$ comes from the fact that $\dim(M) = \mathfrak{c} = \dim(\mathcal{C}(I, \mathbb{R}^2))$, because, by Baire's category theorem, the dimension of any separable infinite dimensional F-space is $\mathfrak{c}$.

Next, let us prove the strong algebrability. As a first step, we construct an appropriate sequence $\{f_n\}_{n \geq 1}$ generating a free algebra in $\mathcal{C}(I, \mathbb{R}^2)$. By Lemma 1.8.9, there exists a Peano curve $\varphi \in CS(I, [-1, 1]^2)$ such that $\varphi(0) = (0,0) = \varphi(1)$. If $T = [a, b] \subset \mathbb{R}$ is an interval, we define $\varphi_T : [a, b] \to \mathbb{R}^2$ as $\varphi_T(t) = \varphi\left(\frac{t-a}{b-a}\right)$, so that $\varphi_T(T) = [-1, 1]^2$ and $\varphi_T(a) = (0,0) = \varphi_T(b)$. Denote by $\mathbb{Q}$ the set of rational numbers and consider the countable set $J = \{\sigma_k\}_{k \geq 1}$ defined as

$$J := \{\sigma = (q_1, q_2, .., q_j, 0, 0, ...) \in (-1, 1)^{\mathbb{N}} : q_1, ..., q_j \in \mathbb{Q} \cap (-1, 1), j \in \mathbb{N}\}.$$

Take a sequence (a_n) with $0 < a_1 < \cdots < a_n < \cdots \to 1$ and consider the sequence of intervals $[a_n, a_{n+1}]$ $(n \geq 1)$. Then we can extract from it infinitely many countable families of sequences of intervals $\{I_{n,k} : k \in \mathbb{N}\}$ $(n \in \mathbb{N})$ such that $I_{n,k} \cap I_{m,l} = \varnothing$ as soon as $(n, k) \neq (m, l)$ and, for every $n \in \mathbb{N}$, the intervals $I_{n,k}$ approach 1 as $k \to \infty$. Split each interval $I_{n,k}$ into three segments of equal length, say

$$I_{n,k} = I_{n,k,1} \cup I_{n,k,2} \cup I_{n,k,3},$$

where $I_{n,k,2}$ is the middle segment.

Fix $n \in \mathbb{N}$ and define the mapping $f_n : I \to \mathbb{R}^2$ as follows. For all $k \in \mathbb{N}$, we set $f_n := k^{-1} \varphi_{I_{n,k,2}}$ on $I_{n,k,2}$ and $f_n := (0,0)$ on $I_{n,k,1} \cup I_{n,k,3}$. If $m \neq n$ then we set $f_n := (k^{-1} q_n, k^{-1} q_n)$ on $I_{m,k,2}$, where q_n is the nth component of the sequence $\sigma_k \in J$. Both components of f_n are defined as affine linear on $I_{m,k,1}$ and $I_{m,k,3}$, with value $(0,0)$ at the left endpoint of $I_{m,k,1}$ and at the right endpoint of $I_{m,k,3}$. Finally, set $f_n := (0,0)$ on $I \setminus \bigcup_{k=1}^{\infty} I_{n,k}$. Each f_n is clearly continuous on $[0, 1)$, while its continuity at $t = 1$ (where f_n takes the value $(0,0)$) is guaranteed by the fact that $\sup_{t \in I_{n,k}} \|f_n(t)\| \leq k^{-1}$ for all $k \in \mathbb{N}$.

Now, let $N \in \mathbb{N}$ and consider a nonzero polynomial P of N variables without constant term, say $P(x_1, \ldots, x_N)$. Without loss of generality, we may assume that x_N appears explicitly in P, so that there is $m \in \mathbb{N}$ and polynomials Q_j $(j = 0, 1, \ldots, N-1)$ of $N-1$ real variables, with $Q_m \not\equiv 0$, such that

$$P(x_1, \ldots, x_N) = \sum_{j=0}^{m} Q_j(x_1, \ldots, x_{N-1}) \, x_N^j.$$

Let $F := P(f_1, \ldots, f_N)$. Our aim is to show that $F \in \mathcal{SF}$ (it must also be proved that $F \not\equiv 0$, but this is unnecessary because $0 \notin \mathcal{SF}$).

Assume first that $Q_m(0, \ldots, 0) \neq 0$. Since Q_m is continuous, there is $r \in (0,1)$ such that $Q_m(x_1, \ldots, x_{N-1}) \neq 0$ for all $(x_1, \ldots, x_{N-1}) \in (-r, r)^{N-1} \setminus \{(0, \ldots, 0)\}$. Taking $p \in \mathbb{N}$ with $1/p < r$ and $q_j := 1/p$ $(j = 1, \ldots, N-1)$, we get the existence of a point $(q_1, \ldots, q_{N-1}) \in (\mathbb{Q} \cap (-1,1))^{N-1}$ such that

$$Q_m(k^{-1}q_1, \ldots, k^{-1}q_{N-1}) \neq 0 \quad \text{for all } k \in \mathbb{N}. \tag{1.6}$$

If, on the contrary, we had $Q_m(0, \ldots, 0) = 0$, then we would get a point $q = (q_1, \ldots, q_{N-1}) \in (\mathbb{Q} \cap (-1,1))^{N-1}$ satisfying (1.6) too. In order to see this, assume, by way of contradiction, that for each point $p = (p_1, \ldots, p_{N-1}) \in (\mathbb{Q} \cap (-1,1))^{N-1}$ there are infinitely many $t \in \mathbb{R}$ with $Q_m(tp_1, \ldots, tp_{N-1}) = 0$. Since the left hand side of the latter equation is a polynomial in the variable t, we would have $Q_m(tp_1, \ldots, tp_{N-1}) = 0$ for all t. Fixing t and taking into account the density of $\mathbb{Q} \cap (-1,1)$ in $(-1,1)$ and the continuity of Q_m, we get $Q_m(tx_1, \ldots, tx_{N-1}) = 0$ for all $(x_1, \ldots, x_N) \in (-1,1)^{N-1}$ and all $t \in \mathbb{R}$, so $Q_m \equiv 0$, a contradiction. Hence there is $p \in (\mathbb{Q} \cap (-1,1))^{N-1}$ such that the set of $t \in \mathbb{R}$ for which $Q_m(tp_1, \ldots, tp_{N-1}) = 0$ is finite. Since 0 is one of such t's, there is $s \in \mathbb{N}$ with $Q_m(tp_1, \ldots, tp_{N-1}) \neq 0$ for all $t \in (0, 1/s]$. Therefore we get (1.6) if we set $q = (s^{-1}p_1, \ldots, s^{-1}p_{N-1})$.

Let $\sigma := (q_1, \ldots, q_{N-1}, 0, 0, \ldots) \in J$, where $(q_1, \ldots, q_{N-1}) \in (\mathbb{Q} \cap (-1,1))^{N-1}$ satisfies (1.6). Then there is $k \in \mathbb{N}$ such that $\sigma = \sigma_k$. Consider the interval $I_{N,k}$ and its subinterval $I_{N,k,2}$. It happens that, for every $t \in I_{N,k,2}$,

$$
\begin{aligned}
F(t) &= P(f_1(t), \cdots, f_N(t)) \\
&= P((k^{-1}q_1, k^{-1}q_1), \ldots, (k^{-1}q_{N-1}, k^{-1}q_{N-1}), k^{-1}\varphi_{I_{N,k,2}}(t)) \\
&= \sum_{j=0}^{m} Q_j((k^{-1}q_1, k^{-1}q_1), \ldots, (k^{-1}q_{N-1}, k^{-1}q_{N-1}))(k^{-1}\varphi_{I_{N,k,2}}(t))^j. \tag{1.7}
\end{aligned}
$$

Recall that, according to the multiplication law defined on $\mathbb{R}^2$, given any polynomial $H(x_1, \ldots, x_N)$, we have that

$$H((a_1, b_1), \cdots, (a_N, b_N)) = (H(a_1, \ldots, a_N), H(b_1, \ldots, b_N)).$$

By the definition of the f_n's, the image $L_2 := F(I \setminus I_{N,k})$ is the union of two piecewise continuously differentiable curves in $\mathbb{R}^2$, so having empty interior L_2^0 (hence $L_2 = \partial L_2$) and Lebesgue measure $\lambda(\partial L_2) = 0$. Thanks to (1.7), the set $L_1 := F(I_{N,k})$ is the image of the square $R := [-1/k, 1/k]^2$ under the C^1-mapping $S : \mathbb{R}^2 \to \mathbb{R}^2$ given by $S(x,y) = (H(x), H(y))$, where H is the nonconstant polynomial

$$H(x) = \sum_{j=0}^{m} \alpha_j \, x^j,$$

with $\alpha_j := Q_j(k^{-1}q_1, \ldots, k^{-1}q_{N-1})$ (it is nonconstant because, from (1.6), $\alpha_m \neq 0$; this also yields $F \not\equiv 0$). Therefore there is a point $x_0 \in (-1/k, 1/k)$ such that $H'(x_0) \neq 0$, so the determinant of the Jacobian matrix $J_S(x,y)$ of the transformation S at $(x_0, x_0) \in R$ is $H'(x_0)^2 \neq 0$. By the inverse mapping theorem, S has a local differentiable (hence continuous) inverse at (x_0, x_0), and so S is locally open at this point, which yields $L_1^0 = (S(R))^0 \neq \varnothing$.

Finally, since L_1 is compact (hence closed) in $\mathbb{R}^2$, one has that $\partial L_1 \subset L_1 = S(R) = S(R^0) \cup S(\partial R)$. Since S is locally open at those points $(x,y) \in R^0$ with $\det J_S(x,y) \neq 0$, we deduce that $\partial L_1 \subset S(C) \cup S(\partial R)$, where

$$C := \{(x,y) \in R^0 : \det J_S(x,y) = 0\}.$$

On one hand, since S is continuously differentiable on $\mathbb{R}^2$, Sard's theorem (see, e.g., [370, p. 47]) tells us that $\lambda(S(C)) = 0$. On the other hand, the continuous differentiability of S on $\mathbb{R}^2$ implies the well known estimation

$$\lambda(S(\partial R)) \leq \int_{\partial R} |\det J_S| \, d\lambda \leq \sup_{\partial R} |\det J_S| \cdot \lambda(\partial R) = 0.$$

Thus, $\lambda(S(\partial R)) = 0$, hence $\lambda(\partial L_1) = 0$. To sum up, we get $(F(I))^0 \supset L_1^0 \neq \varnothing$ and

$$\lambda(\partial F(I)) \leq \lambda((\partial L_1) \cup (\partial L_2)) \leq \lambda(\partial L_1) + \lambda(\partial L_2) = 0.$$

This entails $(F(I))^0 \neq \varnothing$ and $\lambda(\partial F(I)) = 0$. In other words, $F \in \mathcal{SF}$, which finishes the proof. $\qquad \square$

1.9   Exercises

Exercise 1.1. (a) Prove directly that $CS(\mathbb{R}, \mathbb{R})$ is $\mathfrak{c}$-lineable.
(b) Let $N \in \mathbb{N}$. Demonstrate that $CS(\mathbb{R}, \mathbb{R}^N)$ is $\mathfrak{c}$-lineable.
Hint: For (a), consider, for each $r > 0$, the function $\varphi_r(t) := e^{rt} - e^{-rt}$, and show that the set $\{\varphi_r\}_{r>0}$ is linearly independent and that every nonzero element of $\mathrm{span}(\{\varphi_r\}_{r>0})$ is continuous and surjective. Use (a) and the composition given in the second paragraph of Section 1.8 to prove (b).

Exercise 1.2. Complete the details of the final part of the proof of Theorem 1.8.6. Specifically, prove that the functions F_λ's are linearly independent and that each nonzero member of their linear span is continuous and surjective.

Exercise 1.3. Give a proof for Lemma 1.8.8.
Hint: Convergence in supremum norm implies pointwise convergence. For fixed $f \in \overline{\text{span}} \{f_n\}_{n \geq 1}$, there is a sequence of finite linear combinations of the functions f_n's norm-tending to f. Now, if $x_0 \in I$, there is at most one n such that $x_0 \in \{x \in I : f_n(t) \neq 0\}$.

Exercise 1.4. In the case $\mathbb{K} = \mathbb{C}$, prove the following result, that is slightly stronger than Theorem 1.4.1: for each infinite subset $M \subset \mathbb{N}_0$ and each sequence $c_n \subset (0, \infty)$, the set $\{f \in \mathcal{C}^\infty(I) :$ there are infinitely many $n \in M$ such that $|f^{(n)}(x)| > c_n$ for all $x \in I\}$ is residual in $\mathcal{C}^\infty(I)$.
Hint: Replace the function $\sin(bx)$ used in the mentioned theorem by the function $\exp(ibx)$.

Exercise 1.5. In the case $\mathbb{K} = \mathbb{C}$, show that the set $\{f \in \mathcal{C}^\infty(I) : \limsup_{n \to \infty} \inf_{x \in I} \frac{|f^{(n)}(x)|}{c_n} = \infty\}$ is dense-lineable for any sequence $(c_n) \subset (0, \infty)$. As a consequence, derive that the set

$$\left\{ f \in \mathcal{C}^\infty(I) : \limsup_{n \to \infty} \inf_{x \in I} \left(\frac{|f^{(n)}(x)|}{n!} \right)^{1/n} = \infty \right\},$$

which is smaller that $\mathcal{PS}(I)$, is dense-lineable.
Hint: Use Exercise 1.4 together with the approach of Theorem 1.4.2(a).

Exercise 1.6. Let $\{C_j : j \in J\}$ be a family of distinct nonempty sets such that for each $j \in J$, $C_j^\mathbb{N}$ is the space of all sequences of elements of C_j and $\chi_{C_j^\mathbb{N}}$ is the characteristic function of $C_j^\mathbb{N}$. Show that $\{\chi_{C_j^\mathbb{N}} : j \in J\}$ is a linearly independent set in the space of functions from $\mathbb{R}^\mathbb{N}$ to $\mathbb{R}$.
Hint: Check [28, Lemma 4.1] (or [210, Lemma 3] for the solution).

Exercise 1.7. Prove that $\widehat{C}(D)$ in Example 1.7.22 is $\mathfrak{c}$-lineable.
Hint: Replace the projections π_j by continuous functionals. Check [140, Remark 4.2].

1.10   Notes and remarks

Section 1.2. Concerning Weierstrass' monster, a continuous function $f : [0, 1] \to \mathbb{R}$ is called *Besicovitch* if it has a nowhere one-sided derivative (finite or infinite). A.S. Besicovitch [118] discovered in 1925 the existence of such functions. In 2007, Bobok [125] established the spaceability in $\mathcal{C}[0,1]$ of the family of Besicovitch functions (see also [68, 126]), although he did not explicitly use the terms lineability or spaceability.

Section 1.4. Apart from Lerch's function given in the text, other explicit examples of nowhere analytic functions can be seen in [253, 274].

Papers by Morgenstern [314] and Salzmann and Zeller [357] are probably not well known (with the additional handicap that the proof in [314] contained a gap, as noted in [357]), because several authors have published later similar results. Namely, Christensen [163] established in 1971 that the set $\mathcal{S}_0(I) := \{f \in \mathcal{C}^\infty(I) :$ there exists a residual subset $A_f \subset I$ such that $\rho(f, x) = 0$ for all $x \in A_f\}$ is residual in $\mathcal{C}^\infty(I)$, and Darst [180] proved in 1973 the residuality of $\mathcal{S}(I)$ in $\mathcal{C}^\infty(I)$. Note that $\mathcal{PS}(I) \subset \mathcal{S}_0(I) \subset \mathcal{S}(I)$, where the last inclusion derives from the fact that the set of points of I where f is not analytic is closed. The property that $\mathcal{PS}(I)$ is residual was rediscovered in 1991 by Ramsamujh [339], whose main result is covered by a more general one in [70] for the case $\mathbb{K} = \mathbb{C}$. More recently, Bastin, Esser and Nikolay [52] have studied the genericity of $\mathcal{C}^\infty$-functions which are nowhere analytic in a measure-theoretic sense.

Since each function in $\mathcal{PS}(I)$ is everywhere differentiable, we have by the classical result by Gurariy [239] that $\mathcal{PS}(I)$ is not spaceable in $\mathcal{C}(I)$. But it is not known whether $\mathcal{PS}$ is spaceable in $\mathcal{C}^\infty(I)$ or not. In relation to Theorem 1.4.2(d), it is even possible to obtain, by using the general criteria of Chapter 7, the maximal dense-lineability of $\mathcal{PS}(I)$ (if $\mathbb{K} = \mathbb{C}$). The maximal lineability seems to be unknown in the case $\mathbb{K} = \mathbb{R}$. The algebrability –in its diverse degrees– of $\mathcal{PS}(I)$ in $\mathcal{C}(I)$ seems to be an open problem as well. Another unsolved question is the existence or non-existence of large algebras within $\mathcal{PS}(\mathbb{R})$, all of whose members possess infinitely many zeros.

In relation to the results on annulling functions, the reader is referred to [148] for a very accessible work on uniqueness theorems for analytic functions. The definition of the functions generating the algebra in Theorem 1.4.3 is inspired by the construction of Kim and Kwon in [274]. Theorem 1.4.3 will be improved in Theorem 7.5.2 of this book. Very recently, Enflo, Gurariy and Seoane [196] proved that for every infinite dimensional subspace X of $\mathcal{C}(I)$ the subset of its annulling functions contains an infinite dimensional closed subspace; see also Section 6.4. Finally, Conejero, Muñoz, Murillo and Seoane [176] have been able to demonstrate that the family of smooth annulling functions on $\mathbb{R}$ is strongly $\mathfrak{c}$-algebrable. Note that this result is the best possible in terms of dimension, since the set of continuous functions has dimension $\mathfrak{c}$. The functions constructed in [176] vanished on a same set with cardinality $\mathfrak{c}$.

Within the framework of vector-valued analytic functions we refer the interested reader to the paper [292] by López-Salazar, where large vector spaces of entire functions of unbounded type are constructed.

Also, and although not quite related to the previous results, in [5], the authors study the lineability and spaceability of continuous functions on locally compact groups. Namely, and if G denotes a non-compact locally compact group, the authors consider the problem of the spaceability of the set

$$\left(C_0(G) \cap \left(C_0(G) * C_0(G) \right) \right) \setminus C_{00}(G)$$

and show that, for $G = \mathbb{R}^n$, the above set is strongly $\mathfrak{c}$-algebrable (and, therefore, algebrable and lineable) with respect to the convolution product.

Section 1.7. The results of this section related to functions defined in subspaces of the real line are due to Gurariy and Quarta [241]. The paper [241] is investigated in detail in [328], which inspired our presentation and is an interesting reference on the subject (in Portuguese). The extension to higher dimensions was recently obtained by Botelho, Cariello, Fávaro, Pellegrino and Seoane [140].

As mentioned earlier, one of the first results in the topic of lineability was due to Gurariy and Quarta [241]. They considered subsets of continuous functions attaining their maximum at exactly one point. To our surprise, sometimes one cannot achieve lineability for certain nontrivial sets, as the following result shows.

Theorem. *Let* $\mathcal{M} = \{f \in \mathcal{C}([0,1]) : f$ *attains its maximum at exactly one point of* $[0,1]\}$.

(1) $\mathcal{M}$ *is a dense* G_δ *set in* $\mathcal{C}([0,1])$. *In particular,* M *is residual.*

(2) *If* $V \subset \mathcal{M} \cup \{0\}$ *is a vector space, then* $\dim(V) \leq 1$ (*as seen in Theorem 1.7.8*).

In [241], the authors provided a number of partial extensions of the last theorem. For instance, for both $\mathcal{C}(\mathbb{R})$ and $\mathcal{C}_0(\mathbb{R})$ (continuous functions $\mathbb{R} \to \mathbb{R}$ vanishing at $\pm\infty$), there is a 2-dimensional subspace every nonzero element of which attains its maximum at exactly one point of $\mathbb{R}$. In the case of $\mathcal{C}_0(\mathbb{R})$, there is no 3-dimensional subspace having this property ([241]). After all the effort invested in this class of functions, the following problem still remains open (see also [18]):

> *Is there a n-dimensional vector subspace of* $\mathcal{C}(\mathbb{R})$*, with* $n > 2$*, every nonzero element of which attains its maximum at exactly one point of* $\mathbb{R}$?

Related to the above questions, in [337], it was proved that there exists a continuous function on $\mathbb{R}$ with a proper local maximum at each point of a dense subset of $\mathbb{R}$. One could ask whether the set of all functions enjoying this property, denoted $CM(\mathbb{R})$, is lineable. Apparently this is not true, the problem being that the proper local maxima become proper local minima for any negative multiple of f, with $f \in CM(\mathbb{R})$. If f also had a dense set of proper local minima then this problem would not arise. Let us denote by $CMm(\mathbb{R})$ the nonempty (see [193]) set of continuous functions such that both of their sets of proper local minima *and* maxima are dense in $\mathbb{R}$. García, Grecu, Maestre and Seoane [213] proved in 2010 the following result.

Theorem. *There exists an infinite dimensional Banach space of continuous functions on $\mathbb{R}$ all of whose nonzero members have the property that their sets of proper local minima and maxima, respectively, are dense subsets of $\mathbb{R}$. In other words, $CMm(\mathbb{R})$ is maximal spaceable (and, also, algebrable).*

Section 1.8. The proof of Theorem 1.8.1 tells us more; namely, the set $\{f \in \mathcal{C}(\mathbb{R}, \mathbb{R}^2) : f^{-1}(\{(a,b)\})$ is unbounded for each $(a,b) \in \mathbb{R}^2\}$ is dense-lineable and spaceable. In fact, by following the general criteria given in Chapter 7, one can get the maximal dense-lineability of the same set. Analogously, the set $\{f \in \mathcal{C}(\mathbb{R}, \mathbb{C}) : f^{-1}(\{z\})$ is unbounded for each $z \in \mathbb{C}\}$ is algebrable. Moreover, if X is a σ-Peano topological vector space then the set $\{f \in \mathcal{C}(\mathbb{R}, X) : f^{-1}(\{x\})$ is unbounded for each $x \in X\}$ is maximal lineable in $\mathcal{C}(\mathbb{R}, X)$, see [9]. The lineability of the set of continuous surjections $\mathbb{R}^\Lambda \to \mathbb{R}^\Gamma$, where Λ and Γ are infinite sets, is also analyzed in [9].

In the realm of smooth functions, we can also find related results. For instance, Aron, Jaramillo and Randsford [31] have recently obtained the following remarkable theorem.

Theorem. *Suppose that $m > n$. There the set of $\mathcal{C}^\infty$ maps $f : \mathbb{R}^m \to \mathbb{R}^n$ such that, for every proper affine subspace G of $\mathbb{R}^m$, the restriction $f|_G$ is not surjective, is lineable.*

In [99] it is proved that the family of continuous functions $f : I \to \mathbb{R}^2$ such that $(f(I))^0 \neq \varnothing$ is maximal dense-lineable in $\mathcal{C}(I, \mathbb{R}^2)$. Hence the family of functions $f \in \mathcal{C}(I, \mathbb{R}^2)$ satisfying $\lambda(f(I)) > 0$ is maximal dense-lineable as well. Nevertheless, even the mere dense-lineability of $\mathcal{SF}$ seems to be an open problem. Other notions related to space-filling curves can be found in [313] and [373].

Exercise 1.1 is taken from Albuquerque [8], who gave an elegant proof of the maximal lineability of $CS(\mathbb{R}^m, \mathbb{R}^n)$.

Chapter 2

Complex Analysis

In 1884 the Swedish mathematician Mittag-Leffler discovered that for every domain G in the complex plane $\mathbb{C}$ there is at least a holomorphic function $f : G \to \mathbb{C}$ that is not continuable as a holomorphic function to any larger domain. The study of the lineability of this interesting phenomenon will be the first subject of this chapter. Special attention will be paid to the case $G = \mathbb{D} := \{z \in \mathbb{C} : |z| < 1\}$, the open unit disc. Besides, there are a few other kinds of strange functions, such as strongly annular functions, entire functions tending to zero on every straight line but with a speedy overall growth, functions having a wild behavior near the boundary of the domain and infinitely differentiable complex functions on a real interval which are nowhere almost analytic in the sense of Gevrey. All of them will be investigated in this chapter from the point of view of lineability.

2.1 What one needs to know

In the complex plane $\mathbb{C}$ we will consider, as usual, the Euclidean distance $d(z, w) = |z - w|$. With respect to d, $B(z, r)$ will denote the open unit ball with center $z \in \mathbb{C}$ and radius r. A *domain* $G \subset \mathbb{C}$ is a nonempty connected open subset of $\mathbb{C}$. If G is a domain of the complex plane $\mathbb{C}$ then the *compact-open topology* on the space $H(G)$ of holomorphic (or analytic) functions on G is the linear topology for which an open basis is the collection of all sets

$$V(f, K, \varepsilon) := \{g \in H(G) : |g(z) - f(z)| < \varepsilon \text{ for all } z \in K\},$$

where $f \in H(G)$, $\varepsilon > 0$ and $K \subset G$ is compact. Then $H(G)$ is an *F-space*, that is, a completely metrizable topological vector space, meaning that its topology is generated by a complete distance. In fact, if (K_n) is a sequence of compact subsets of G such that $\bigcup_{n \geq 1} K_n = G$ and $K_n \subset K_{n+1}^0$ (A^0 stands for the interior of A) for all $n = 1, 2, ...$, then

$$\rho(f, g) := \sum_{n \geq 1} \frac{1}{2^n} \frac{\max\{|f(z) - g(z)| : z \in K_n\}}{1 + \max\{|f(z) - g(z)| : z \in K_n\}}$$

defines one of such distances on $H(G)$; see, e.g., [2, Chapter 5] or [177, Chapter 7]. Furthermore, we have that $f_n \to f$ in $H(G)$ for this topology if and only if $f_n \to f$ uniformly on each compact subset of G. In fact, ρ is a complete distance on the bigger space $C(G)$ of continuous functions $G \to \mathbb{C}$, and Weierstrass convergence theorem asserts that $H(G)$ is a closed subspace of $C(G)$; in other words, if f is a function defined on G and $f_n \to f$ uniformly on compacta in G and each $f_n \in H(G)$, then $f \in H(G)$. Since $C(G)$ is a separable metrizable space, we get that $H(G)$ is also separable.

Note that if $\{K_n : n \in \mathbb{N}\}$ is the sequence of compact sets considered in the preceding paragraph then each compact subset of G is contained in some K_n. Moreover, such a sequence can be constructed so as to satisfy the following property: each connected component of the complement of every K_n contains some connected component of the complement of G (see [177, Chapter 7]). In other words, each "hole" of K_n contains at least one "hole" of G.

By a *Jordan domain* we understand as usual the bounded component of the complement of a Jordan curve, and a Jordan curve is a topological image in $\mathbb{C}$ of the unit circle $\mathbb{T} := \{z \in \mathbb{C} : |z| = 1\}$. A domain $G \subset \mathbb{C}$ is called *simply connected* if it "lacks holes," that is, the set $\mathbb{C}_\infty \setminus G$ is connected, where $\mathbb{C}_\infty$ is the one-point compactification of $\mathbb{C}$. We recall that the one-point compactification of a domain G is the set $G_* := G \cup \{\omega\}$, where ω is a point to which the whole boundary ∂G "collapses." Then G_* is a compact metrizable topological space if one defines the neighborhoods of the "infinity point" ω as those subsets of G_* containing the complement of some compact set in G. A domain $G \subset \mathbb{C}$ is said to be *regular* whenever $G = \overline{G}^0$. Any Jordan domain is simply connected and regular, but no other implication is true. For instance, $\{1 < |z| < 2\}$ is regular but not simply connected, and $\{z : |z - 1| < 1, |z - \frac{1}{2}| > 1/2\}$ is a regular simply connected non-Jordan domain.

For the statements in this paragraph and the following one, the reader is referred, for instance, to the books of Rudin [351] and Gaier [205]. A very useful result in complex analysis is Runge's approximation theorem, which can be stated as follows. Let K be a compact subset of $\mathbb{C}$, $\varepsilon > 0$, f be a function that is holomorphic in some open set containing K and $A \subset \mathbb{C}_\infty$ be a set containing exactly one point of each connected component of $\mathbb{C}_\infty \setminus K$. Then there exists a rational function R, with poles only in A, such that $|R(z) - f(z)| < \varepsilon$ for all

$z \in K$. As a special instance, we get that the set of polynomials is dense in $H(G)$ provided that G is simply connected. Mergelyan's theorem is a stronger result, stating that if $K \subset \mathbb{C}$ is a compact set with $\mathbb{C} \setminus K$ connected then the set of polynomials is dense in $A(K)$. Here $A(K)$ denotes the Banach space –under the maximum norm– of all continuous functions $K \to \mathbb{C}$ that are holomorphic in the interior of K, that is, $A(K) = \mathcal{C}(K) \cap H(K^0)$.

For unbounded closed sets, Arakelian's approximation theorem comes to our help. Its formulation requires some terminology. If G is a domain of $\mathbb{C}$, a nonempty relatively closed subset $F \subset G$ is said to be an *Arakelian subset* of G whenever $G_* \setminus F$ is both connected and locally connected. Arakelian's theorem asserts that, given $\varepsilon > 0$, an Arakelian subset F of G and $f \in A(K)$, there is $g \in H(G)$ such that $|g(z) - f(z)| < \varepsilon$ for all $z \in F$. Nersesjan's theorem establishes a "tangential" approximation by imposing the additional restriction of "lacking large islands" to the set F, namely: (A) for every compact subset $K \subset G$ there exists a neighborhood V of ω in G_* such that no component of F^0 intersects both K and V. Nersesjan's theorem asserts that if $G \subset \mathbb{C}$ is a domain, F is an Arakelian subset of G satisfying (A), $\varepsilon : F \to (0, +\infty)$ is a continuous function and $f \in A(K)$, then there exists a function $g \in H(G)$ such that $|g(z) - f(z)| < \varepsilon(z)$ for all $z \in F$. Finally, tangential approximation can be achieved in $\mathbb{C}$ for certain functions $\varepsilon(\cdot)$ without imposing the "large island condition". Specifically, we have the following Arakelian's tangential approximation theorem: Assume that $F \subset \mathbb{C}$ is an Arakelian set, that $\varepsilon(t)$ is a continuous and positive function for $t \geq 0$ and that $\int_1^\infty t^{-3/2} \log(1/\varepsilon(t)) \, dt < \infty$. Then for every $g \in A(F)$ there exists an entire function f such that $|f(z) - g(z)| < \varepsilon(|z|)$ for all $z \in F$.

Faber series are an extension of Taylor series to domains more general than discs. For the basic results on Faber series and Faber transforms we refer the reader to [205, 206, 268, 363], among others. Riemann's isomorphism theorem (see, e.g., [2, Chapter 6]) asserts that, given a simply connected domain $\Omega \subset \mathbb{C}$ with $\Omega \neq \mathbb{C}$ and $z_0 \in \Omega$, there exists a unique holomorphic bijection $\varphi : \Omega \to \mathbb{D}$ satisfying $\varphi(z_0) = 0 < \varphi'(z_0)$. From here it can be easily proved that if $G \subset \mathbb{C}$ is a Jordan domain then there is a unique one-to-one function $g \in H(\{w : |w| > 1\})$ such that $g(\{w : |w| > 1\}) = \mathbb{C} \setminus \overline{G}$ and has expansion $g(w) = c \, w + c_0 + c_1 \, w^{-1} + c_2 \, w^{-2} + \cdots$ (with $c > 0$) in a neighborhood of ∞. The Faber polynomials associated with G are the polynomials Φ_n ($n \in \mathbb{N}_0$) determined by the generating function relationship $\frac{g'(w)}{g(w)-z} = \sum_{k=0}^{\infty} \frac{\Phi_k(z)}{w^{k+1}}$. The operator $\mathcal{F}$ that takes a function $f(w) := \sum_{k=0}^{\infty} c_k w^k \in H(\mathbb{D})$ and maps it to the (formal) Faber series $(\mathcal{F}f)(z) := \sum_{k=0}^{\infty} c_k \Phi_k(z)$ ($z \in G$) is called the Faber transform. If the boundary of G is an analytic curve then the series $\mathcal{F}f$ converges uniformly on compact subsets of G to a function $F \in H(G)$ and, in addition, the function g can be holomorphically and univalently continued to some domain $\{|w| > r_0\}$ for some $r_0 \in (0, 1)$. In this case, the Faber transform $\mathcal{F} : f \in H(\mathbb{D}) \mapsto F \in H(G)$ is (linear and) continuous.

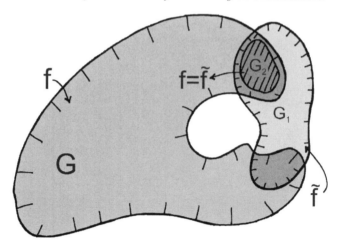

FIGURE 2.1: A holomorphically extendable function.

Osgood–Carathéodory's theorem is an improved version of Riemann's theorem where boundaries are considered. It can be found in e.g. [246], and it asserts that if G is a Jordan domain then there exists a homeomorphism φ from the closure of G onto $\overline{\mathbb{D}}$ whose restriction on G is a holomorphic isomorphism from G onto $\mathbb{D}$.

2.2   Nonextendable holomorphic functions: genericity

We want to analyze the property of nonextendability from the generic point of view, in both topological and algebraic aspects. This section deals with the topological genericity. We start with the precise definitions. As usual, ∂A will denote the boundary of the subset A of a topological space.

Definition 2.2.1. Let G be a domain of $\mathbb{C}$ with $G \neq \mathbb{C}$. A function $f \in H(G)$ is said to be *holomorphically nonextendable* beyond ∂G provided that there do not exist two domains G_1 and G_2 in E and $\widetilde{f} \in H(G_1)$ such that $G_2 \subset G \cap G_1$, $G_1 \not\subset G$ and $\widetilde{f} = f$ on G_2.

A function $f \in H(G)$ that is *extendable* is represented in Figure 2.1. The definition of holomorphic nonextendability is complicated because it must take into account the possibility that ∂G could intersect itself; see Exercise 2.1 for a maybe slightly more workable equivalent definition. Along the pertinent literature, the fact that an $f \in H(G)$ is holomorphically nonextendable beyond ∂G is phrased with several synonymous sentences, namely: f is *holomorphic exactly on* G, G is the *domain of holomorphy* of f, ∂G is the *natural boundary* of f. The set of these functions will be denoted by $H_e(G)$.

According to the Mittag-Leffler theorem, $H_e(G) \neq \varnothing$ for any domain $G \subset \mathbb{C}$. It is easy to see that $f \in H_e(G)$ if and only if $R(f, a) = \text{dist}\,(a, \partial G)$ for all $z \in G$, where $R(f, z)$ stands for the radius of convergence of the Taylor series of f at z, that is,

$$R(f, z) = \frac{1}{\limsup_{n \to \infty} \left| \frac{f^{(n)}(z)}{n!} \right|^{1/n}}.$$

If $f \in H_e(G)$ then f has no holomorphic extension to any domain containing G strictly, but the converse is not true (consider, for instance, $G := \mathbb{C} \backslash (-\infty, 0]$ and $f :=$ the principal branch of $\log z$). Both properties are equivalent if G is a Jordan domain, in particular if $G = \mathbb{D}$.

In 1933, Kierst and Szpilrajn [273] showed that the family $H_e(\mathbb{D})$ is residual in $H(\mathbb{D})$. Their approach can be adapted to any domain G. In 2000, Kahane [262] was able to extend this result to subspaces X of $H(G)$. In fact, the same proof given in [262] allows to weaken the hypotheses on X, so as to obtain (see [77]) the following assertion.

Theorem 2.2.2. *Let* $G \subset \mathbb{C}$ *be a domain with* $G \neq \mathbb{C}$ *and* X *be a Baire topological vector space with* $X \subset H(G)$ *such that the next conditions hold:*

(a) *All evaluation functionals* $f \in X \mapsto f^{(k)}(a) \in \mathbb{C}$ *(*$a \in G$; $k \in \mathbb{N}_0$*) are continuous.*

(b) *For every* $a \in G$ *and every* $r > \text{dist}(a, \partial G)$ *there exists* $f \in X$ *such that*
$$R(f, a) < r.$$

Then $X \cap H_e(G)$ *is residual in* X.

Proof. Our set $X \cap H_e(G)$ is the set $\{f \in X : R(f, a) = \text{dist}(a, \partial G) \text{ for all } a \in G\}$. According to Exercise 2.1, we can replace G by a dense subset D in order to estimate the convergence radii. Let us take as D a countable dense set in G. Then

$$X \cap H_e(G) = \{f \in X : R(f, a) = \text{dist}(a, \partial G) \text{ for all } a \in D\}.$$

According to a well known theorem due to Abel, the radius of convergence of a power series $\sum_{n=0}^{\infty} c_n z^n$ equals $\sup\{r > 0 : \text{the sequence } \{|c_n| r^n\}_{n \geq 1} \text{ is}$

bounded}. Hence we have

$$X \cap H_e(G) = \bigcap_{a \in D} \{f \in X : \sup\{r > 0 : \{|f^{(n)}(a)/n!| \, r^n\}_{n \geq 1} \text{ is bounded}\}$$

$$= \text{dist}(a, \partial G)\}$$

$$= \bigcap_{a \in D} \{f \in X : \{|f^{(n)}(a)/n!| \, r^n\}_{n \geq 1} \text{ is unbounded for all }$$

$$r > \text{dist}(a, \partial G)\}$$

$$= \bigcap_{a \in D} \{f \in X : \{|f^{(n)}(a)/n!| \, \delta(a, k)^n\}_{n \geq 1} \text{ is unbounded for all }$$

$$k \in \mathbb{N}\}$$

$$= \bigcap_{a \in D} \bigcap_{k \in \mathbb{N}} G(a, k),$$

where $\delta(a, k) := \text{dist}(a, \partial G) + \frac{1}{k}$ and $G(a, k) := \{f \in X : \{|f^{(n)}(a)/n!|\delta(a, k)^n\}_{n \geq 1} \text{ is unbounded}\}$ for $k \geq 1$.

Define, for each $a \in G$ and each $n \in \mathbb{N}$, the functional

$$\Phi_{a,n} : f \in X \mapsto f^{(n)}(a) \in \mathbb{C}.$$

According to hypothesis (a), every $\Phi_{a,n}$ is continuous. Now, consider the sets $G(a, k, l) := \{f \in X : \text{exists } n \in \mathbb{N} \text{ such that } |f^{(n)}(a)/n!| \, \delta(a, k)^n > l\}$ ($a \in G$ and $k, l \in \mathbb{N}$). It is evident that

$$G(a, k, l) = \bigcup_{n \in \mathbb{N}} \Phi_{a,n}^{-1}(\{z \in \mathbb{C} : |z| > \frac{n!l}{\delta(a, k)^n}\}),$$

so the continuity of the $\Phi_{a,n}$'s guarantees that every $G(a, k, l)$ is an open set. But plainly $G(a, k) = \bigcap_{l \in \mathbb{N}} G(a, k, l)$ and $X \cap H_e(G) = \bigcap_{\substack{a \in D \\ k, l \in \mathbb{N}}} G(a, k, l)$. This proves that $X \cap H_e(G)$ is a G_δ subset of X.

Since X is a Baire space, it suffices to show that each set $G(a, k)$ is dense in X. To this end, we use hypothesis (b). Fix a nonempty open subset V of X. Then there are $f \in X$ and a neighborhood U of the origin in X such that $V = f + U$. If $f \in G(a, k)$ then, trivially, $G(a, k) \cap V \neq \varnothing$. If, on the contrary, $f \notin G(a, k)$, we have that there is $\alpha \in (0, \infty)$ satisfying $|f^{(n)}(a)/n!| \, \delta(a, k)^n \leq \alpha$ for all $n \geq 1$. Select a function $g \in X$ with $R(g, a) < \delta(a, k)$. Then the sequence $\{|g^{(n)}(a)/n!| \, \delta(a, k)^n\}_{n \geq 1}$ is unbounded. From the triangle inequality it is derived that the sequence $\{\frac{|f^{(n)}(a) + \varepsilon g^{(n)}(a)|}{n!} \delta(a, k)^n\}_{n \geq 1}$ is unbounded for every $\varepsilon > 0$. Therefore $f + \varepsilon g \in G(a, k)$. Since multiplication by scalars is continuous on a topological vector space, we obtain the existence of an $\varepsilon > 0$ with $\varepsilon g \in U$, and so $f + \varepsilon g \in V$. Consequently, $G(a, k) \cap V \neq \varnothing$, and we are done. □

Of course, the case $X = H(G)$ is included in Theorem 2.2.2. As for other spaces to which the last theorem applies, consider the domain $G = \mathbb{D} = \{|z| < 1\}$. For $0 < p < \infty$ the *Hardy space* H^p and the *Bergman space* B^p are defined as the set $\{f \in H(\mathbb{D}) : \|f\|_p < \infty\}$, where $\|f\|_p := \sup_{0<r<1} \left(\int_0^{2\pi} |f(re^{i\theta})|^p \frac{d\theta}{2\pi} \right)^{1/p}$ for $f \in H^p$, $\|f\|_p := \left(\int \int_{\mathbb{D}} |f(z)|^p \frac{dA(z)}{\pi} \right)^{1/p}$ for $f \in B^p$ and $dA(z)$ denotes the normalized area measure on $\mathbb{D}$. They become separable F-spaces with the distance $d(f,g) = \|f - g\|_p^{\alpha(p)}$, where $\alpha(p) = 1$ if $p \geq 1$ (= p if $p < 1$, resp.). In fact, they are Banach spaces if $p \geq 1$. A singular family is H^∞, the space of all bounded holomorphic functions in $\mathbb{D}$, endowed with the supremum norm $\|f\|_\infty := \sup\{|f(z)| : z \in \mathbb{D}\}$. It is a nonseparable Banach space. If $G \subset \mathbb{C}$ is a domain, consider the space $A^\infty(G)$ of holomorphic functions on G having highly boundary-regular behavior, that is,

$$A^\infty(G) := \{f \in H(G) : f^{(k)} \text{ extends continuously to } \overline{G} \text{ for all } k \geq 0\}.$$

It is also an F-space when it is endowed with the topology of uniform convergence of functions and all their derivatives on each compact set $K \subset \overline{G}$. That is, in $A^\infty(G)$, one has that $f_n \to f$ if and only if, for every compact set $K \subset \overline{G}$ and every $k \in \mathbb{N}_0$, $f_n^{(k)} \to f^{(k)}$ uniformly on K. Then $X \cap H_e(\mathbb{D})$ is residual in X if $X = H^p$, B^p, H^∞ or $A^\infty(\mathbb{D})$, because these spaces satisfy (a) and (b); see Exercise 2.3. The Cauchy estimates together with some elementary manipulation of Taylor coefficients yield that

$$A^\infty(\mathbb{D}) = \left\{ f(z) = \sum_{n=0}^\infty a_n z^n : \{n^N a_n\}_{n=0}^\infty \text{ is bounded for all } N \in \mathbb{N} \right\}.$$

An explicit example of a nonextendable, very well-behaved function in the boundary is given in [351, Chap. 16]: the function

$$f(z) := \sum_{n=0}^\infty a_n \exp(-\sqrt{n}) z^n, \quad \text{where } a_n = \begin{cases} 1 & \text{if } n \text{ is a power of 2} \\ 0 & \text{otherwise,} \end{cases}$$

belongs to $A^\infty(\mathbb{D}) \cap H_e(\mathbb{D})$. In fact, this example is a special instance of the Hadamard lacunary theorem, which asserts that if the power series $\sum_{k\geq 1} c_k z^{n_k}$ (with $n_1 < n_2 < \cdots$) has radius of convergence 1 and $\liminf_{n\to\infty} \frac{n_{k+1}}{n_k} > 1$ then the series defines a function $f \in H_e(\mathbb{D})$.

More generally, let $G \subset \mathbb{C}$ be a *regular* domain. In 1980, Chmielowski [162] discovered –as a consequence of an N-dimensional result– that $A^\infty(G) \cap H_e(G) \neq \emptyset$ for every such domain. This yields, trivially, condition (b) of Theorem 2.2.2, and (a) is evident because convergence in $A^\infty(G)$ implies pointwise convergence of all derivatives. Then we obtain that $A^\infty(G) \cap H_e(G)$ is residual in $A^\infty(G)$.

2.3   Vector spaces of nonextendable functions

Plainly, the set $H_e(G)$ is not a linear space, so the study of its lineability makes sense. Aron, García and Maestre [25] settled the problem for any space $H(G)$.

Theorem 2.3.1. *Let $G \subset \mathbb{C}$ be a domain with $G \neq \mathbb{C}$. Then $H_e(G)$ is dense-lineable. Moreover, there is a closed infinitely generated algebra contained in $H(G) \cup \{0\}$. In particular, $H_e(G)$ is spaceable and algebrable in $H(G)$.*

Proof. As seen in Section 2.1, we can select an increasing sequence $\{K_n : n \in \mathbb{N}\}$ of compact sets such that each compact subset of G is contained in some K_n and, in addition, each hole of every K_n contains some hole of G. Choose also a sequence $\{a_n : n \in \mathbb{N}\}$ of distinct points of G such that it has no accumulation point in G and each point of ∂G is an accumulation point of the sequence "through any way of approximation." More precisely, the sequence $\{a_n\}$ should have the following property: for every $a \in G$ and every $r > \mathrm{dist}(a, \partial G)$, the intersection of $\{a_n\}$ with the connected component of $B(a, r) \cap G$ containing a is infinite. An example of the required sequence may be defined as follows. Let $A = \{\alpha_k\}$ be a dense countable subset of G. For each $k \in \mathbb{N}$ choose $b_k \in \partial G$ such that $|b_k - \alpha_k| = \mathrm{dist}(\alpha_k, \partial G)$. For every $k \in \mathbb{N}$ let $\{a_{k,l} : l \in \mathbb{N}\}$ be a sequence of points of the line interval joining α_k with the corresponding point b_k such that

$$|a_{k,l} - b_k| < \frac{1}{k+l} \quad (k, l \in \mathbb{N}).$$

Each one-fold sequence $\{a_n\}$ (without repetitions) consisting of all distinct points of the set $\{a_{k,l} : k, l \in \mathbb{N}\}$ has the required property. Since only finitely many points a_n live in a given compact subset of G, it can be assumed without loss of generality that, for every $j \in \mathbb{N}$, $a_n \notin K_j$ for all $n \geq j$.

Fix $j \in \mathbb{N}$. The function $\frac{-1}{z - a_j}$ is holomorphic in a neighborhood of K_j. By Runge's theorem and the property of the holes of K_j, there exists a rational function g_j with poles outside G (hence $f_j \in H(G)$) such that

$$\left| g_j(z) + \frac{1}{z - a_j} \right| < \frac{1}{2^j \max\{|w - a_j| : w \in K_j\}} \quad \text{for all } z \in K_j.$$

Let $f_j(z) := (z - a_j) g_j(z) + 1$. Then $f_j \in H(G)$, $f(a_j) = 1$ and $|f_j(z)| < 2^{-j}$ for all $z \in K_j$. For each $k \in \mathbb{N}$, consider the infinite product

$$h_k(z) := \prod_{j=k}^{\infty} (1 - f_j(z)).$$

For fixed $j_0 \in \mathbb{N}$, we have $\sup_{K_{j_0}} |f_j| \leq \sup_{K_j} |f_j| < 2^{-j}$ for all $j \geq j_0$, which entails the convergence of the series $\sum_j \sup_K |f_j|$ for every compact set $K \subset G$. This implies (see, e.g., [2, Chapter 5]) that the infinite product $\prod_{j \geq 1}(1 - f_j)$ converges uniformly on compacta to a holomorphic function in G. In particular, the remainders h_k converge uniformly on compacta to 1. Moreover, they satisfy $h_k(a_j) = 0$ for all $j \geq k$. Define

$$M_0 := \{f \in H(G) : f \text{ vanishes on all but a finite number of points } a_j\}.$$

Plainly, M_0 is a vector space. It is easy to show that M_0 is dense in $H(G)$: given $g \in H(G)$ and a compact set $K \subset G$, we have

$$h_k g \to 1 \cdot g = g \quad \text{as} \quad k \to \infty \quad \text{uniformly on} \quad K,$$

and the functions $h_k g$ belong to M_0. Hence $H_e(G)$ is dense-lineable. The particular construction of the sequence (a_n) and the Identity Principle for analytic functions entail that, for any boundary point ξ_0, each nonzero $f \in M_0$ is not holomorphically extendable to boundary point ξ_0 (see Exercise 2.1), so $f \in H_e(G)$. Therefore $M_0 \subset H_e(G) \cup \{0\}$ and $H_e(G)$ is dense-lineable.

Now, the fact that convergence in $H(G)$ is stronger than pointwise convergence implies that the set

$$M_1 := \{f \in H(G) : f \text{ vanishes on all } a_j\}$$

is a closed subalgebra of $H(G)$. Arguing as in the preceding paragraph, we have $M_1 \subset H_e(G) \cup \{0\}$. Finally, let us show that the algebra M_1 is infinitely generated. For this, select a function $h \in H(G)$ with zeros exactly at the points a_n and being all of multiplicity 1. The existence of such an h is guaranteed by the Weierstrass interpolation theorem; see [351, Chapter 15]. The functions $z^n h(z)$ $(n \geq 0)$ live in M_1 and are linearly independent; hence $\dim(M_1) \geq \aleph_0$. Assume, by way of contradiction, that M_1 is generated by finitely many functions $g_1, \ldots, g_N$. Then M_1 is generated as a vector space by countably many functions, namely, the products of powers $g_1^{m_1} \cdots g_N^{m_N}$ with $(m_1, \ldots, m_N) \in \mathbb{N}_0^N \setminus \{(0, \ldots, 0)\}$. Therefore $\dim(M_1) = \aleph_0$, which is not possible because the subspace M_1 is infinite dimensional and closed, see Exercise 2.4. This is the desired contradiction, which concludes the proof.   □

Concerning subspaces of $H(G)$, it is easy to get lineability if not much more than mere nonvacuousness is assumed, as Theorem 2.3.2 below shows. In addition, by using the Faber transform one can obtain large closed manifolds of nonextendable boundary-regular function if ∂G enjoys a soft structure. Recall that a closed curve $\Gamma \subset \mathbb{C}$ is called analytic provided that there is a periodic function $\varphi : \mathbb{R} \to \mathbb{C}$ such that $\varphi(\mathbb{R}) = \Gamma$ and, for each $t_0 \in \mathbb{R}$, there are a $\delta > 0$ and a sequence $\{a_n\}_{n \geq 0} \subset \mathbb{C}$ such that $\varphi(t) = \sum_{n \geq 0} a_n(t - t_0)^n$ for all $t \in (t_0 - \delta, t_0 + \delta)$.

Theorem 2.3.2. *Let $G \subset \mathbb{C}$ be a domain with $G \neq \mathbb{C}$. We have:*

(1) *Assume that ∂G does not contain isolated points and X be a vector space over $\mathbb{C}$ with $X \subset H(G)$ satisfying $X \cap H_e(G) \neq \varnothing$ and $\{\varphi f : f \in X\} \subset X$ for some nonconstant function $\varphi \in H(\overline{G})$. Then $X \cap H_e(G)$ is lineable. In particular, $A^\infty(G) \cap H_e(G)$ is lineable if G is regular.*

(2) *If G is a Jordan domain with analytic boundary then $A^\infty(G) \cap H_e(G)$ is spaceable in $A^\infty(G)$.*

(3) *If G is regular, $\mathbb{C} \setminus \overline{G}$ is connected and G is length-finite, that is, there is $M \in (0, \infty)$ such that for any $a, b \in G$ there exists a curve $\gamma \subset G$ joining a to b for which $\mathrm{length}\,(\gamma) \leq M$, then $A^\infty(G) \cap H_e(G)$ is dense-lineable in $A^\infty(G)$.*

Proof. (1) Choose $f \in X \cap H_e(G)$ and consider the function φ provided in the hypothesis, so that $\varphi \in H(\Omega)$ for some domain $\Omega \supset \overline{G}$ and, in addition, $\varphi^n f \in X$ $(n \in \mathbb{N}_0)$. Therefore $(P \circ \varphi)f \in X$ for every (holomorphic) polynomial P because X is a vector space. Let us define

$$M := \{(P \circ \varphi)f : \ P \text{ is a polynomial}\}.$$

It is clear that M is a linear subspace of X. Let us show that M has infinite dimension. For this, since M is the linear span of the functions $\varphi^n f$ $(n \in \mathbb{N}_0)$, it is enough to prove that such functions are linearly independent. Assume, by way of contradiction, that this is not the case. Then there exists a nonzero polynomial P with $(P \circ \varphi)f \equiv 0$ on G. Since $H(G)$ is an integral ring –and, clearly, $f \not\equiv 0$– we get $P \circ \varphi \equiv 0$ on G. Due to the Open Mapping theorem (see, e.g., [2]) and the fact that φ is not constant, the set $\varphi(G)$ is (nonempty and) open, and P vanishes on it. From the Analytic Continuation Principle, $P \equiv 0$, which is absurd.

To conclude the proof, it must be shown that each function $F \in M \setminus \{0\}$ is in $H_e(G)$. Indeed, for such a function F there exists a nonzero polynomial P with $F = (P \circ \varphi)f$. Suppose, again by way of contradiction, that $F \notin H_e(G)$. Let us denote by S_{z_0} the sum of the Taylor series of F with center at z_0. Then there are a point $a \in G$ and a number $r > \mathrm{dist}(a, \partial G)$ such that $S_a \in H(B(a, r))$. Of course, $S_a = (P \circ \varphi)f$ in $B(a, |a - b|)$, where b is a point on ∂G such that

$$|a - b| = \mathrm{dist}(a, \partial G).$$

Therefore there are a point $c \in \partial G$ and a number $\varepsilon > 0$ with $B(c, \varepsilon) \subset \Omega \cap B(a, r)$ and $P(\varphi(z)) \neq 0$ for all $z \in B(c, \varepsilon)$; indeed, $\Omega \cap B(a, r)$ is a neighborhood of b, the point b is not isolated in ∂G and the set of zeros of $P \circ \varphi$ in Ω is discrete in Ω. Now, take a point $\zeta \in B(c, \varepsilon/2) \cap G$. Then

$$B(\zeta, \varepsilon/2) \subset B(c, \varepsilon) \subset B(a, r)$$

and $P(\varphi(z)) \neq 0$ for all $z \in B(\zeta, \varepsilon/2)$. The function S_ζ equals F in a neighborhood of ζ, whence $S_\zeta/(P \circ \varphi)$ equals f in a neighborhood of ζ. But

$S_\zeta \in H(B(\zeta, \varepsilon/2))$, hence also $S_\zeta/(P \circ \varphi) \in H(B(\zeta, \varepsilon/2))$. Finally, we get from the non-extendability of f that

$$\frac{\varepsilon}{2} > \mathrm{dist}(\zeta, c) \geq \mathrm{dist}(\zeta, \partial G) = R(f, \zeta) = R\left(\frac{S_\zeta}{P \circ \varphi}, \zeta\right) \geq \frac{\varepsilon}{2},$$

which is the sought-after contradiction. The second part of (1) follows from Chmielowsky's result given in Section 2.2 and the trivial fact that if φ is the identity $\varphi(z) = z$ and $f \in A^\infty(G)$ then $\varphi f \in A^\infty(G)$.

(2) For this we will make use of part (5) of Theorem 2.4.1 (which, or course, will be proved independently) asserting the spaceability of $A^\infty(\mathbb{D}) \cap H_e(\mathbb{D})$ in $A^\infty(\mathbb{D})$. We need, in addition, the following two auxiliary results on Faber series, which can be found, respectively, in [259, Theorem 1] and [268, Section 3]:

- Let G be a Jordan domain with analytic boundary and J be a subarc of ∂G. Let $f \in H(\mathbb{D})$ and consider its Faber transform $F = \mathcal{F}f \in H(G)$. Then F has an analytic continuation across J if and only if f has an analytic continuation across $g^{-1}(J)$. Here $g : \{|z| > 1\} \to \mathbb{C} \setminus \overline{G}$ is the one-to-one analytic function associated to G given in Section 2.1.

- Let G be a Jordan domain with analytic boundary. Then the Faber transform $\mathcal{F} : H(\mathbb{D}) \to H(G)$ is a topological isomorphism such that $\mathcal{F}(A^\infty(\mathbb{D})) = A^\infty(G)$ and the restriction map $\mathcal{F} : A^\infty(\mathbb{D}) \to A^\infty(G)$ is also a topological isomorphism.

From here, the proof is easy: simply choose an infinite dimensional space $M_0 \subset (A^\infty(\mathbb{D}) \cap H_e(\mathbb{D})) \cup \{0\}$ that is closed in $A^\infty(\mathbb{D})$, and define $M := \mathcal{F}(M_0)$. Since $\mathcal{F}$ is an isomorphism, M is an infinite dimensional closed vector subspace of $A^\infty(G)$. Finally, given a function $F \in M \setminus \{0\}$, there is a (unique) function $f \in M_0 \setminus \{0\}$ such that $F = \mathcal{F}f$. Suppose, by way of contradiction, that $F \notin H_e(G)$. Therefore there would exist a subarc $J \subset \partial G$ with the property that F has an analytic continuation across J. Hence the function f would have an analytic continuation across the subarc $g^{-1}(J) \subset \partial\mathbb{D}$, which is absurd since $M_0 \setminus \{0\} \subset H_e(\mathbb{D})$.

(3) As can be seen in [268], the set of polynomials is dense in $A^\infty(G)$ (see Exercise 2.6). From the proof of (1), if we select any $f \in A^\infty(G)$ and the function $\varphi(z) := z$, we get that the set

$$M = \{(P \circ \varphi)f : P \text{ is a polynomial}\} = \{Pf : P \text{ is a polynomial}\}$$

is a vector space contained –except for zero– in $A^\infty(G) \cap H_e(G)$. It is enough to show that M itself is dense in $A^\infty(G)$. Note that, since G is length-finite, it must be bounded. Hence by adding, if necessary, an adequate constant to f, one can assume that f is zero-free on G, so $1/f \in A^\infty(G)$. Fix $g \in A^\infty(G)$. Then g/f belongs to $A^\infty(G)$. By the density of polynomials, there exists a sequence (P_n) of polynomials with $P_n^{(j)} \to (g/f)^{(j)}$ $(n \to \infty)$ uniformly on G for each $j \in \mathbb{N}_0$. Fix $N \in \mathbb{N}_0$. By Leibniz's rule, we have (under the convention $\binom{N}{0} = 1$ even if $N = 0$) that, as $n \to \infty$,

$$(P_n f)^{(N)} = \sum_{j=0}^{N} \binom{N}{j} P_n^{(j)} f^{(N-j)} \longrightarrow \sum_{j=0}^{N} \binom{N}{j} (g/f)^{(j)} f^{(N-j)}$$

$$= ((g/f) f)^{(N)} = g^{(N)}$$

uniformly on G, which proves the density of M. □

2.4   Nonextendability in the unit disc

In the special domain $\mathbb{D}$, a number of additional results on lineability of $H_e(G)$ have recently been obtained. In order to analyze them, we consider properties (a)–(b) of Theorem 2.2.2 (for $G = \mathbb{D}$), as well as the following ones, where X is a (topological) vector space contained in $H(G)$:

(c) For every $f(z) = \sum_{n=0}^{\infty} a_n z^n \in X$, the function $\sum_{n \in Q} a_n z^n \in X$ for every $Q \subset \mathbb{N}_0$.

(d) Some denumerable subset of $H(\overline{\mathbb{D}})$ is a dense subset of X.

(e) $X \not\subset H(\overline{\mathbb{D}})$.

Here $H(\overline{\mathbb{D}})$ stands for the space of functions $f \in H(\mathbb{D})$ having holomorphic extension to some open set $\Omega_f \supset \overline{\mathbb{D}}$. Observe that properties (c) and (e) do not require any topological or algebraic structure on X. Parts (1)–(5) of the following theorem are due to Bernal [77], while (6) was proved by Aron, García and Maestre [25].

Theorem 2.4.1. *Assume that X is a topological vector space with $X \subset H(\mathbb{D})$. We have:*

(1) *If X is Baire metrizable and satisfies* (a), (b), (c) *and* (d), *then $X \cap H_e(\mathbb{D})$ is dense-lineable in X.*

(2) *If X is metrizable, X satisfies* (d) *and there is a subset of X for which* (c) *and* (e) *hold, then $X \cap H_e(\mathbb{D})$ is dense-lineable in X.*

(3) *If X is Baire and satisfies* (a), (b) *and* (c), *then $X \cap H_e(\mathbb{D})$ is spaceable in X.*

(4) *If X satisfies* (a) *and there is a subset of X for which* (c) *and* (e) *hold, then $X \cap H_e(\mathbb{D})$ is spaceable in X.*

(5) *The set $X \cap H_e(\mathbb{D})$ is dense-lineable and spaceable in X for $X = A^\infty(\mathbb{D})$, H^p or B^p $(0 < p < \infty)$.*

(6) *The set $H^\infty \cap H_e(\mathbb{D})$ contains, except for zero, an infinitely generated algebra that is nonseparable and closed in H^∞. In particular, $H^\infty \cap H_e(\mathbb{D})$ is spaceable and algebrable.*

Proof. Some notation will be needed. If $f(z) := \sum_{n=0}^\infty a_n z^n \in H(\mathbb{D})$ then the *support* of f (or of the sequence $\{a_n\}_{n=0}^\infty$) is the set $\mathrm{supp}\,(f) = \{n \in \mathbb{N}_0 : a_n \neq 0\}$. If $Q \subset \mathbb{N}_0$ then we denote by $H_Q(\mathbb{D})$ the space of all $f \in H(\mathbb{D})$ with gaps outside Q, that is, such that $\mathrm{supp}\,(f) \subset Q$. The symbol P_Q will stand for the natural projection

$$P_Q : \sum_{n=0}^\infty a_n z^n \in H(\mathbb{D}) \mapsto \sum_{n \in Q} a_n z^n \in H_Q(\mathbb{D}).$$

We first prove the following property:

(P) Suppose that X is a topological vector space with $X \subset H(\mathbb{D})$ satisfying (c) and that $F \in X \setminus H(\overline{\mathbb{D}})$. Then $H_{\mathrm{supp}\,(F)}(\mathbb{D}) \cap H_e(\mathbb{D}) \cap X$ is lineable.

Indeed, fix $F \in X \setminus H(\overline{\mathbb{D}})$. Then we can write $F(z) := \sum_{n=0}^\infty a_n z^n$, where the radius of convergence of the power series is 1. By the Cauchy-Hadamard formula, we have $\limsup_{n\to\infty} |a_n|^{1/n} = 1$. Therefore there exists a strictly increasing sequence $\{n(k) : k \in \mathbb{N}\} \subset \mathbb{N}$ such that

$$\lim_{k\to\infty} |a_{n(k)}|^{1/n(k)} = 1. \tag{2.1}$$

We can extract a sequence $\{m(1) < m(2) < \cdots\} \subset \{n(k) : k \in \mathbb{N}\}$ with

$$m(k+1) > 2m(k) \quad (k \in \mathbb{N}). \tag{2.2}$$

Now we divide the sequence $\{m(k) : k \in \mathbb{N}\}$ into infinitely many strictly increasing sequences $A_j = \{p(j,k) : k \in \mathbb{N}\}$ $(j \in \mathbb{N})$ so that they are pairwise disjoint. By property (c), each series $F_j(z) = \sum_{k=1}^\infty a_{p(j,k)} z^{p(j,k)}$ defines a function belonging to X. But from (2.1) we have clearly that

$$\lim_{k\to\infty} |a_{p(j,k)}|^{1/p(j,k)} = 1 \quad (j \in \mathbb{N}) \tag{2.3}$$

whereas by inequality (2.2) we obtain that every F_j possesses Hadamard gaps. Consider the linear span $L(F) := \mathrm{span}\,\{F_j : j \in \mathbb{N}\}$. Then, obviously, $L(F)$ is a vector subspace of X. Moreover, $L(F)$ is infinite dimensional because the functions F_j $(j \in \mathbb{N})$ are linearly independent due to the fact that $\mathrm{supp}\,(F_j) \cap \mathrm{supp}\,(F_l) \subset A_j \cap A_l = \emptyset$ whenever $j \neq l$. Furthermore, it is evident that if

$$h := \sum_{j=1}^N c_j F_j \in L(F) \quad (c_j \in \mathbb{C}, j = 1, ..., N) \tag{2.4}$$

then $\mathrm{supp}\,(h) \subset \bigcup_{j=1}^{N} \mathrm{supp}\,(F_j) \subset \mathrm{supp}\,(F)$, hence $h \in H_{\mathrm{supp}\,(F)}(\mathbb{D})$.

Finally, assume that $h \in L(F) \setminus \{0\}$. Without loss of generality, we can suppose that h has the expression (2.4) with $c_N \neq 0$. By (2.3), the radius of convergence of the power series defining $c_N F_N$ is 1. But the same is true for h because the corresponding radii for $c_j F_j$ $(j = 1, ..., N-1)$ are ≤ 1 and the supports of the $c_j F_j$ $(j = 1, ..., N)$ are pairwise disjoint. On the other hand, if $\mathrm{supp}\,(h) = \{p(1) < p(2) < \cdots\}$ $(\subset \{m(k) : k \in \mathbb{N}\})$ then from (2.2) we have that $p(k+1) > 2p(k)$ for all $k \in \mathbb{N}$. Thus the Hadamard lacunary theorem asserts that $h \in H_e(\mathbb{D})$. This concludes the proof of (P).

(1)–(2) Let us denote by d a distance on X which is translation-invariant and compatible with the topology of X. If the assumptions of (1) are satisfied then we can apply Theorem 2.2.2 on $G = \mathbb{D}$ to obtain that $X \cap H_e(\mathbb{D})$ is residual in X. In particular, such subset is nonempty and we can pick a function $F \in X \cap H_e(\mathbb{D})$, hence $F \in X \setminus H(\overline{\mathbb{D}})$. If the hypotheses of (2) are assumed, then, by the property (e), we obtain the existence of a function $F \in Y \setminus H(\overline{\mathbb{D}})$ for some subset $Y \subset X$ satisfying, in addition, stability under projections. Thus, we may start in both cases with a function $F \in X \setminus H(\overline{\mathbb{D}})$ whose all projections $P_Q(F)$ $(Q \subset \mathbb{N}_0)$ are in X. Moreover, due to (d), there is a sequence $\{g_n : n \in \mathbb{N}\} \subset H(\overline{\mathbb{D}}) \cap X$ that is dense in X.

Consider the vector space $L(F) = \mathrm{span}\,\{F_n : n \in \mathbb{N}\}$ provided in the proof of the property (P). Recall that by construction we had in fact that

$$F_n = P_{A_n}(F) \quad (n \in \mathbb{N})$$

for certain sets $A_n \subset \mathbb{N}$. Then $F_n \in X$ for all $n \in \mathbb{N}$. Let us fix an $n \in \mathbb{N}$. The continuity of the multiplication by scalars in the topological vector space X gives the existence of a constant $\varepsilon_n > 0$ for which $d(\varepsilon_n F_n, 0) < 1/n$. Now we define

$$f_n := g_n + \varepsilon_n F_n \quad \text{and} \quad M := \mathrm{span}\,\{f_n : n \in \mathbb{N}\}.$$

We have that $f_n \in X$ for all n because $g_n, F_n \in X$, whence M is a vector space contained in X. Furthermore, the translation-invariance of d implies $d(f_n, g_n) = d(\varepsilon_n F_n, 0) < 1/n$, so $d(f_n, g_n) \to 0$ as $n \to \infty$. This and the density of $\{g_n : n \in \mathbb{N}\}$ imply the density of $\{f_n : n \in \mathbb{N}\}$, which in turn implies, trivially, that M is dense in X.

Finally, take a function $f \in M \setminus \{0\}$. Then there exist $N \in \mathbb{N}$ and complex constants $c_1, ..., c_N$ with $c_N \neq 0$ such that $f = c_1 g_1 + \cdots + c_N g_N + h$, where

$$h := \sum_{j=1}^{N} c_j \varepsilon_j F_j \in L(F) \setminus \{0\}.$$

Due to (P), $h \in H_e(\mathbb{D})$. But the function $g := c_1 g_1 + \cdots + c_N g_N$ is holomorphically continuable on $B(0, R)$ for some $R > 1$ (in fact, for $R = \min_{1 \leq n \leq N} R_n$, where R_n is the radius of convergence of the Taylor series of g_n). Consequently, the sum $f = g + h$ can be holomorphically continued beyond *no* point of $\partial\mathbb{D}$, that is, $f \in H_e(\mathbb{D})$, as required.

(3)–(4) We get as in the first part of the proof of (1)–(2) the existence of a function $F \in X \setminus H(\overline{\mathbb{D}})$. Let us follow the same notation as that in the proof of property (P). It is clear that the sequence $\{n(k) : k \in \mathbb{N}\}$ selected there may be chosen to satisfy $a_{n(k)} \neq 0$ for all $k \in \mathbb{N}$. In addition, let us denote $Q := \bigcup_{j\in\mathbb{N}} A_j$ and consider again the infinite dimensional vector space $L(F) = \operatorname{span}\{F_n : n \in \mathbb{N}\}$ constructed in (P). Then its closure $M := \overline{L(F)}$ in X is a closed infinite dimensional vector subspace. All that should be proved is $M \setminus \{0\} \subset H_e(\mathbb{D})$.

To this end, we observe that the conclusion will follow as soon as we demonstrate the following three properties:

(i) The set $\Lambda := \left\{f(z) = \sum_{n\in Q} c_n z^n \in X : \text{there exists } \{\lambda_j\}_{j=1}^\infty \subset \mathbb{C} \text{ such that } c_{p(j,k)} = \lambda_j a_{p(j,k)} \text{ for all } j, k \in \mathbb{N}\right\}$ contains $L(F)$.

(ii) Λ is closed in X.

(iii) $\Lambda \setminus \{0\} \subset H_e(\mathbb{D})$.

Indeed, (i) together with (ii) would imply that $M \subset \Lambda$, whence $M \setminus \{0\} \subset \Lambda \setminus \{0\} \subset H_e(\mathbb{D})$ by (iii), and we would be done.

Property (i) is trivial: it suffices to choose $\lambda_j = 0$ $(j > N)$ for each prescribed function $f = \sum_{j=1}^N \lambda_j F_j \in L(F)$. As for (ii), assume that

$$\left\{f_\alpha(z) := \sum_{n\in Q} c_n^{(\alpha)} z^n\right\}_{\alpha\in I} \subset \Lambda$$

is a net with $f_\alpha \to f$ in X. It must be shown that $f \in \Lambda$. Suppose that f has a Taylor expansion $f(z) = \sum_{n=0}^\infty c_n z^n$ $(z \in \mathbb{D})$. Due to (a), we have that $f_\alpha^{(n)}(0) \to f^{(n)}(0)$ for each $n \in \mathbb{N}_0$, so $c_n^{(\alpha)} \to c_n$. Then $c_n = 0$ for all $n \notin Q$ and $f(z) = \sum_{n\in Q} c_n z^n$. Moreover, for every $\alpha \in I$ there exists a sequence $\{\lambda_j^{(\alpha)}\}_{j=1}^\infty \subset \mathbb{C}$ such that $c_{p(j,k)}^{(\alpha)} = \lambda_j^{(\alpha)} a_{p(j,k)}$ for all $j, k \in \mathbb{N}$. Again by (a), we get $c_{p(j,k)}^{(\alpha)} \to c_{p(j,k)}$, hence

$$\lambda_j^{(\alpha)} \longrightarrow \frac{c_{p(j,k)}}{a_{p(j,k)}} \quad \text{for all } j, k \in \mathbb{N}.$$

But by the uniqueness of the limit, there must be constants $\lambda_j \in \mathbb{C}$ $(j \in \mathbb{N})$ satisfying $\lambda_j = c_{p(j,k)}/a_{p(j,k)}$, or equivalently, $c_{p(j,k)} = \lambda_j a_{p(j,k)}$ for all $j, k \in \mathbb{N}$. Then $f \in \Lambda$.

Finally, assume that $f \in \Lambda \setminus \{0\}$ and that f has a Taylor expansion about the origin as in the definition of Λ; see (i). Then there exists $J \in \mathbb{N}$ with $\lambda_J \neq 0$. Of course, $\limsup_{n\to\infty} |c_n|^{1/n} \leq 1$. But by (2.1),

$$\lim_{k\to\infty} |c_{p(J,k)}|^{1/p(J,k)} = \lim_{k\to\infty} |\lambda_J|^{1/p(J,k)} \cdot \lim_{k\to\infty} |a_{p(J,k)}|^{1/p(J,k)} = 1.$$

Therefore $\limsup_{n\to\infty} |c_n|^{1/n} = 1$, that is, the radius of convergence of the Taylor expansion of f is 1. On the other hand, the set Q consisted of the integers of the sequence $\{m(1) < m(2) < \cdots \}$, which had Hadamard gaps by virtue of (2.2). Hence (again) Hadamard's lacunary theorem comes to our aid yielding $f \in H_e(\mathbb{D})$. This shows (iii) and finishes the proof of (3)–(4).

(5) The spaces $A^\infty(\mathbb{D})$, H^p and B^p are F-spaces, so they are Baire and metrizable. That they satisfy (a) and (b) has been already mentioned. Since

$$A^\infty(\mathbb{D}) = \left\{ f(z) = \sum_{n=0}^{\infty} a_n z^n : \{n^N a_n\}_{n=0}^{\infty} \text{ is bounded for all } N \in \mathbb{N} \right\},$$

condition (c) is fulfilled by this space. Moreover, the set of polynomials is dense in it, because $\mathbb{D}$ is length-finite and $\mathbb{C} \setminus \overline{\mathbb{D}}$ is connected, see Exercise 2.7. Hence there is a countable dense subset in $A^\infty(\mathbb{D})$ that is contained in $H(\overline{\mathbb{D}})$, namely, the set of polynomials whose coefficients have rational real and imaginary parts. Therefore $A^\infty(\mathbb{D})$ satisfies (d), so (1) applies and $A^\infty(\mathbb{D}) \cap H_e(\mathbb{D})$ is dense-lineable in it. But observe that $A^\infty(\mathbb{D})$ is contained in B^p and in H^p, having its topology stronger than those of the last two spaces. This implies the dense-lineability part of (5). Now (3) and the properties above entail the spaceability of $A^\infty(\mathbb{D}) \cap H_e(\mathbb{D})$ in $A^\infty(\mathbb{D})$. Finally, we have already seen that (c) holds for the subset $Y := A^\infty(\mathbb{D})$ of H^p and B^p. Note that (e) also holds for Y, because $A^\infty(\mathbb{D}) \cap H_e(\mathbb{D}) \neq \varnothing$. From (4), we get the spaceability of $X \cap H_e(\mathbb{D})$ in X for $X = H^p$, B^p.

(6) Here we need to invoke the Carleson–Newman interpolation theorem, from which a consequence is the following (see, e.g., [252]): if $(z_n) \subset \mathbb{D}$ is a sequence such that

$$\sup_{n \in \mathbb{N}} \frac{1 - |z_n|}{1 - |z_{n-1}|} < 1 \quad (n \in \mathbb{N})$$

then it is H^∞-interpolating, that is, given a bounded sequence $(\alpha_n) \subset \mathbb{C}$, there is a function $f \in H^\infty$ such that $f(z_n) = \alpha_n$ for all $n \geq 1$. It is not difficult to find a sequence $(z_n) \subset \mathbb{D}$ satisfying the above property and such that, in addition, the set of accumulation points of $\{z_{2n} : n \geq 1\}$ is the unit circle $\mathbb{T}$ (it is enough that, for every rational m/n, there is a subsequence of (z_{2n}) converging to $e^{\frac{2\pi i m}{n}}$).

Now, consider the set

$$M := \{f \in H^\infty : f(z_{2n}) = 0 \text{ for every } n \in \mathbb{N}\}.$$

Plainly, M is a closed algebra in H^∞. As in the proof of Theorem 2.3.1, each nonzero member of M belongs to $H_e(\mathbb{D})$ and M is infinitely generated. Hence, we only need to show that M is non-separable. Indeed, let J be an arbitrary set of odd positive integers. Select an element $f_J \in M$ such that $f_J(z_n) = 1$ for $n \in J$ and $f_J(n) = 0$ for all other n. Since the infinite subsets of $\mathbb{N}$ form an uncountable set, we have just found an uncountable family $\{f_J\}_J \subset M$ such that $\|f_J - f'_J\|_\infty \geq 1$ as soon as $J \neq J'$. Hence the $\|\cdot\|_\infty$-balls $B(f_J, 1/2)$ are pairwise disjoint and, consequently, there cannot be a denumerable dense subset in M. In other words, M is non-separable, as required. $\square$

The spaces H^p, B^p and $A^\infty(\mathbb{D})$ considered above have, in some sense, moderated growth near the boundary $\mathbb{T}$. From an opposite point of view, one way of being nonextendable is to grow fast near the boundary. In this vein, an interesting family in $H(\mathbb{D})$ is the one defined below.

Definition 2.4.2. A function $f \in H(\mathbb{D})$ is called *strongly annular* provided that

$$\limsup_{r \to 1} \min\{|f(z)| : |z| = r\} = \infty.$$

The set of these functions will be denoted by $\mathcal{SA}$.

Note that $\mathcal{SA} \subset H_e(\mathbb{D})$. In 1975, Bonar and Carroll [127] established the residuality of $\mathcal{SA}$. In fact, a "weighted" improvement is possible, as Theorem 2.4.3 shows, which in turn will be useful to state algebraic genericity for $\mathcal{SA}$; see Theorem 2.4.5 below. The content of these two theorems can be found in [86]. Prior to establishing them, it is convenient to introduce some new notation. Denote by Σ the set of all strictly increasing sequences $\sigma = \{r_n\}_{n \geq 1} \subset (0, 1)$ with $r_n \to 1$. If $\sigma = \{r_n\}_{n \geq 1} \in \Sigma$, we set $C(\sigma) := \bigcup_{n=1}\{z : |z| = r_n\}$. If $\varphi : \mathbb{D} \to (0, +\infty)$ is continuous and $\sigma \in \Sigma$, we define

$$\mathcal{SA}(\varphi) := \left\{ f \in H(\mathbb{D}) : \limsup_{r \to 1} \min\{\frac{|f(z)|}{\varphi(z)} : |z| = r\} = \infty \right\} \quad \text{and}$$

$$\mathcal{SA}(\varphi, \sigma) := \{ f \in H(\mathbb{D}) : \lim_{\substack{|z| \to 1 \\ z \in C(\sigma)}} \frac{|f(z)|}{\varphi(z)} = \infty \}.$$

Then it is plain that $\mathcal{SA}(\varphi) = \bigcup_{\sigma \in \Sigma} \mathcal{SA}(\varphi, \sigma)$ and that $\mathcal{SA}(1) = \mathcal{SA}$.

Theorem 2.4.3. *Let be prescribed a continuous function* $\varphi : \mathbb{D} \to (0, +\infty)$ *and a sequence* $\sigma \in \Sigma$. *Then the set* $\mathcal{SA}(\varphi, \sigma)$ *is residual in* $H(\mathbb{D})$. *Consequently,* $\mathcal{SA}(\varphi)$ *is also residual in* $H(\mathbb{D})$.

Proof. Let $\sigma = (r_n)$, so that $0 < r_1 < r_2 < \cdots \to 1$. For every pair $m, n \in \mathbb{N}$ we denote $\mathcal{S}_{m,n} := \{f \in H(\mathbb{D}) : |f(z)| > n\varphi(z) \text{ for all } z \in C_m\}$, where $C_m := \{z : |z| = r_m\}$. If we set $\mathcal{S}_n = \bigcup_{m \geq n} \mathcal{S}_{m,n}$ $(n \in \mathbb{N})$ then one can express

$$\mathcal{SA}(\varphi, \sigma) = \bigcap_{n=1}^{\infty} \mathcal{S}_n.$$

For each compact set $K \subset \mathbb{D}$ and each continuous function f on $\mathbb{D}$ we set $\|f\|_K := \sup\{f(z)| : z \in K\}$ and $m(f, K) := \min\{|f(z)| : z \in K\}$. Recall that a basic open neighborhood of a function $g \in H(\mathbb{D})$ has the form $V(g, K, \varepsilon) = \{h \in H(\mathbb{D}) : \|h - g\|_K < \varepsilon\}$, where $\varepsilon > 0$ and K is a compact subset of $\mathbb{D}$.

Fix $m, n \in \mathbb{N}$. If $g \in \mathcal{S}_{m,n}$ then $\delta := m(|g| - n\varphi, C_m) > 0$. If $h \in V(g, C_m, \delta)$ then we have for all $z \in C_m$ that $-|h(z)| + |g(z)| \leq |h(z) - g(z)| < m(|g| - n\varphi, C_m)$, so

$$|h(z)| > |g(z)| - m(|g| - n\varphi, C_m) \geq |g(z)| - |g(z)| + n\varphi(z) = n|\varphi(z)|.$$

Hence $V(g, C_m, \delta) \subset \mathcal{S}_{m,n}$, which proves that $\mathcal{S}_{m,n}$ is open. Therefore every $\mathcal{S}_n$ is open. By Baire's theorem it is enough to show that each $\mathcal{S}_n$ is dense. To this end, fix a basic open set $V(g, K, \varepsilon)$. Choose $m \geq \max\{n, 3\}$ such that $K \subset B(0, r_{m-2})$. Since $r_{m-2} < r_{m-1} < r_m$, we can select $p \in \mathbb{N}$ satisfying

$$\left(\frac{r_{m-2}}{r_{m-1}}\right)^p < \frac{\varepsilon}{\|\varphi\|_{C_m}} \quad \text{and} \quad \left(\frac{r_m}{r_{m-1}}\right)^p > n + \frac{\|g\|_{C_m}}{\|\varphi\|_{C_m}}.$$

Define $f(z) := g(z) + (z/r_{m-1})^p \|\varphi\|_{C_m}$. Then

$$\|f - g\|_K \leq (r_m/r_{m-1})^p \|\varphi\|_{C_m} < \varepsilon,$$

so $f \in V(g, K, \varepsilon)$. Furthermore, for all $z \in C_m$,

$$|f(z)| > (r_m/r_{m-1})^p \|\varphi\|_{C_m} - |g(z)| > n\|\varphi\|_{C_m} + \|g\|_{C_m} - |g(z)| \geq n\varphi(z).$$

Thus, $f \in V(g, K, \varepsilon) \cap \mathcal{S}_n$, which proves the density of $\mathcal{S}_n$. $\square$

The following auxiliary result is needed in the proof of Theorem 2.4.5.

Lemma 2.4.4. *Assume that $\varphi : \mathbb{D} \to (0, +\infty)$ is a continuous function satisfying*

$$\lim_{|z| \to 1} \frac{\log \varphi(z)}{\log \frac{1}{1-|z|}} = \infty.$$

If $f \in \mathcal{SA}(\varphi)$ and $g \in H(\overline{\mathbb{D}}) \setminus \{0\}$ then $fg \in \mathcal{SA}$.

Proof. Choose a connected open set with $G \supset \overline{\mathbb{D}}$ and $g \in H(G)$. From the Analytic Continuation Principle one derives that there are only finitely many zeros of g on $\overline{\mathbb{D}}$. Hence we can assume that g possesses zeros $z_1, \ldots, z_p$ in $\mathbb{D}$ and zeros $w_1, \ldots, w_q$ on $\mathbb{T} = \{z : |z| = 1\}$, with respective multiplicities $m_1, \ldots, m_p, n_1, \ldots, n_q$ (other cases are easier to handle). Then $g = PQh$, where $h \in H(G)$, h lacks zeros in $\overline{\mathbb{D}}$ and $P(z) := \prod_{k=1}^p (z - z_k)^{m_k}$, $Q(z) := \prod_{k=1}^q (z - w_k)^{n_k}$. By hypothesis, $f \in \mathcal{SA}(\varphi, \sigma)$ for some sequence $\sigma = (r_n) \in \Sigma$. Let n_0 be such that $r_n > \max\{|z_1|, \ldots, |z_p|\}$ for all $n \geq n_0$, and choose $\alpha, \beta > 0$ with $|h(z)| > \alpha$ ($z \in \mathbb{D}$) and $|z - z_k| > \beta$ ($|z| = r_n$, $n \geq n_0$; $k = 1, \ldots, p$). If $z \in \bigcup_{n \geq n_0} r_n \mathbb{T} =: A$ we have

$$|f(z)g(z)| = |h(z)P(z)Q(z)| \cdot \varphi(z) \cdot \frac{|f(z)|}{\varphi(z)}$$

$$> \alpha \prod_{k=1}^p |z - z_k|^{m_k} \prod_{k=1}^q (1 - |z|)^{n_k} \varphi(z) \cdot \frac{|f(z)|}{\varphi(z)}$$

$$> \alpha \beta^{\text{degree}(P)} (1 - |z|)^{\text{degree}(Q)} \varphi(z) \cdot \frac{|f(z)|}{\varphi(z)}.$$

By hypothesis, $\lim_{|z|\to 1}(1-|z|)^N \varphi(z) = \infty$ for all $N \in \mathbb{N}$. But recall that $\lim_{\substack{|z|\to 1 \\ z \in A}} \frac{|f(z)|}{\varphi(z)} = \infty$. Therefore $\lim_{\substack{|z|\to 1 \\ z \in A}} |f(z)g(z)| = \infty$, that is, $fg \in \mathcal{SA}$. $\square$

Theorem 2.4.5. *$\mathcal{SA}$ is maximal dense-lineable and strongly algebrable in* $H(\mathbb{D})$.

Proof. Fix a translation-invariant distance d on $H(\mathbb{D})$ generating its topology, and let (P_n) be an enumeration of the complex polynomials whose coefficient have rational real and imaginary parts. Then (P_n) is a dense subset of $H(\mathbb{D})$. Assign to each $n \in \mathbb{N}$ any fixed $\beta_n \in [n-1, n)$. Consider the function

$$\varphi(z) := \exp \frac{1}{1-|z|} \quad (z \in \mathbb{D}).$$

According to Theorem 2.4.3 we can select a function $f_0 \in \mathcal{SA}(\varphi)$. Define the functions $e_\alpha(z) := \exp(\alpha z)$ $(\alpha > 0)$. Due to the continuity of the multiplication by scalars in a topological vector space, there is $\varepsilon_n > 0$ such that $d(0, \varepsilon_n e_{\beta_n} f_0) < 1/n$.

Consider the set

$$M := \mathrm{span}\{P_n + \varepsilon_n e_\alpha f_0 : \alpha \in [n-1, n), \, n \in \mathbb{N}\}.$$

It is clear that M is a vector subspace of $H(\mathbb{D})$. In addition, it is dense, because it contains the set $\{P_n + \varepsilon_n e_{\beta_n} f_0 : n \geq 1\}$, which in turn is dense since (P_n) is dense and

$$d(P_n, P_n + \varepsilon_n e_{\beta_n}) = d(0, \varepsilon_n e_{\beta_n} f_0) < \frac{1}{n} \to 0 \quad \text{as} \quad n \to \infty.$$

Moreover, $\dim(M) = \mathfrak{c}$. Indeed, since $\mathrm{card}\,[0, \infty) = \mathfrak{c}$, it is enough to prove the linear independence of the family $\{P_n + \varepsilon_n e_\alpha f_0 : \alpha \in [n-1, n), \, n \in \mathbb{N}\}$. To this end, consider a nontrivial linear combination

$$F := \sum_{j=1}^{N} a_j (P_j + \varepsilon_j e_{\alpha_j} f_0) = 0,$$

where $\alpha_j \in [j-1, j)$ and, without loss of generality, $a_N \neq 0$. Denote by Q the polynomial $\sum_{j=1}^{N} a_j P_j$, so that $Q + f_0 \sum_{j=1}^{N} a_j \varepsilon_j e_{\alpha_j} = 0$. Note that if $\sum_{j=1}^{N} a_j \varepsilon_j e_{\alpha_j} = 0$ on $\mathbb{D}$ then, by the analytic continuation principle, $\sum_{j=1}^{N} a_j \varepsilon_j e_{\alpha_j} = 0$ on $\mathbb{C}$ and, in particular,

$$a_N = -\varepsilon_N^{-1} \sum_{j=1}^{N-1} a_j \varepsilon_j e^{(\alpha_j - \alpha_N)x} \quad \text{for all } x \in \mathbb{R}.$$

Letting $x \to +\infty$, we have $e^{(\alpha_j - \alpha_N)x} \to 0$ $(j = 1, \ldots, N-1)$, hence $a_N = 0$, a contradiction. Hence the function $g := \sum_{j=1}^{N} a_j \varepsilon_j e_{\alpha_j}$ belongs to $H(\overline{\mathbb{D}}) \setminus \{0\}$.

By Lemma 2.4.4, $f_0 \, g \in \mathcal{SA}$ and, since Q is bounded on $\mathbb{D}$, $F = Q + f_0 g \in \mathcal{SA}$, which is absurd because $F = 0$. This shows the required linear independence. But the same argument proves that $M \setminus \{0\} \subset \mathcal{SA}$. Consequently, $\mathcal{SA}$ is $\mathfrak{c}$-lineable.

Let us now face the algebrability of $\mathcal{SA}$. For $f \in H(\mathbb{D})$ the standard notation $M(f, r) := \max\{|f(z)| : |z| = r\}$ $(0 < r < 1)$ will be used. In addition, set $m(f, r) := \min\{|f(z)| : |z| = r\}$.

We start with a function $f_1 \in \mathcal{SA}$. Then there is a sequence of radii $\sigma = (r_n) \in \Sigma$ such that $\lim_{\substack{|z| \to 1 \\ z \in C(\sigma)}} |f_1(z)| = \infty$, where $C(\sigma) = \bigcup_{n=1}^{\infty} r_n \mathbb{T}$. Hence $f_1 \in \mathcal{SA}(\varphi_0, \sigma)$, where $\varphi_0 \equiv 1$. Let $\varphi_1(z) := \exp M(f_1, |z|)$. According to Theorem 2.4.3, we can select a function $f_2 \in \mathcal{SA}(\varphi_1, \sigma)$. By induction, assume that for some $N \geq 2$ the functions $f_1, \ldots, f_{N-1}, \varphi_0, \ldots, \varphi_{N-2}$ have been already determined. Then we define $\varphi_{N-1}(z) := \exp M(f_{N-1}, |z|)$ and, again by Theorem 2.4.3, one can choose a function $f_N \in \mathcal{SA}(\varphi_{N-1}, \sigma)$. Therefore we obtain a sequence of functions $(f_n) \subset H(\mathbb{D})$ such that $f_n \in \mathcal{SA}(\varphi_{n-1}, \sigma)$ $(n \geq 1)$, where $\varphi_0 \equiv 1$ and $\varphi_j(z) := \exp M(f_j, |z|)$ $(j \geq 1)$. Define M as the algebra generated by the functions f_k $(k \in \mathbb{N})$. Our task is to show that M is a free algebra such that $M \setminus \{0\} \subset \mathcal{SA}$. To achieve this, it is enough to show that, for any $N \in \mathbb{N}$, any nonzero polynomial $P(w_1, \ldots, w_N)$ of N complex variables without constant term and any collection of different indexes $i_1, \ldots i_N \in \mathbb{N}$, the function $F := P(f_{i_1}, \ldots, f_{i_N})$ belongs to $\mathcal{SA}$.

We proceed by induction. Let $P(w) = c_1 w + \cdots + c_m w^m$ be a nonzero polynomial with $c_m \neq 0$, and let f be one of the functions f_k. Then $|P(w)| \to \infty$ if $|w| \to \infty$ and $m(f, r_n) \to \infty$ as $n \to \infty$. Therefore, given $\alpha > 0$, there exists $\beta > 0$ such that $|P(w)| > \alpha$ whenever $|w| > \beta$. Moreover, there exists $n_0 \in \mathbb{N}$ such that $|f(z)| > \beta$ for all $z \in \bigcup_{n \geq n_0} r_n \mathbb{T}$. Hence $|P(f(z))| > \beta$ for all these z. In other words, $m(P(f), r_n) \to \infty$ as $n \to \infty$, so $F = P(f) \in \mathcal{SA}$ and the desired property is proved in the case $N = 1$. Assume that the property is proved for $N - 1$, where $N \geq 2$. Let $F := P(f_{i_1}, \ldots, f_{i_N})$, where $P(w_1, \ldots, w_N)$ is as in the preceding paragraph. Without loss of generality, one can assume that $i_1 < \cdots < i_N$ and that the polynomial P has the form

$$P(w_1, \cdots, w_n) = \sum_{j=1}^{m} P_j(w_1, \ldots, w_{N-1}) \, w^j$$

with P_j $(j = 1, \ldots, m)$ polynomials of $N - 1$ variables and $P_m \not\equiv 0$. Then either P_m is a nonzero constant or, by the induction hypothesis, it satisfies $P_m(g_1, \ldots, g_N) \in \mathcal{SA}$, where we have denoted $g_j := f_{i_j}$. In any case, one gets the existence of a constant $\alpha > 0$ such that $m(P_m(g_1, \ldots, g_{N-1}), r_n) > \alpha$ for n large enough. Observe that, since each function $P_j(g_1, \ldots, g_{N-1})$ is a finite sum of finite products of powers of $g_1, \ldots, g_{N-1}$, the selection of the functions $f_1, f_2, \ldots$ allows to assert that $M(P_j(g_1, \ldots, g_{N-1}), r_n)/m(r_n, g_N) \to 0$ as $n \to \infty$. Hence

$$\lim_{n \to \infty} M(P_j(g_1, \ldots, g_{N-1}), r_n) \, m(g_N, r_n)^{-k} = 0$$

for all $k \in \mathbb{N}$. Finally, the triangle inequality and the definition of $m(\cdot, r)$ and $M(\cdot, r)$ leads us to

$$
\begin{aligned}
m(F, r_n) &\geq \min_{|z|=r_n} \left(|g_N(z)|^N \, P_m(g_1(z), \dots, g_{N-1}(z)) | \right. \\
&\quad \left. - \sum_{j=1}^{m-1} |P_j(g_1(z), \dots, g_{N-1}(z))| \, |g_N(z)|^j \right) \\
&\geq \min_{|z|=r_n} |g_N(z)|^N \left[m(P_m(g_1, \dots, g_{N-1}), r_n) \right. \\
&\quad \left. - \sum_{j=1}^{m-1} |P_j(g_1(z), \dots, g_{N-1}(z))| \, |g_N(z)|^{j-m} \right] \\
&\geq m(g_N, r_n)^N \left(\alpha - \sum_{j=1}^{m-1} M(P_j(g_1, \dots, g_{N-1}), r_n) \, m(g_N, r_n)^{j-m} \right) \\
&\longrightarrow \infty \quad (n \to \infty).
\end{aligned}
$$

This completes induction and shows that $M \setminus \{0\} \subset \mathcal{SA}(1, \sigma) \subset \mathcal{SA}$, which had to be proved. $\qquad\square$

In fact, weights can be imposed on the growth of functions; see [86]. Turning to subspaces X of $H(\mathbb{D})$, recall that some criteria for lineability properties of $X \cap H_e(\mathbb{D})$ have been shown in Theorem 2.4.1. This raises the question of the lineability of the smaller set $X \cap \mathcal{SA}$. But this set is *empty* (hence non-lineable) for important spaces $X \subset H(\mathbb{D})$. For instance, it is evident that no $f \in \mathcal{SA}$ has a continuous extension to $\partial \mathbb{D}$; and $\mathcal{SA} \cap H^p = \varnothing$ $(p > 0)$ thanks to the Fatou theorem asserting the existence of finite radial limit almost everywhere on $\partial \mathbb{D}$ for every $f \in H^p$; see [194].

2.5 Tamed entire functions

One of the most beautiful results in complex analysis is Liouville's theorem (see, e.g., [2]): every bounded entire function is constant. In spite of this, Mittag-Leffler [306] gave in 1920 an example of an entire function tending to 0 (hence bounded) along every (straight) line through the origin. In 1924, K. Grandjot [230] modified Mittag-Leffler's function to get an entire function $\mathbb{C} \to \mathbb{C}$ receding to 0 along any algebraic curve, so along any straight line. This surprising result has been improved in several ways, for instance, adding boundedness to all derivatives or imposing integrability on every line; see, e.g., [14, 386].

The question naturally arises as to whether this kind of "tamed" (or "anti-

Liouville") entire function enjoys some sort of lineability. The answer is affirmative, as the following theorem shows.

Theorem 2.5.1. *Let* $\varphi : [0, \infty) \to (0, \infty)$ *be an increasing function. There exists a vector subspace* M *which is dense in* $H(\mathbb{C})$ *such that* $\lim_{\substack{z \to \infty \\ z \in L}} f(z) = 0$

for every line L *and every* $f \in M$, *and* $\lim_{r \to +\infty} \dfrac{\max\{|f(z)| : |z| = r\}}{\varphi(r)} = \infty$

for every $f \in M \setminus \{0\}$.

Proof. The main idea is to apply Arakelian's tangential approximation theorem, see Section 2.1. Let $\{P_n\}_{n \geq 1}$ be an enumeration of all polynomials having rational real and imaginary parts. Then $\{P_n\}_{n \geq 1}$ is a dense subset of $H(\mathbb{C})$. Consider the closed balls $B_n := \overline{B}(0, n)$ $(n \geq 1)$. Consider also the curve $y = x\,e^{-x}$ $(x > 0)$, which is positive, attains its maximum at $x = 1$ and tends to 0 when $x \to +\infty$. Then it is evident that

$$A := \mathbb{C} \setminus \{z = x + iy \in \mathbb{C} : x > 0 \ \text{and} \ 0 < y < x\,e^{-x}\}$$

is an Arakelian subset of $\mathbb{C}$ and that each line L is contained in A except for a bounded subset (see Figure 2.2). We can select a sequence $\{z_n = x_n + iy_n : n \geq 1\}$ satisfying $|z_n| = n$ and $z_n \notin A$ for all $n \geq 1$. Therefore, for each $n \in \mathbb{N}$, the set $A_n := B_n \cup (A \cap \{z : |z| \geq n+1\}) \cup \{z_k\}_{k \geq 1}$ is an Arakelian set as well: note that B_n does not introduce new holes, and the sequence $\{z_k\}_{k \geq 1}$ makes no influence on the connectedness or the local connectedness of $\mathbb{C}_\infty \setminus A_n$. For fixed $n \in \mathbb{N}$, let $\varepsilon(t) := \frac{1}{n+t}$. This is a continuous positive function on $[0, \infty)$ satisfying, clearly, that

$$\int_1^\infty t^{-3/2} \log(1/\varepsilon(t))\, dt < \infty.$$

For every entire function f, we use the notation $M(f, r) := \max\{|f(z)| : |z| = r\}$ $(r > 0)$. The maximum modulus principle guarantees that $M(f, r) = \max\{|f(z)| : |z| \leq r\}$ and that $M(f, \cdot)$ is an increasing function of r.

   With all these ingredients at hand, we are ready to use the mentioned Arakelian theorem. We define recursively the functions $f_n : A_n \to \mathbb{C}$ and the entire functions g_n $(n = 1, 2, \dots)$ by following the order $f_1, g_1, f_2, g_2, f_3, g_3, \dots$. For this, we let

$$f_n(z) := \begin{cases} P_n(z) & \text{if} \quad z \in B_n \\ 0 & \text{if} \quad z \in A \cap \{w : |w| \geq n+1\} \\ \gamma_{k,n} & \text{if} \quad z = z_k \ (k \geq n+1), \end{cases}$$

where $\gamma_{k,n} := (1+k)\big(1 + \sum_{j=1}^{n-1} M(g_j, n)\big)(1 + \varphi(k+1))$ and the sum $\sum_{j=1}^{n-1}$ is defined as 0 if $n = 1$. It is evident that each f_n belongs to $A(A_n)$. Once one has defined $f_1, g_1, \dots, f_{n-1}$ and g_{n-1}, the function g_n is selected so as to satisfy

$$|g_n(z) - f_n(z)| < \frac{1}{n + |z|} \quad \text{for all } z \in A_n. \tag{2.5}$$

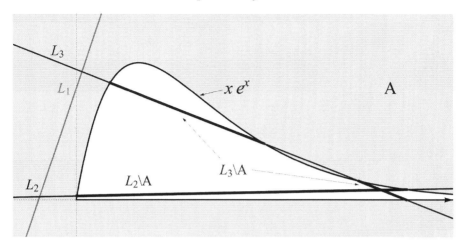

FIGURE 2.2: Any line L is almost contained in the set A.

Define $M := \operatorname{span}\{g_n : n = 1, 2, \ldots\}$. From (2.5), we get that

$$|g_n(z) - P_n(z)| < 1/n \text{ for all } z \in B_n.$$

Since (P_n) is a dense sequence in $H(\mathbb{C})$ and the balls B_n form an exhaustive sequence of compact subsets of $\mathbb{C}$, we obtain that the set (g_n) is also dense. As $M \supset (g_n)$, we get that M is a dense vector subspace of $H(\mathbb{C})$. Again by (2.5), we have that $|g_n(z)| < \frac{1}{1+|z|}$ for all n and all $z \in A \cap \{w : |w| \geq n+1\}$. But, for each n, the set $A \cap \{w : |w| \geq n+1\}$ contains each given line L except for a bounded subset, because A does. Hence $\lim_{\substack{z \to \infty \\ z \in L}} g_n(z) = 0$ for every such L and, by linearity, $\lim_{\substack{z \to \infty \\ z \in L}} g(z) = 0$ for all $g \in M$.

It remains to show that every function $g \in M \setminus \{0\}$ grows more rapidly than φ. Without loss of generality, we can assume that there are $N \in \mathbb{N}$ and scalars $c_1, \ldots, c_N$ such that $c_N \neq 0$ and $g = \sum_{j=1}^{N} c_j g_j$. In the following, the sums $\sum_{j=1}^{N-1}$ should be interpreted as 0 if $N = 1$. From the triangle inequality, we get

$$M(g, r) \geq |c_N| M(g_N, r) - \sum_{j=1}^{N-1} |c_j| M(g_j, r)$$

$$= |c_N| M(g_N, r) \left(1 - \sum_{j=1}^{N-1} \frac{|c_j| M(g_j, r)}{|c_N| M(g_N, r)}\right)$$

for all $r > 0$. From (2.5) and the definition of the f_n's, one derives that

$$M(g_N, n) \geq |g_N(z_n)| > (n+1)\,\varphi(n+1) \quad \text{and}$$

$$(n+1)\,M(g_j, n) < |g_N(z_n)| \leq M(g_N, n)$$

as soon as $n \geq N+1$ and $j \in \{1, \ldots, N-1\}$. Let $r \geq N+1$ and denote by $[x]$ the integer part of every $x \in \mathbb{R}$. We can assume $\varphi(r) \geq 1$ for all $r > 0$. Then

$$\frac{M(g,r)}{\varphi(r)} \geq \frac{M(g,[r])}{\varphi([r]+1)} \geq \frac{|c_N| M(g_N,[r])}{\varphi([r]+1)} \left(1 - \sum_{j=1}^{n-1} \frac{|c_j|}{|c_N|} \frac{M(g_j,[r])}{M(g_N,[r])}\right)$$

$$\geq |c_N|([r]+1) \left(1 - \sum_{j=1}^{N-1} \frac{|c_j|}{|c_N|} \frac{M(g_j,[r])}{M(g_N,[r])}\right)$$

$$\geq |c_N|([r]+1) \left(1 - \sum_{j=1}^{N-1} \frac{|c_j|}{|c_N|} \frac{1}{[r]+1}\right).$$

Hence $\lim\limits_{r \to +\infty} \dfrac{M(g,r)}{\varphi(r)} = \infty$, which had to be shown. $\qquad\square$

Notice that one obtains entire functions being small and big simultaneously. Additional properties enjoying lineability as well can be found in [16, 71, 72, 133].

2.6   Wild behavior near the boundary

In this section we deal with holomorphic functions in a plane domain G having, in some sense, wild behavior near all points of ∂G. The formal definitions are as follows. Recall that a subset A of a topological space X is said to be relatively compact if its closure $\overline{A}$ is a compact subset of X.

Definition 2.6.1. Let $G \subset G$ be a domain with $G \neq \mathbb{C}$, $f \in H(G)$ and A be a subset of G that is not relatively compact in G. The *cluster set* of f along A is defined as the set

$$C(f,A) := \{w \in \mathbb{C}_\infty : \exists (z_n) \subset A \text{ such that } (z_n) \text{ tends to}$$
$$\text{some point of } \partial G \text{ and } f(z_n) \to w\}.$$

If $\xi \in \overline{A} \cap \partial G$, the *cluster set* of f along A at ξ is the set

$$C(f,A,\xi) := \{w \in \mathbb{C}_\infty : \exists (z_n) \subset A \text{ such that } z_n \to \xi \text{ and } f(z_n) \to w\}.$$

The *cluster set* of f at ξ is $C(f,\xi) := C(f,G,\xi)$. Any of these cluster sets is said to be *maximal* whenever it equals $\mathbb{C}_\infty$.

Observe that all these cluster sets are nonempty and closed in $\mathbb{C}_\infty$ and that $C(f,A) = \overline{\bigcup_{\xi \in \overline{A} \cap \partial G} C(f,A,\xi)}$, where the closure is taken on $\mathbb{C}_\infty$. In 1933,

Kierst and Szpilrajn [273] established, for a Jordan domain G, the residuality of the family of functions f holomorphic in G such that $C(f, \xi)$ is maximal for every $\xi \in \partial G$. In 2004, Bernal, Calderón and Prado-Bassas [95] gave in 2004 a strong linear version of this result. This will be established in Theorem 2.6.2 but, prior to this, recall that a (continuous) curve $\gamma : [0, 1) \to G$ is said to tend to the boundary of G provided that, given a compact set $K \subset G$, there exists $t_0 = t_0(K) \in [0, 1)$ such that $\gamma(t) \notin K$ for all $t > t_0$. By abuse of language, we sometimes identify $\gamma = \gamma([0, 1))$.

Theorem 2.6.2. *Assume that $G \subset \mathbb{C}$ is a Jordan domain. Then the set $CM(G) := \{f \in H(G) : C(f, \gamma)$ is maximal for every curve $\gamma \subset G$ tending to ∂G whose closure does not contain $\partial G\}$ is dense-lineable in $H(G)$.*

Proof. By the Osgood-Carathéodory theorem there exists an homeomorphism $\varphi : \overline{G} \to \overline{\mathbb{D}}$ whose restriction on G is a holomorphic isomorphism from G onto $\mathbb{D}$. Then if $\mathcal{D}$ were the dense vector subspace contained in $CM(\mathbb{D}) \cup \{0\}$ then the set

$$\mathcal{D}_1 := \{f \circ \varphi : f \in \mathcal{D}\}$$

would be the desired dense vector subspace in $CM(G) \cup \{0\}$. This is easy. Suffice it to say that: the composition operator $C_\varphi f = f \circ \varphi$ is linear; for every compact subset $K \subset G$ the image $\varphi(K)$ is a compact subset of $\mathbb{D}$ and for every curve $\gamma \subset G$ tending to the boundary ∂G with $\overline{\gamma} \not\supset \partial G$ we have that $\varphi(\gamma) \subset \mathbb{D}$ is also a curve tending to the boundary $\partial \mathbb{D}$ with $\overline{\varphi(\gamma)} \not\supset \partial \mathbb{D}$. The reader can easily complete the remaining details.

Hence we may suppose that $G = \mathbb{D}$ from now on. Assume that $\{P_n^*\}_{n=1}^\infty$ is a countable dense subset of $H(\mathbb{D})$ (for instance, an enumeration of the holomorphic polynomials having coefficients with rational real and imaginary parts). Then we consider a sequence $\{P_n\}_{n=1}^\infty$ where each P_n^* occurs infinitely many times. We also fix two sequences $\{r_n\}, \{s_n\}$ of positive real numbers satisfying

$$r_1 < s_1 < r_2 < s_2 < \cdots < r_n < s_n < \cdots$$

and $\lim_{n\to\infty} r_n = 1 = \lim_{n\to\infty} s_n$. Let us divide $\mathbb{N}$ into infinitely many strictly increasing sequences $\{p(n, j) : j = 1, 2, \dots\}$ $(n \in \mathbb{N})$. For fixed $n \in \mathbb{N}$ we consider the set $F_n \subset \mathbb{D}$ given by the disjoint union

$$F_n = \overline{B}\left(0, \frac{n}{n+1}\right) \cup \bigcup_{j=J(n)}^\infty K_j,$$

where $J(n) := \min\{j \in \mathbb{N} : r_j > \frac{n}{n+1}\}$ and each K_j is the spiral compact set

$$K_j = \left\{\left(r_j + \frac{s_j - r_j}{4\pi}\theta\right)\exp(i\theta) : \theta \in [0, 4\pi]\right\}.$$

Observe that each K_j has connected complement and that the sequence $\{K_j\}_{j=1}^\infty$ goes to $\mathbb{T}$. Note also that every F_n is closed in $\mathbb{D}$. Consider the one-point compactification $\mathbb{D}_*$ of $\mathbb{D}$, with ω standing for its infinity point. A simple glance reveals that $\mathbb{D}_* \setminus F_n$ is connected (indeed, $\mathbb{D} \setminus F_n$ is connected and $\mathbb{D} \setminus F_n \subset \mathbb{D}_* \setminus F_n \subset$ the closure in $\mathbb{D}_*$ of $\mathbb{D} \setminus F_n$) and locally connected at ω (by a similar reason). In addition, F_n satisfies the following property: for every compact subset $K \subset \mathbb{D}$ there exists a neighborhood V of ω in $\mathbb{D}_*$ such that no component of the interior F_n^0 of F_n intersects both K and V; indeed, $F_n^0 = B(0, \frac{n}{n+1})$ and for any K we can choose $V := \{\omega\} \cup \{\frac{n}{n+1} < |z| < 1\}$. Under these three topological conditions, Nersesjan's theorem (see Section 2.1) asserts the existence of a function $f_n \in H(\mathbb{D})$ approaching a given continuous function $g_n : F_n \to \mathbb{C}$ with g_n holomorphic in F_n^0 within a prescribed continuous error function $\varepsilon(z)$. Let $\{q_j\}_{j=1}^{\infty}$ be any fixed dense sequence in $\mathbb{C}$. If we select $\varepsilon(z) := \frac{1-|z|}{n}$ then we obtain

$$|f_n(z) - g_n(z)| < \frac{1 - |z|}{n} \qquad (z \in F_n), \qquad (2.6)$$

where $g_n : F_n \to \mathbb{C}$ is the function defined as

$$g_n(z) = \begin{cases} P_n(z) & \text{if } z \in \overline{B}(0, \frac{n}{n+1}) \\ q_j & \text{if } z \in K_{p(n,j)} \text{ and } p(n,j) \geq J(n) \\ 0 & \text{if } z \in K_{p(k,j)} \ (k \neq n) \text{ and } p(k,j) \geq J(n). \end{cases}$$

Observe that, trivially, g_n is continuous on F_n and holomorphic in F_n^0, so Nersesjan's theorem applies properly. Let us define $\mathcal{D}$ as the linear span

$$\mathcal{D} = \text{span} \{f_n : n \in \mathbb{N}\}.$$

Of course, $\mathcal{D}$ is a linear submanifold of $H(\mathbb{D})$, and $\mathcal{D}$ is dense because $\{f_n\}_{n=1}^{\infty}$ is. Indeed, from (2.6) we have that

$$|f_n(z) - P_n(z)| < \frac{1}{n} \quad \text{for all } z \in \overline{B}(0, \frac{n}{n+1}).$$

Then if we fix a function P_m^* there exists a sequence $n_1 < n_2 < \cdots$ with $P_{n_j} = P_m^*$ for all $j \in \mathbb{N}$. Now if $K \subset \mathbb{D}$ is compact then there is $j_0 \in \mathbb{N}$ such that $K \subset \overline{B}(0, \frac{n_j}{n_j+1})$ for every $j > j_0$. Therefore

$$|f_{n_j}(z) - P_m^*(z)| < \frac{1}{n_j} \quad \text{for all } z \in K \text{ and all } j > j_0,$$

so $f_{n_j} \to P_m^*$ $(j \to \infty)$ uniformly on compacta in $H(\mathbb{D})$. Hence the closure of $\{f_n : n \in \mathbb{N}\}$ in $H(\mathbb{D})$ contains the dense set $\{P_m^* : m \in \mathbb{N}\}$, which proves the density of $\{f_n\}_{n=1}^{\infty}$.

It remains to show that $\mathcal{D} \setminus \{0\} \subset \mathcal{CM}(\mathbb{D})$. That is, it should be proved that, for every prescribed curve $\gamma \subset \mathbb{D}$ escaping towards $\mathbb{T}$ with $\overline{\gamma} \not\supset \mathbb{T}$, and for every function $f \in \mathcal{D} \setminus \{0\}$, we have $C(f, \gamma) = \mathbb{C}_{\infty}$. Note that for such a function f there exist $N \in \mathbb{N}$ and complex scalars $\lambda_1, \ldots, \lambda_N$ such that $\lambda_N \neq 0$ and

$$f = \lambda_1 f_1 + \cdots + \lambda_N f_N.$$

By the conditions on γ, this curve must intersect all spirals K_j except finitely many of them; indeed, if this were not the case then the shape of K_j's together with the continuity of γ would force γ to make infinitely many windings around the origin while approaching $\mathbb{T}$, which would contradict the hypothesis $\overline{\gamma} \not\supset \mathbb{T}$. Therefore there exists $j_0 \in \mathbb{N}$ such that $p(k, j_0) \geq J(N)$ ($k = 1, \ldots, N$) and $\gamma \cap K_{p(N,j)} \neq \emptyset$ ($j \geq j_0$). Choose points $z_j \in \gamma \cap K_{p(N,j)}$ ($j \geq j_0$). Then by (2.6) we obtain, for every $j \geq j_0$,

$$|f_N(z_j) - q_j| = |f_N(z_j) - g_N(z_j)| < \frac{1 - |z_j|}{N} \leq 1 - |z_j| \leq 1 - r_j \quad \text{and}$$

$$|f_n(z_j)| = |f_n(z_j) - g_n(z_j)| < \frac{1 - |z_j|}{n} \leq 1 - r_j \quad (n = 1, \ldots, N - 1).$$

Hence we get

$$
\begin{aligned}
|f(z_j) - \lambda_N q_j| &= |\lambda_1 f_1(z_j) + \cdots + \lambda_N f_N(z_j) - \lambda_N q_j| \\
&\leq |\lambda_N| \cdot |f_N(z_j) - q_j| + \sum_{n=1}^{N-1} |\lambda_n f_n(z_j)| \\
&< \left(\sum_{n=1}^{N} |\lambda_n| \right) (1 - r_j) \to 0 \quad (j \to \infty).
\end{aligned}
$$

But since $\lambda_N \neq 0$ the sequence $\{\lambda_N q_j : j \in \mathbb{N}\}$ is dense in $\mathbb{C}$, so for given $\alpha \in \mathbb{C}$ there is a sequence $\{j_1 < j_2 < \cdots\} \subset \mathbb{N}$ with $\lambda_N q_{j_k} \to \alpha$ as $k \to \infty$. Now we can select a sequence $\{k(1) < k(2) < \cdots\} \subset \mathbb{N}$ and a point $t \in \mathbb{T}$ with $w_l := z_{j_{k(l)}} \to t$ ($l \to \infty$). Then $\{w_l\}_{l=1}^{\infty} \subset \gamma$ and $f(w_l) = f(w_l) - \lambda_N q_{j_{k(l)}} + \lambda_N q_{j_{k(l)}} \to \alpha$ ($l \to \infty$), so $\alpha \in C(f, \gamma)$. In other words, $C(f, \gamma) = \mathbb{C}$, as required. $\qquad \square$

Note that, in particular, the set of functions $f \in H(\mathbb{D})$ such that the images $\{f(re^{i\theta}) : 0 \leq r < 1\}$ ($\theta \in [0, 2\pi]$) of all rays from the origin are dense in $\mathbb{C}$ forms a dense-lineable subset of $H(\mathbb{D})$.

2.7 Nowhere Gevrey differentiability

Throughout this section we shall focus on a very particular class of functions, the so-called nowhere Gevrey differentiable functions. Here, $\mathcal{C}^{\infty}([0,1])$

denotes the Fréchet space of the (real or complex) functions of class $\mathcal{C}^\infty$ on $[0,1]$, endowed with the sequence $(p_k)_{k\in\mathbb{N}_0}$ of semi-norms defined by

$$p_k(f) = \sup_{j\le k}\ \sup_{x\in[0,1]}\ |f^{(j)}(x)|$$

or, equivalently, with the distance d defined by

$$d(f,g) = \sum_{k=0}^{\infty} 2^{-k}\,\frac{p_k(f-g)}{1+p_k(f-g)}.$$

Let us recall the following definition of a Gevrey differentiable function; see, e.g., [165].

Definition 2.7.1. For a real number $s>0$ and an open subset Ω of $\mathbb{R}$ an infinitely differentiable function f in Ω is said to be *Gevrey differentiable* of order s at $x_0 \in \Omega$ if there exist a compact neighborhood I of x_0 and constants $C, h > 0$ such that

$$\sup_{x\in I} |f^{(n)}(x)| \le Ch^n(n!)^s \quad \text{for all } n\in\mathbb{N}_0.$$

Clearly, if a function is Gevrey differentiable of order s at x_0, it is also Gevrey differentiable of any order $s' > s$ at x_0 (the case $s = 1$ corresponds to analyticity). On the other hand, we have the following notion of a nowhere Gevrey differentiable function.

Definition 2.7.2. A function f is said to be *nowhere Gevrey differentiable* (NG from now on) on $\mathbb{R}$ if f is not Gevrey differentiable of order s at x_0, for every $s > 1$ and every $x_0 \in \mathbb{R}$.

Recall that (following [255]) a Borel set B in a complete metric linear space E is said to be *shy* if there exists a Borel probability measure μ on E with compact support such that $\mu(B + x) = 0$ for any $x \in E$. A set is said to be *prevalent* if it is the complement of a shy set. It is evident that every nowhere Gevrey differentiable function is, in particular, nowhere analytic. The set of nowhere analytic functions in $\mathcal{C}^\infty([0,1])$ is known to be prevalent ([52]), residual ([314]), lineable ([82]) and even algebraic ([174]). In [52] it was also shown that the set of nowhere Gevrey differentiable functions in $\mathcal{C}^\infty([0,1])$ is

(i) a prevalent subset of $\mathcal{C}^\infty([0,1])$ and

(ii) a residual subset of $\mathcal{C}^\infty([0,1])$.

Thus (in [52]) the authors obtained genericity in both the measure-theoretical and the topological senses. On the other hand nothing was known until nowadays about the algebraic structure of the set NG. One might think that since NG enjoys such a rich Borel structure, it might also contain large algebraic structures (linear spaces, algebras, etc.). Of course, and as we already know, this is (in general) not true.

This section will settle this question for the set of nowhere Gevrey differentiable functions. We shall follow the results by Bastin, Conejero, Esser and Seoane [51], and provide the details of their proof of the maximal-dense-lineability of NG in $\mathcal{C}^\infty([0,1])$. Following the steps from [51], we shall use *any* nowhere Gevrey differentiable function (see for example [52] for an explicit construction).

However, to tackle the problem of algebrability, a more precise knowledge of a very particular "key" function in NG is needed. Although we will not enter into the details of this latter part (due to the highly complex calculations involved), we can say that following some ideas from [165, 174] the authors were able to build a (real valued) infinitely differentiable nowhere Gevrey differentiable function and, once they had that, they also proved the maximal-dense-algebrability of the set of nowhere Gevrey differentiable functions in $\mathcal{C}^\infty([0,1])$.

In order to show that the set NG is maximal-dense-lineable, we need the assistance of the following result, which we shall leave as an exercise to the reader (see Exercise 2.10, page 116).

Proposition 2.7.3. *For every $\alpha \in \mathbb{R}$, let $e_\alpha(x) = \exp(\alpha x)$, $x \in \mathbb{R}$. If f is nowhere Gevrey differentiable on $\mathbb{R}$, if $a_1, \ldots, a_N \in \mathbb{C}$ are not all equal to 0 and if $\alpha_1 < \cdots < \alpha_N$ are real numbers, then the function*

$$g = \sum_{j=1}^{N} a_j e_{\alpha_j} \circ f = \left(\sum_{j=1}^{N} a_j e_{\alpha_j} \right) \circ f$$

is nowhere Gevrey differentiable on $\mathbb{R}$. It follows that NG is strongly algebrable.

Of course, as an immediate corollary, we would have:

Corollary 2.7.4. NG *is lineable in* $\mathcal{C}^\infty([0,1])$.

Lemma 2.7.5. *If $\mathcal{P}$ denotes the set of polynomials, then $\mathcal{P} + \mathrm{NG} \subset \mathrm{NG}$.*

Proof. Let us consider $g \in \mathrm{NG}$ and P a polynomial. We proceed by contradiction. Assume that $g + P$ is Gevrey differentiable of order $s > 0$ at $x_0 \in \mathbb{R}$. Since P is analytic at x_0, P is also Gevrey differentiable of order s at x_0 and the same holds for $g = (g + P) - P$, hence a contradiction. $\square$

In order to obtain the dense-lineability of NG in $\mathcal{C}^\infty([0,1])$ we need to recall the following proposition, that is contained in [26, Theorem 2.2 and Remark 2.5] (see also Section 7.3).

Proposition 2.7.6. *Let X be a metrizable topological vector space and consider two subsets A, B of X such that A is lineable and B is dense-lineable in X. If $A + B \subset A$, then A is dense-lineable in X.*

With this result at hand, we can now infer that the set NG is dense-lineable in $C^\infty([0,1])$. Indeed, it follows directly from Corollary 2.7.4, Lemma 2.7.5 and Proposition 2.7.6. Next, let us show that the lineability dimension of NG is the largest possible one, that is, $\mathfrak{c}$. In order to do that, let us fix a function $f \in$ NG. As earlier, we consider

$$\mathcal{D} = \operatorname{span}\{f e_\alpha : \alpha \in [0,1]\},$$

where $e_\alpha(x) = \exp(\alpha x)$. From Proposition 2.7.3, we just have to show that $\dim \mathcal{D} = \mathfrak{c}$. For this, it suffices to show that the functions $f e_\alpha$, $\alpha \in [0,1]$, are linearly independent. Let us assume that it is not the case. Then there exist $c_1, \cdots, c_N \in \mathbb{C}$ not all zero, and $\alpha_1 < \cdots < \alpha_N$ in $[0,1]$ such that $c_1 f e_{\alpha_1} + \cdots + c_N f e_{\alpha_N} = 0$ on $[0,1]$, i.e. $f(c_1 e_{\alpha_1} + \cdots + c_N e_{\alpha_N}) = 0$ on $[0,1]$. Since the functions $e_{\alpha_1}, \cdots, e_{\alpha_N}$ are linearly independent (see [82, Theorem 3.1]), there exists $x \in [0,1]$ such that $c_1 e_{\alpha_1}(x) + \cdots + c_N e_{\alpha_N}(x) \neq 0$. By continuity, there exists a subinterval $J \subset [0,1]$ such that $c_1 e_{\alpha_1} + \cdots + c_N e_{\alpha_N}$ does not vanish on J. It follows that $f = 0$ on J, which is impossible since f is nowhere Gevrey differentiable.

2.8 Exercises

Exercise 2.1. Let G be a domain of $\mathbb{C}$ with $G \neq \mathbb{C}$, $\xi_0 \in \partial G$ and D be a dense subset of G. A function $f \in H(G)$ is said to be *holomorphically extendable through* ξ_0 if there exist $r > 0$, $g \in H(B(\xi_0, r))$ and a connected component A of $G \cap B(\xi_0, r)$ such that $f = g$ in A. Fix $f \in H(G)$. Prove the following properties:

(a) f is holomorphically extendable through ξ_0 if and only if $R(f, a) > \operatorname{dist}(a, \xi_0)$.

(b) $f \in H_e(G)$ if and only if, for any $\xi_0 \in \partial G$, f is *not* holomorphically extendable through ξ_0.

(c) $f \in H_e(G)$ if and only if $R(f, a) = \operatorname{dist}(a, \partial G)$ for all $a \in G$.

(d) $f \in H_e(G)$ if and only if $R(f, a) = \operatorname{dist}(a, \partial G)$ for all $a \in D$.

Hint: Take into account that $\operatorname{dist}(a, \partial G)$ is always reached at some point of ∂G, that we always have $R(f, a) \geq \operatorname{dist}(a, \partial G)$ and that if f and g are as in the definition given at the beginning of this exercise then for many points $a \in A$ one has $R(f, a) = R(g, a) > \operatorname{dist}(a, \xi_0) \geq \operatorname{dist}(a, \partial G)$.

Exercise 2.2. Adapt the proof of Theorem 2.2.2 to any domain of $\mathbb{C}^N$. It is assumed that $X \cap H_e(G) \neq \varnothing$.
Hint: Replace discs by polydiscs and total derivatives $g^{(n)}(a)$ ($n \in \mathbb{N}_0$) by partial derivatives $\partial^\alpha g(a)$ ($\alpha = (\alpha_1, \ldots, \alpha_N) \in \mathbb{N}_0^N$).

Exercise 2.3. Prove that $X \cap H_e(\mathbb{D})$ is residual in X if $X = A^\infty(\mathbb{D})$, H^∞, H^p or B^p $(0 < p < \infty)$.

Hint: Apply Theorem 2.2.2 with $G = \mathbb{D}$. Observe that condition (b) in this theorem is trivially satisfied if $X \cap H_e(\mathbb{D}) \neq \varnothing$, and this holds for $X = A^\infty(\mathbb{D})$ (see the example given in Section 2.2). Note now that both H^p and B^p contain $A^\infty(\mathbb{D})$. Finally, condition (a) is evident for $X = A^\infty(\mathbb{D})$, while for H^p and B^p it is a consequence of the following inequalities, which can be, respectively, found in [194, Chapter 3] and [189, page 13]:

$$|f(z)| \leq 2^{1/p}\|f\|_p(1 - |z|)^{-1/p} \quad \text{and} \quad |f(z)| \leq C\|f\|_p(1 - |z|)^{-2/p}.$$

They are valid for all $p \in (0, \infty)$, all $z \in \mathbb{D}$ and, respectively, for all $f \in H^p$ and $f \in B^p$. Here C is a constant depending only on p.

Exercise 2.4. Complete the details of the final part of the proof of Theorem 2.3.1. Specifically:

(a) Verify that the space $M_1 = \{f \in H(G) : f \text{ vanishes on all } a_j\}$ defined in the proof Theorem 2.3.1 is infinite dimensional.

(b) Let X be an F-space that is also an algebra, and let $A \subset X$ be a closed subalgebra. Prove that if A is infinite dimensional as a vector space, then A is an infinitely generated algebra.

Hint: For (a), select any function $h \in M_1$ and show, by using the Identity Principle, that the functions $z^n h(z)$ $(n \in \mathbb{N}_0)$ are in M_1 and form a linearly independent family. For (b), proceed by way of contradiction by assuming that A is a closed algebra generated by finitely many elements $u_1, \ldots, u_N$. Then the vector space A is generated by the countably many products $u_1^{m_1} \cdots u_N^{m_N}$, with $m_1, \ldots, m_N \in \mathbb{N}_0$ and not all are zero. These products can be ordered as a sequence $v_1, v_2, \ldots, v_n, \ldots$. Then the sets $A_n := \operatorname{span}\{v_1, \ldots, v_n\}$ are increasing finite dimensional vector spaces such that $A = \bigcup_n A_n$. Since A is closed, it is also an F-space, so Baire's theorem is at our disposal.

Exercise 2.5. Let $G \subset \mathbb{C}$ be a domain with $G \neq \mathbb{C}$ and $\varphi : G \to (0, +\infty)$ be a given function. Prove that the set $\mathcal{S}_\varphi$ defined in the Notes and Remarks section is dense-lineable in $H(G)$.

Hint: Fix sequences $(a_n) \subset G$ and (K_n) as in the proof of Theorem 2.3.1 as well as a dense sequence $(g_n) \subset H(G)$. Define $F_n := K_n \cup \{a_k : k \in \mathbb{N}\}$ $(n = 1, 2, \ldots)$. Verify that each F_n is an Arakelian subset of G, and apply Arakelian's approximation theorem so as to obtain a function $f_n \in H(G)$ satisfying $|f_n(z) - g_n(z)| < 1/n$ for all $z \in K_n$ and $|f_n(a_k) - k(1 + \varphi(a_k))| < 1$ for all $k \in \mathbb{N}$ such that $a_k \notin K_n$. Finally, consider $M := \operatorname{span}\{f_n : n \geq 1\}$.

Exercise 2.6. Prove that if $G \subset \mathbb{C}$ is a length-finite domain such that $\mathbb{C} \setminus \overline{G}$ is connected then the set of holomorphic polynomials is dense in the space $A^\infty(G)$.

Hint: Fix a derivation order $N \in \mathbb{N}_0$, a function $g \in A^\infty(G)$ and an $\varepsilon > 0$, and apply appropriately Mergelyan's theorem together with Cauchy's formula for derivatives.

Exercise 2.7. Use the characterization of $A^\infty(\mathbb{D})$ via the boundedness of the sequences $(n^N|a_n|)$ $(N \in \mathbb{N})$ to give a proof of the density of the sets of polynomials in $A^\infty(\mathbb{D})$.

Exercise 2.8. Given an infinite subset $Q \subset \mathbb{N}_0$, demonstrate that the set $H_Q(\mathbb{D}) \cap H_e(\mathbb{D})$ is lineable.
Hint: Proceed as in the proof of the property (P) given inside the proof of Theorem 2.4.1.

Exercise 2.9. Prove that the set $\mathcal{SA}$ of strongly annular functions is, in fact, densely strongly algebrable in $H(\mathbb{D})$.
Hint: Sharpen the second part of the proof of Theorem 2.4.5 so as to get that the sequence $\{f_n\}_{n \geq 1}$ be dense. Use Theorem 2.4.3.

Exercise 2.10. Using the fact that the composition of Gevrey functions is still Gevrey (see [384]) and following the lines of the proof of [45, Theorem 5.10], as well as the functions given in the statement of this Theorem, show Proposition 2.7.3.

2.9   Notes and remarks

Section 2.1. The notions of domain and compact-open topology and the properties of the space $H(G)$ given at the beginning of this chapter can be easily extended, mutatis mutandis, to domains in the space $\mathbb{C}^N = \mathbb{C} \times \cdots \times \mathbb{C}$ (N-fold, with $N \in \mathbb{N}$), which is a metric space under the distance $d(z, w) = \left(\sum_{k=1}^N |z_k - w_k|^2 \right)^{1/2}$, where $z = (z_1, \ldots, z_N)$ and $w = (w_1, \ldots, w_N)$.

Section 2.2. In contrast with the case $N = 1$, not every domain in $\mathbb{C}^N$ (with $N \geq 2$) is the domain of holomorphy of some function, that is, we may have $H_e(G) = \varnothing$. In fact, the Cartan–Thullen theorem asserts that $H_e(G) \neq \varnothing$ if and only if G is holomorphically convex, that is, for every compact subset K of G, the set $\widehat{K} := \{x \in G : |f(x)| \leq \sup_K |f| \text{ for all } f \in H(G)\}$ satisfies $\text{dist}(\widehat{K}, E \setminus G) > 0$ (see for instance [271]).

The proof of Theorem 2.2.2 applies when G is a domain of $\mathbb{C}^N$ ($N \geq 2$) if one assumes that $X \cap H_e(G) \neq \varnothing$; see Exercise 2.2.

For a thorough study of Hardy spaces and Bergman spaces the reader can consult the books by Duren [194] and Hedenmalm *et al.* [244], respectively.

Section 2.3. The proof of Theorem 2.3.1 given in [25] holds for any domain of holomorphy $G \subset \mathbb{C}^N$. An alternative proof of the dense-lineability of $H_e(G)$

(for $N = 1$) can be found in [77]. Valdivia [374] has shown that the dense subspace contained in $H_e(G)$ (with $G \subset \mathbb{C}^N$, $N \in \mathbb{N}$) can be chosen to be nearly-Baire. A locally convex space E is called nearly-Baire if, given a sequence (A_j) of sum-absorbing balanced closed subsets covering E, there is j_0 such that A_{j_0} is a neighborhood of 0; and a subset $A \subset E$ is said to be sum-absorbing whenever there is $\lambda > 0$ such that $\lambda(A + A) \subset A$.

The use of the Arakelian approximation theorem (see [205] or Section 2.1) leads us to an extension of Theorem 2.3.1 in which the growth of f near each boundary point is as fast as prescribed. This extension can be found in [79] and establishes that if $G \subset \mathbb{C}$ is a domain and $\varphi : G \to (0, +\infty)$ is any prescribed function then the set

$$\mathcal{S}_\varphi := \Big\{ f \in H_e(G) : \limsup_{\substack{z \to \xi \\ z \in A}} \frac{|f(z)|}{\varphi(z)} = \infty \ \text{ for all } A \in \mathcal{A}(\xi, r),$$

$$\text{all } \xi \in \partial G \text{ and all } r > 0 \Big\}$$

is spaceable and maximal dense-lineable in $H(G)$, where $\mathcal{A}(\xi, r)$ denotes the collection of all connected components of $G \cap B(\xi, r)$. In particular, $H_e(G)$ is always maximal dense-lineable.

The contents of Theorem 2.3.2 can be found in [93]. Concerning its part (3), another proof –which is based on a result about universality of sequences of operators, see Theorem 4.3.5– is furnished in [93]. Moreover, Bernal and Ordóñez [104] have shown, under the same conditions on G specified in the theorem, the *maximal* dense-lineability of $A^\infty(G) \cap H_e(G)$. This can be done by using a more general technique of search of lineability; see Chapter 7. Under the only assumption of regularity for G, Valdivia [375] was able to prove that the (bigger) family $A^\infty(G) \cap \widetilde{H}_e(G)$ is dense-lineable in $A^\infty(G)$. We have denoted $\widetilde{H}_e(G) := \{ f \in H(G) :$ there is no domain $\Omega \supsetneq G$ to which f extends holomorphically$\}$ (recall that $\widetilde{H}_e(G) = H_e(G)$ if G is a Jordan domain). In fact, there is a dense vector subspace E in $A^\infty(G)$ such that $E \setminus \{0\} \subset \widetilde{H}_e(G)$ and E is nearly-Baire. In [375] the problem is posed as to whether "nearly-Baire" can be replaced by "Baire."

In the infinite dimensional setting, Alves has recently proved that if G is a domain of a complex separable Banach space X and $H_e(G) \neq \varnothing$ then $H_e(G)$ is $\mathfrak{c}$-lineable and algebrable (see [11]), and even closely strongly algebrable (hence spaceable) and densely strongly algebrable (hence dense-lineable) (see [12]) when $H(G)$ is endowed with the topology of uniform convergence on compacta. In this respect, a study of the family $H(G)$ of holomorphic functions $G \to \mathbb{C}$ can be found in Mujica's book [317].

Section 2.4. Together with H^p, B^p and $A^\infty(\mathbb{D})$, further subspaces of $H(\mathbb{D})$ are studied in [77] in connection with topological genericity or lineability of their respective subsets of nonextendable functions.

There is an extensive literature dealing with strongly annular functions. The interested reader is referred to the references contained in [86].

Concerning the negative fact $H^p \cap \mathcal{SA} = \varnothing$, we have, on the contrary, the following positive result provided by Redett [341]: $B_\alpha^p \cap \mathcal{SA} \neq \varnothing$ for every $p \in (0, \infty)$ and every $\alpha \in (-1, \infty)$, where B_α^p denotes the α-weighted Bergman space, that is, the class of functions $f \in H(\mathbb{D})$ for which

$$\|f\|_{p,\alpha} := \left(\int \int_{\mathbb{D}} |f(z)|^p (1 - |z|)^\alpha \frac{dA(z)}{\pi} \right)^{\min\{1, 1/p\}} < \infty.$$

It becomes a separable F-space under the F-norm $\| \cdot \|_{p,\alpha}$. Note that $B_0^p = B^p$, the classical Bergman space. By exploiting Redett's approach in [341] it was obtained in [86] that the set $\mathcal{SA} \cap B_\alpha^p$ is, in fact, dense-lineable in B_α^p. Whether $X \cap \mathcal{SA}$ is spaceable in X for subspaces X of $H(\mathbb{D})$ (including the largest space $X = H(\mathbb{D})$) seems to be an open problem.

Section 2.5. The task of searching lineability in the family of anti-Liouville functions was started in the papers [71, 72], whose statements were improved by Armitage [16] and Bonilla [133]. In fact, these authors proved the following result.

Theorem. *Let* $\alpha \in (0, \infty)$ *and let* $\varphi : [0, \infty) \to (0, \infty)$ *be an increasing function. There exists a vector subspace* M *which is dense in* $H(\mathbb{C})$ *such that* $\lim\limits_{\substack{z \to \infty \\ z \in S}} \exp(|z|^\alpha) f(z) = 0$ *for every strip or unbounded algebraic curve* S *and every* $f \in M$, *and* $\lim\limits_{r \to +\infty} \dfrac{\max\{|f(z)| : |z| = r\}}{\varphi(r)} = \infty$ *for every* $f \in M \setminus \{0\}$.

Additional properties of integrability along rays can be added. A harmonic version is discussed in [132], where, in addition, it is proved the dense-lineability in $\mathcal{H}(\mathbb{B})$ of the family of harmonic functions f on the Euclidean unit ball $\mathbb{B}$ of $\mathbb{R}^N$ having zero nontangential limit at every point of $\partial\mathbb{B}$.

Section 2.6. An explicit, rather elementary example of an "anti-Liouville function" is the following one, which was furnished by Newman [326]: the function $F(z) := \frac{g(z+i\pi)}{g(iz+i\pi)}$, where $g(z) := \int_0^\infty t^{-t} e^{zt} \, dt$, is entire, nonconstant and $F(z) \to 0$ as $z \to \infty$ along any ray starting from the origin.

An extension of Theorem 2.6.2 to L-analytic functions on domains in $\mathbb{R}^N$, where L is an elliptic operator, can be found in [87]. The assertion given in the mentioned theorem can be completed so as to include universality properties: see Sections 4.4 and 4.5 and their corresponding Notes and Remarks.

Section 2.7. In connection with Gevrey differentiable functions, other related classes of real and complex C^∞-functions have been studied. For instance, Esser [199] has recently analyzed topological genericity, prevalence and lineability of certain subfamilies within the set of the so-called Denjoy–Carleman ultradifferentiable functions, which are also "near" to being analytic.

Chapter 3

Sequence Spaces, Measure Theory and Integration

The study of lineability and spaceability (although not with these terms) within the context of measure spaces can be traced back to, at least, the 1950's. In fact, from a famous theorem due to Grothendieck (see [352, Chapter 6]) we can conclude that if $0 < p < \infty$ and μ is finite, then $L_\infty(X, \mu)$ is not spaceable in $L_p(X, \mu)$. Also, if one looks at Rosenthal's work (e.g., [348, 349]) we essentially find the proof of the spaceability of $\ell_\infty \setminus c_0$. This chapter deals with lineability and spaceability results in the lines of the latter ones, in the framework of Measure Theory and Sequence Spaces. Since the results presented here also encompass ℓ_p spaces for $0 < p < 1$ and, in this case, the ℓ_p spaces are not Banach spaces, we present some basic facts from the theory of quasi-Banach spaces (which are *not easy to find* in Functional Analysis textbooks).

We would like to emphasize that, due to the vast existing material related to lineability within the context of Measure Theory, it was imperative to make a selection of just some illustrative results. In the final section of Notes and Remarks we shall attempt to present a sketch of a more complete panorama of the state of the art of this line of research.

3.1   What one needs to know

The reader of this particular chapter will need certain knowledge of basic Functional Analysis, Topology and, of course, Measure Theory. For instance, the Banach–Steinhaus, Hahn–Banach and closed graph theorems will be needed as well as the notion of basic sequences and the Banach–Grumblum Criterion (sometimes called Grumblum–Nikolskii's Theorem or simply Nikolskii's Theorem). We shall state these notions and results when needed in this chapter. First, let $\mathbb{K} = \mathbb{R}$ or $\mathbb{C}$ and let (X, Σ, μ) be a measure space and $0 < p < \infty$. The set of all measurable maps from X to $\mathbb{K}$ such that

$$\|f\|_p := \left(\int_X |f|^p \, d\mu \right)^{\frac{1}{p}} < \infty$$

will be denoted by $\mathcal{L}_p(X, \Sigma, \mu)$.

For $p = \infty$ we define $\mathcal{L}_\infty(X, \Sigma, \mu)$ as the set of measurable maps bounded μ-almost everywhere, i.e., there is a set $N \in \Sigma$ and there is a real number $M > 0$ such that $\mu(N) = 0$ and $|f(x)| \leq M$ for all $x \notin N$. We define

$$S_f(N) := \sup \{ |f(x)| : x \notin N \}$$

and

$$\|f\|_\infty := \inf \{ S_f(N) : N \in \Sigma \text{ and } \mu(N) = 0 \} .$$

The next results, due to J. Rogers and O. Hölder (Theorem 3.1.1) and H. Minkowski (Theorem 3.1.2), are both essential when studying L_p-spaces. They are, respectively, known as *Hölder's inequality for integrals* and *Minkowski's inequality for integrals*.

Theorem 3.1.1. *Let $p, q > 1$ such that $\frac{1}{p} + \frac{1}{q} = 1$ and let (X, Σ, μ) be a measure space. If $f \in \mathcal{L}_p(X, \Sigma, \mu)$ and $g \in \mathcal{L}_q(X, \Sigma, \mu)$, then $fg \in \mathcal{L}_1(X, \Sigma, \mu)$ and*

$$\|fg\|_1 \leq \|f\|_p \cdot \|g\|_q .$$

Above, it is very common to write $q = p^*$ whenever $\frac{1}{p} + \frac{1}{q} = 1$.

Theorem 3.1.2. *Let $1 \leq p < \infty$ and (X, Σ, μ) be a measure space. If $f, g \in \mathcal{L}_p(X, \Sigma, \mu)$, then $f + g \in \mathcal{L}_p(X, \Sigma, \mu)$ and*

$$\|f + g\|_p \leq \|f\|_p + \|g\|_p .$$

In general $\| \cdot \|_p$ is not a norm in $\mathcal{L}_p(X, \Sigma, \mu)$, since it may happen that $\|f\|_p = 0$ for $f \neq 0$. Thus, if (X, Σ, μ) is a measure space, we introduce an equivalence relation in $\mathcal{L}_p(X, \Sigma, \mu)$ in the following way: $f, g \colon X \longrightarrow \mathbb{K}$ are equivalent if $f = g$ μ-a.e., i.e., there is a set $A \in \Sigma$ such that $\mu(A) = 0$ and

$f(x) = g(x)$ for all $x \notin A$. Denoting the equivalence class of f by $[f]$, it is straightforward to see that, in

$$L_p(X, \Sigma, \mu) := \{[f] : f \in \mathcal{L}_p(X, \Sigma, \mu)\},$$

the operations

$$[f] + [g] := [f + g] \quad \text{and} \quad c[f] := [cf]$$

are well defined and, thus, $L_p(X, \Sigma, \mu)$ is a vector space. If we define

$$\|[f]\|_p := \|f\|_p,$$

we have a norm in $L_p(X, \Sigma, \mu)$ for $1 \leq p \leq \infty$. It is a simple task to check that $L_p(X, \Sigma, \mu)$ is a Banach space (although the case $p = \infty$ is a little bit more subtle to study).

For the sake of simplicity, and as usual in Functional Analysis and Measure Theory, the elements of $L_p(X, \Sigma, \mu)$ shall be denoted by f (instead of $[f]$).

If $\mu(X) < \infty$ (for instance if $X = [0, 1]$ with the Lebesgue measure) then $L_q(X, \Sigma, \mu) \subset L_p(X, \Sigma, \mu)$ whenever $p > q$. In fact, and using Hölder's inequality,

$$\left(\int |f|^q \, d\mu\right)^{\frac{1}{q}} = \left(\int |f|^q \cdot 1 \, d\mu\right)^{\frac{1}{q}} \leq \left(\left(\int (|f|^q)^{\frac{p}{q}} \, d\mu\right)^{\frac{1}{\left(\frac{p}{q}\right)}} \cdot \left(\int 1^{\left(\frac{p}{q}\right)^*} \, d\mu\right)^{\frac{1}{\left(\frac{p}{q}\right)^*}}\right)^{\frac{1}{q}}$$

and thus

$$\left(\int |f|^q \, d\mu\right)^{\frac{1}{q}} \leq \left(\int |f|^p \, d\mu\right)^{\frac{1}{p}} \mu(X)^{\frac{1}{\left(\frac{p}{q}\right)^* q}}.$$

Sometimes, for the sake of simplicity $L_p(X, \Sigma, \mu)$ is simply denoted by $L_p(X, \mu)$, and when $X = [0, 1]$ and μ is the Lebesgue measure over the Borel sigma-algebra Σ we simply write $L_p[0, 1]$. For $0 < p < \infty$, let us recall that

$$\ell_p := \left\{ (a_j)_{j=1}^{\infty} : a_j \in \mathbb{K} \text{ for all } j \in \mathbb{N} \text{ and } \sum_{j=1}^{\infty} |a_j|^p < \infty \right\}.$$

Denoting by $\mathcal{P}(\mathbb{N})$ the set of all subsets of $\mathbb{N}$, and considering the counting measure μ_c in $\mathcal{P}(\mathbb{N})$ we can realize that ℓ_p is precisely $L_p(\mathbb{N}, \mathcal{P}(\mathbb{N}), \mu_c)$, by identifying, of course, the sequences with functions defined in the positive integers. In this case, for $p \geq 1$ (note that here we exclude the case $0 < p < 1$) the norm $\| \cdot \|_p$ is

$$\|(a_j)_{j=1}^{\infty}\|_p = \left(\sum_{j=1}^{\infty} |a_j|^p\right)^{\frac{1}{p}}.$$

So, since L_p-spaces for $p \geq 1$ are Banach spaces, we conclude that for $1 \leq p < \infty$ the spaces ℓ_p are Banach spaces with the usual operations and with the norm $\| \cdot \|_p$. Restricted to sequence spaces, Hölder's and Minkowski's inequalities for integrals become *Hölder's inequality for sequences* and *Minkowski's inequality for sequences*, and they read, respectively, as stated in the following two theorems.

Theorem 3.1.3. *Let $n \in \mathbb{N}$ and $p, q > 1$ such that $\frac{1}{p} + \frac{1}{q} = 1$. Then*

$$\sum_{j=1}^{n} |a_j b_j| \leq \left(\sum_{j=1}^{n} |a_j|^p \right)^{\frac{1}{p}} \cdot \left(\sum_{j=1}^{n} |b_j|^q \right)^{\frac{1}{q}}$$

for all positive integers n and all scalars $a_1, \ldots, a_n, b_1, \ldots, b_n$.

Theorem 3.1.4. *For $p \geq 1$, we have*

$$\left(\sum_{k=1}^{n} |a_j + b_j|^p \right)^{\frac{1}{p}} \leq \left(\sum_{k=1}^{n} |a_j|^p \right)^{\frac{1}{p}} + \left(\sum_{k=1}^{n} |b_j|^p \right)^{\frac{1}{p}}$$

for all $n \in \mathbb{N}$ and all scalars $a_1, \ldots, a_n, b_1, \ldots, b_n$.

For $p = \infty$, we define ℓ_∞ as the space of all bounded sequences of scalars, i.e.,

$$\ell_\infty := \left\{ (a_j)_{j=1}^{\infty} : a_j \in \mathbb{K} \text{ for all } j \in \mathbb{N} \text{ and } \sup_{j \in \mathbb{N}} |a_j| < \infty \right\}.$$

Since

$$\ell_\infty = L_\infty(\mathbb{N}, \mathcal{P}(\mathbb{N}), \mu_c),$$

we conclude that ℓ_∞ is a Banach space. The set $c_0 := \{(a_j)_{j=1}^{\infty} : a_j \in \mathbb{K}$ for all j and $\lim_{j \to \infty} a_j = 0\}$, called the *space of null sequences*, is a vector subspace of ℓ_∞ which becomes a Banach space under the norm $\|\cdot\|_\infty$. Clearly, $\ell_p \subset c_0 \subset \ell_\infty$ for all $p \in (0, \infty)$.

It is interesting to note that as p increases the spaces ℓ_p also increases, i.e., if $0 < p < q$, then $\ell_p \subset \ell_q$. The case $1 \leq p < q$ can be found in books of Functional Analysis. If $0 < p < 1$ the inclusion is still valid: let $(a_j) \in \ell_p$, and let $b_j = a_j^p$. Then $(b_j) \in \ell_1 \subset \ell_{\frac{q}{p}}$. Thus $\sum |b_j|^{\frac{q}{p}} < \infty$, i.e., $\sum |a_j|^q < \infty$.

When dealing with L_p and ℓ_p spaces for $0 < p < 1$ the panorama may change radically. The results from the case $p \geq 1$ are not transferred in general to the case $0 < p < 1$. For instance the function

$$\left\| (a_j)_{j=1}^{\infty} \right\|_p = \left(\sum_{j=1}^{\infty} |a_j|^p \right)^{\frac{1}{p}}$$

is not a norm for $0 < p < 1$. Let us now recall the notion of quasi-Banach space. Let X be real or complex linear space. A *quasi-norm* in X is a map $\|\cdot\| : X \to [0, \infty)$ satisfying

(Q1) $\|x\| = 0$ implies $x = 0$.

(Q2) $\|x + y\| \leq C (\|x\| + \|y\|)$, where $C \geq 1$ does not depend on x, y.

(Q3) $\|\lambda x\| = |\lambda| \|x\|$ for all $x \in X$ and $\lambda \in \mathbb{K}$.

Every quasi-norm on a linear space X generates a metrizable topology. If this topology is complete, we call X a quasi-Banach space.

If $0 < p < 1$, the sequence spaces ℓ_p are quasi-Banach spaces with quasi-norms given by

$$\|(x_n)_{n=1}^{\infty}\|_p = \left(\sum_{n=1}^{\infty} |x_n|^p\right)^{\frac{1}{p}}.$$

In fact (Q2) is the only nontrivial item to be proved. To this end, using that $(s+t)^p \leq s^p + t^p$ for all $s, t \geq 0$ and $0 < p < 1$, we can verify that

$$\|x + y\|_p \leq \left(\|x\|_p^p + \|y\|_p^p\right)^{\frac{1}{p}}. \tag{3.1}$$

By the Hölder inequality (Theorem 3.1.3) in (3.1) with $\frac{1}{1/p} + \frac{1}{\frac{1}{1-p}} = 1$ and regarding $\|x\|_p^p + \|y\|_p^p$ as $1 \cdot \|x\|_p^p + 1 \cdot \|y\|_p^p$, we have

$$\|x + y\|_p \leq \left(\left(1^{\frac{1}{1-p}} + 1^{\frac{1}{1-p}}\right)^{1-p}\left(\left(\|x\|_p^p\right)^{\frac{1}{p}} + \left(\|y\|_p^p\right)^{\frac{1}{p}}\right)^p\right)^{\frac{1}{p}}$$

$$= 2^{\frac{1-p}{p}}\left(\|x\|_p + \|y\|_p\right).$$

Similarly, the function spaces $L_p[0,1]$ for $0 < p < 1$ are quasi-Banach spaces. Quasi-Banach spaces satisfying $\|x + y\| \leq (\|x\|^p + \|y\|^p)^{\frac{1}{p}}$ for all x, y and for a certain $0 < p \leq 1$ are called p-Banach spaces. Thus, a 1-Banach space is a Banach space.

The behavior of quasi-Banach spaces (or, more generally, metrizable complete topological vector spaces, called F-spaces) is sometimes quite different from the behavior of Banach spaces. For instance, in 1940 an astonishing result was proved by M.M. Day: if $0 < p < 1$ and $\varphi : L_p[0,1] \to \mathbb{R}$ is linear and continuous then $\varphi = 0$. In general, the extension of lineability/spaceability arguments from Banach to quasi-Banach spaces is not straightforward. For instance, in [381, Section 6] it is essentially proved (with a different terminology) that if Y is a closed infinite-codimensional linear subspace of the Banach space X, then $X \setminus Y$ is spaceable (see Theorem 7.4.1 in Chapter 7). A counterexample due to Kalton [264, Theorem 1.1] shows that this result is not valid for quasi-Banach spaces (there exists a quasi-Banach space K with an 1-dimensional subspace that is contained in all closed infinite dimensional subspaces of K). Besides, the search for closed infinite dimensional subspaces of quasi-Banach spaces is a quite delicate issue. Even fundamental questions sometimes are still without answer; for instance the following problem is still open (cf. [265, Problem 3.1]):

The atomic space problem: Does every (infinite dimensional) quasi-Banach space have a proper closed infinite dimensional subspace?

Thus, it seems interesting to look for lineability and spaceability techniques that also cover the case of quasi-Banach spaces. For more details on quasi-Banach spaces we refer to [265].

3.2   Lineability and spaceability in sequence spaces

Sequence spaces and function spaces are a very suitable environment to study lineability and spaceability. As a matter of fact, we can find results related to spaceability in the framework of sequence spaces at least since the late sixties. For instance, as we already commented in this chapter, the fact that $\ell_\infty \setminus c_0$ is spaceable can be derived from Rosenthal's papers [348, 349].

If $p < q$, as we already know, the sequence space ℓ_p is strictly contained in ℓ_q, so a natural question is: Is $\ell_q \setminus \ell_p$ lineable, spaceable? Using general techniques (see Theorem 7.4.1 and Theorem 7.4.2) it can be observed that for $p > q \geq 1$ the set $\ell_p \setminus \ell_q$ and even the "smaller" set $\ell_p \setminus \bigcup_{1 \leq q < p} \ell_q$ is spaceable for every $p > 1$. In this section we present results showing that even for the case $0 < p < 1$ (and here ℓ_p is non-locally convex) similar results hold.

The abstract concept of invariant sequence spaces is very useful to deal with lineability and spaceability in the framework of sequence spaces. If X is a Banach space and $x \in X^{\mathbb{N}}$, let x^0 be defined as $x^0 = 0$ if x has only finitely many non-zero coordinates; otherwise, $x^0 = (x_j)_{j=1}^\infty$ where x_j is the jth non-zero coordinate of x.

Definition 3.2.1. An *invariant sequence space* over X is an infinite dimensional Banach or quasi-Banach space E of X-valued sequences with the usual operations enjoying the following conditions:

(I1) If $x \in X^{\mathbb{N}}$ is such that $x^0 \neq 0$, then $x \in E$ if and only if $x^0 \in E$, and in this case $\|x\|_E \leq C\|x^0\|_E$ for some constant C depending only on E.

(I2) For every $j \in \mathbb{N}$, we have $\|x_j\|_X \leq \|x\|_E$.

To simplify the terminology, an invariant sequence space will be an invariant sequence space over some Banach space.

The first example below, and a couple of results in this monograph, depend on the Banach–Steinhaus theorem. It is one of the cornerstones of Functional Analysis and can be found in any book dedicated to this area (see for instance [178, Theorem 14.6]). It can be stated as follows: Let X be a Banach space, Y be a normed vector space and $\mathcal{F}$ be a collection of continuous linear operators from X to Y. If for all x in X one has $\sup_{u \in \mathcal{F}} \|u(x)\| < \infty$ then $\sup_{u \in \mathcal{F}} \|u\| < \infty$. In other words, a pointwise-bounded family of continuous linear operators from a Banach space to a normed space is uniformly bounded. Here, as usual,

$\|u\|$ is the operator norm of u, i.e., the supremum of $\|u(x)\|$ for all x in the unit ball of X.

If X is a Banach space, then we denote by $\ell_p(X)$ the vector space of all sequences $(x_j)_{j \geq 1} \subset X$ such that $\sum_{j=1}^{\infty} \|x_j\|^p < \infty$. Analogously, $c_0(X)$ stands for the vector space of all sequences $(x_j)_{j \geq 1} \subset X$ with $\|x_j\| \to 0$ as $j \to \infty$. We next present two interesting vector subspaces of $\ell_p(X)$.

Example 3.2.2. $\ell_p^w(X)$ *(weakly p-summable X-valued sequences). If $0 < p < \infty$ and X is a Banach space, a sequence $(x_j)_{j=1}^{\infty}$ belongs to $\ell_p^w(X)$ if*

$$\sum_{j=1}^{\infty} |\varphi(x_j)|^p < \infty$$

for all continuous linear operators $\varphi : X \to \mathbb{K}$. As a simple consequence of the closed graph theorem we conclude that

$$\sup_{\|\varphi\| \leq 1} \sum_{j=1}^{\infty} |\varphi(x_j)|^p < \infty.$$

The function

$$\left\| (x_j)_{j=1}^{\infty} \right\|_{w,p} := \sup_{\|\varphi\| \leq 1} \left(\sum_{j=1}^{\infty} |\varphi(x_j)|^p \right)^{\frac{1}{p}}$$

is a norm in $\ell_p^w(X)$ if $p \geq 1$ and a p-norm if $0 < p < 1$. In all these cases $\ell_p^w(X)$ is complete and one can verify that it is an invariant sequence space over X.

Example 3.2.3. $\ell_p^u(X)$ *(unconditionally p-summable X-valued sequences). If $0 < p < \infty$ and X is a Banach space, a sequence $(x_j)_{j=1}^{\infty}$ belongs to $\ell_p^u(X)$ if $(x_j)_{j=1}^{\infty} \in \ell_p^w(X)$ and*

$$\lim_{n \to \infty} \left\| (x_j)_{j=n}^{\infty} \right\|_{w,p} = 0.$$

It is not difficult to verify that $\ell_p^u(X)$ is an invariant sequence space over X. Note that the sequence of canonical vectors $(e_j)_{j=1}^{\infty}$ belongs to $\ell_2^w(\ell_2) \setminus \ell_2^u(\ell_2)$.

For $p = 1$ the space $\ell_1^u(X)$ is precisely the space of unconditionally summable sequences in X, i.e., the sequences in X such that its sum is convergent regardless of the choice of the order of the sum. The question of whether $\ell_1^u(X) \setminus \ell_1(X)$ is non void whenever X is infinite dimensional played a fundamental role in the development of the Banach Space Theory. This question appears as Problem 122 in the Scottish Book [300] and also in Banach's monograph [41, page 40]. The complete answer was given in 1950 by Dvoretzky and Rogers [195] but as a matter of fact a partial answer given by MacPhail in

1947 has contributed to the definitive solution. This result rapidly attracted the curiosity and interest of Grothendieck and, as usual to his genius, he soon obtained his own proof (see, for instance [185]); the result can be seen as the beginning of the classical theory of absolutely summing operators (see [186]). As we shall see in Corollary 3.2.8, it has been proved that $\ell_1^u(X) \setminus \ell_1(X)$ is spaceable but, despite much progress, in similar cases, the maximal spaceability of $\ell_1^u(X) \setminus \ell_1(X)$ is still an open problem.

It is also true that $\ell_p^u(X) \setminus \ell_p(X)$ is non void whenever X is infinite dimensional, for all $0 < p < \infty$. This result is usually known as the *weak Dvoretzky–Rogers theorem*.

Now we shall begin to prove lineability results in the context of sequence spaces. First, it is crucial to know that $\ell_p \setminus \bigcup_{0<q<p} \ell_q$ is non void for all $p > 0$, as we shall see in the next result. We present a short proof that seems to be folklore and relies on the Banach-Steinhaus theorem:

Proposition 3.2.4. *The set $\ell_p \setminus \bigcup_{0<q<p} \ell_q$ is non void for all $p > 0$.*

Proof. Since $\left(\frac{1}{\sqrt{n}}\right) \notin \ell_2$ and $\left(\frac{1}{\sqrt{n}}\right) \in \ell_r$ for all $r > 2$, for each $(y_n) \in \ell_q$, $1 < q < 2$, it follows from Hölder's inequality (Theorem 3.1.3) that

$$\sum_{n=1}^{\infty} \left| \frac{1}{\sqrt{n}} y_n \right| \leq \left\| \left(\frac{1}{\sqrt{n}}\right)_{n=1}^{\infty} \right\|_{q^*} \left\| (y_n)_{n=1}^{\infty} \right\|_q < \infty.$$

Supposing that $\ell_2 = \bigcup_{1<q<2} \ell_q$ we have $\sum_{n=1}^{\infty} \left| \frac{1}{\sqrt{n}} y_n \right| < \infty$ for every $(y_n) \in \ell_2$. So, consider, for each positive integer k, the continuous linear functional ℓ_2 defined by $T_k\left((y_n)\right) = \sum_{n=1}^{k} \frac{1}{\sqrt{n}} y_n$. From the Banach-Steinhaus theorem we conclude that $T\left((y_n)_{n=1}^{\infty}\right) := \sum_{n=1}^{\infty} \frac{1}{\sqrt{n}} y_n$ defines a continuous linear functional on ℓ_2 and it follows that $\left(\frac{1}{\sqrt{n}}\right)_{n=1}^{\infty} \in \ell_2$ - a contradiction. Here we have used the fact (sometimes known as Riesz Representation theorem) that when $T : \ell_2 \to \mathbb{K}$ is a continuous linear operator and $T((x_j)_{j=1}^{\infty}) = \sum_{j=1}^{\infty} x_j y_j$ then $(y_j)_{j=1}^{\infty} \in \ell_2$. So there is a $x \in \ell_2 \setminus \bigcup_{1<q<p} \ell_2$ and thus

$$\left(|x_n|^{\frac{2}{p}} \right)_{n=1}^{\infty} \in \ell_p \setminus \bigcup_{1<q<p} \ell_q = \ell_p \setminus \bigcup_{0<q<p} \ell_q.$$

Hence $\ell_p \setminus \bigcup_{0<q<p} \ell_q$ is non void, as required.   □

The following simple lemma will be useful later.

Lemma 3.2.5. *If X is a p-Banach space and $\sum_{j=1}^{\infty} \|x_j\|^p < \infty$, then $\sum_{j=1}^{\infty} x_j$ converges in X.*

Proof. Let $S_m = \sum\limits_{j=1}^{m} x_j$. Then, given $\varepsilon > 0$, there is a positive integer N_0 such that

$$\|S_m - S_n\|^p = \left\| \sum_{j=n+1}^{m} x_j \right\|^p \le \sum_{j=n+1}^{m} \|x_j\|^p < \varepsilon$$

whenever $m > n \ge N_0$. We thus conclude that $(S_n)_{n=1}^{\infty}$ is a Cauchy sequence, and therefore converges. $\square$

Theorem 3.2.6. *Let E be an invariant sequence space over a Banach space X. Then we have:*

(a) *For every $\Gamma \subseteq (0, \infty]$, $E \setminus \bigcup_{q \in \Gamma} \ell_q(X)$ is either empty or spaceable.*

(b) *The set $E \setminus c_0(X)$ is either empty or spaceable.*

Proof. Let us define $A = \bigcup_{q \in \Gamma} \ell_q(X)$ in (a) and $A = c_0(X)$ in (b). If $E \setminus A$ is non-empty, let $x \in E \setminus A$. Since $x \notin A$, we have $\|x\|_q = \infty$ for all $q \in \Gamma$ in the case (a) and $\lim x_j \ne 0$ in the case (b). In both cases, $x^0 \notin A$. We thus have $x^0 := (x_j) \in E \setminus A$, with $x_j \ne 0$ for all positive integers j. Now we separate $\mathbb{N}$ into countably many infinite pairwise disjoint subsets $(\mathbb{N}_i)_{i=1}^{\infty}$. For instance, we can define $\mathbb{N}_1$ as the set of primes (including 1), $\mathbb{N}_2$ as the set of products of exactly two primes, and so on. It is convenient to use the following unusual notation for the elements of the subsets of $\mathbb{N}_i$:

$$\mathbb{N}_i = \{i_1 < i_2 < \ldots\}.$$

Now we define

$$y_i = \sum_{j=1}^{\infty} x_j e_{i_j} \in X^{\mathbb{N}}$$

and observe that $y_i^0 = x^0$ for every i and thus $0 \ne y_i^0 \in E$ for every i. Hence, from (I1), each $y_i \in E$. But one can easily realize that $y_i \notin A$. In fact, in the case (a) this happens because

$$\|y_i\|_r = \|x^0\|_r = \|x\|_r$$

for every $0 < r \le \infty$, and in (b) it occurs since

$$\lim_{j \to \infty} \|x_j\|_X \ne 0.$$

Let C be the constant of condition (I1) and define $\tilde{s} = 1$ if E is a Banach space and $\tilde{s} = s$ if E is an s-Banach space, $0 < s < 1$. For $(a_i)_{i=1}^{\infty} \in \ell_{\tilde{s}}$,

$$\sum_{i=1}^{\infty} \|a_i y_i\|_E^{\tilde{s}} = \sum_{i=1}^{\infty} |a_i|^{\tilde{s}} \|y_i\|_E^{\tilde{s}} \le C^{\tilde{s}} \sum_{i=1}^{\infty} |a_i|^{\tilde{s}} \|y_i^0\|_E^{\tilde{s}}$$

$$= C^{\tilde{s}} \|x^0\|_E^{\tilde{s}} \sum_{i=1}^{\infty} |a_i|^{\tilde{s}} = C^{\tilde{s}} \|x^0\|_E^{\tilde{s}} \|(a_i)_{i=1}^{\infty}\|_{\tilde{s}}^{\tilde{s}} < \infty.$$

Thus

$$\sum_{i=1}^{\infty} \|a_i y_i\|_E < \infty$$

if E is a Banach space, and

$$\sum_{i=1}^{\infty} \|a_i y_i\|_E^s < \infty$$

if E is an s-Banach space, $0 < s < 1$. In both cases, from the previous lemma we conclude that the series $\sum_{i=1}^{\infty} a_i y_i$ converges in E. Therefore, the (linear and injective) operator

$$T \colon \ell_{\tilde{s}} \longrightarrow E \ , \quad T((a_i)_{i=1}^{\infty}) = \sum_{i=1}^{\infty} a_i y_i$$

is well defined. Thus $\overline{T(\ell_{\tilde{s}})}$ is a closed infinite dimensional subspace of E. All that is left to prove is that $\overline{T(\ell_{\tilde{s}})} \setminus \{0\} \subset E \setminus A$. Let $z = (z_n)_{n=1}^{\infty} \in \overline{T(\ell_{\tilde{s}})}$, $z \neq 0$; so, there are sequences $\left(a_i^{(k)}\right)_{i=1}^{\infty} \in \ell_{\tilde{s}}$, $k \in \mathbb{N}$, such that $z = \lim_{k \to \infty} T\left(\left(a_i^{(k)}\right)_{i=1}^{\infty}\right)$ in E. Note that, for each $k \in \mathbb{N}$,

$$T\left(\left(a_i^{(k)}\right)_{i=1}^{\infty}\right) = \sum_{i=1}^{\infty} a_i^{(k)} y_i = \sum_{i=1}^{\infty} a_i^{(k)} \sum_{j=1}^{\infty} x_j e_{i_j} = \sum_{i=1}^{\infty} \sum_{j=1}^{\infty} a_i^{(k)} x_j e_{i_j}.$$

Fix $r \in \mathbb{N}$ such that $z_r \neq 0$. Since $\mathbb{N} = \bigcup_{j=1}^{\infty} \mathbb{N}_j$, there are (unique) $m, t \in \mathbb{N}$ such that $e_{m_t} = e_r$. Thus, for each $k \in \mathbb{N}$, the r-th coordinate of $T\left(\left(a_i^{(k)}\right)_{i=1}^{\infty}\right)$ is the number $a_m^{(k)} x_t$. From (I2) we know that convergence in E implies coordinatewise convergence, and thus

$$z_r = \lim_{k \to \infty} a_m^{(k)} x_t = x_t \cdot \lim_{k \to \infty} a_m^{(k)}.$$

Therefore $x_t \neq 0$ and $\lim_{k \to \infty} |a_m^{(k)}| = \frac{\|z_r\|_X}{\|x_t\|_X} \neq 0$. Defining

$$\alpha_m = \frac{\|z_r\|_X}{\|x_t\|_X} \neq 0,$$

we obtain

$$\lim_{k \to \infty} \|a_m^{(k)} x_j\|_X = \lim_{k \to \infty} |a_m^{(k)}| \|x_j\|_X = \|x_j\|_X \cdot \lim_{k \to \infty} |a_m^{(k)}| = \alpha_m \|x_j\|_X$$

for every $j \in \mathbb{N}$. On the other hand, since we have coordinatewise convergence, we also have

$$\lim_{k \to \infty} \|a_m^{(k)} x_j\|_X = \|z_{m_j}\|_X.$$

Thus from the above equalities we have

$$\|z_{m_j}\|_X = \alpha_m \|x_j\|_X$$

for each $j \in \mathbb{N}$. Observe that m, which depends on r, is fixed, so the natural numbers $(m_j)_{j=1}^\infty$ are pairwise distinct (we shall not forget that $\mathbb{N}_m = \{m_1 < m_2 < \ldots\}$). Now we finish the proof of (a) and (b) separately.

(a) Since $x^0 \notin A$, we have $\|x^0\|_q = \infty$ for all $q \in \Gamma$. Note that

$$\|z\|_q^q = \sum_{n=1}^\infty \|z_n\|_X^q \geq \sum_{j=1}^\infty \|z_{m_j}\|_X^q = \sum_{j=1}^\infty \alpha_m^q \cdot \|x_j\|_X^q = \alpha_m^q \cdot \|x^0\|_q^q = \infty$$

for every q in Γ, $q \neq \infty$. In the case that ∞ belongs to Γ, then

$$\|z\|_\infty = \sup_n \|z_n\|_X \geq \sup_j \|z_{m_j}\|_X = \alpha_m \cdot \sup_j \|x_j\|_X = \alpha_m \|x^0\|_\infty = \infty,$$

proving that $z \notin \bigcup_{q \in \Gamma} \ell_q(X)$.

(b) Since $x^0 \notin A$, we have $\lim \|x_j\|_X \neq 0$. Since $(\|z_{m_j}\|_X)_{j=1}^\infty$ is a subsequence of $(\|z_n\|_X)_{n=1}^\infty$, and $\|z_{m_j}\|_X = \alpha_m \|x_j\|_X$ for every j and $\alpha_m \neq 0$, it is plain that $\lim \|z_n\|_X \neq 0$ and hence $z \notin c_0(X)$.

Hence, the proof is done, because we have proved that $z \notin A$ in both cases, that is, $\overline{T(\ell_{\tilde{s}})} \setminus \{0\} \subset E \setminus A$. $\qquad\qquad \square$

Corollary 3.2.7. *Let E be an invariant sequence space over $\mathbb{K}$.*

(a) *If $0 < p \leq \infty$ and ℓ_p is strictly contained in E, then $E \setminus \ell_p$ is spaceable.*

(b) *If c_0 is strictly contained in E, then $E \setminus c_0$ is spaceable.*[s]

Since $\ell_p(X)$ is strictly contained in $\ell_p^w(X)$ and in $\ell_p^u(X)$ whenever X is infinite dimensional, using Proposition 3.2.4 and Theorem 3.2.6 we have the following consequences.

Corollary 3.2.8. *Let X be an infinite dimensional Banach space. Then*

(a) *$\ell_p^u(X) \setminus \ell_p(X)$ is spaceable for every $p > 0$.*

(b) *$\ell_p^w(X) \setminus \ell_p(X)$ is spaceable.*

(c) *$\ell_p \setminus \bigcup_{0 < q < p} \ell_q$ is spaceable for every $p > 0$.*

3.3   Non-contractive maps and spaceability in sequence spaces

The aim of this section is to continue the ongoing research provided in the previous one. We still consider Banach spaces over $\mathbb{K} = \mathbb{R}$ or $\mathbb{C}$.

Definition 3.3.1. Let X and Y be Banach spaces, E be an invariant sequence space over X, $\Gamma \subseteq (0, \infty]$ and $f \colon X \longrightarrow Y$ be a function. The following definitions play an important role in this section:

$$C(E, f, \Gamma) = \left\{ (x_j)_{j=1}^{\infty} \in E : (f(x_j))_{j=1}^{\infty} \notin \bigcup_{q \in \Gamma} \ell_q(Y) \right\},$$

$$C^w(E, f, \Gamma) = \left\{ (x_j)_{j=1}^{\infty} \in E : (f(x_j))_{j=1}^{\infty} \notin \bigcup_{q \in \Gamma} \ell_q^w(Y) \right\} \quad \text{and}$$

$$C(E, f, 0) = \left\{ (x_j)_{j=1}^{\infty} \in E : (f(x_j))_{j=1}^{\infty} \notin c_0(Y) \right\}.$$

The results from the previous section are just the case of $C(E, f, \Gamma)$ and $C(E, f, 0)$ with f being the identity on X. From now on when we refer to spaceability of $C(E, f, \Gamma)$ or $C^w(E, f, \Gamma)$ or $C(E, f, 0)$ what we mean is being spaceable in E.

Definition 3.3.2. A function $f \colon X \longrightarrow Y$ between normed spaces is:

(a) *Non-contractive* if $f(0) = 0$ and for every scalar $\alpha \neq 0$ there is a constant $K(\alpha) > 0$ such that

$$\|f(\alpha x)\|_Y \geq K(\alpha) \cdot \|f(x)\|_Y$$

for every $x \in X$.

(b) *Strongly non-contractive* if $f(0) = 0$ and for every scalar $\alpha \neq 0$ there is a constant $K(\alpha) > 0$ such that

$$|\varphi(f(\alpha x))| \geq K(\alpha) \cdot |\varphi(f(x))|$$

for all $x \in X$ and all continuous linear functionals $\varphi \colon Y \to \mathbb{R}$.

Note that strongly non-contractive functions are non-contractive; this is a simple consequence of the Hahn–Banach theorem.

Remark 3.3.3. *Observe that a non non-contractive map p is not necessarily a contraction in the usual sense. For instance, the mapping $f \colon \mathbb{R} \to \mathbb{R}$ given by*

$$f(x) = \begin{cases} 0 & \text{if } x \in \mathbb{Q} \\ x^2 & \text{if } x \notin \mathbb{Q} \end{cases}$$

is non non-contractive but it is not a contraction.

Example 3.3.4. *Continuous and non continuous linear operators are strongly non-contractive (thus also non-contractive) with $K(\alpha) = |\alpha|$ for every $\alpha \neq 0$. Besides, homogeneous polynomials (continuous or not) are strongly contractive (hence contractive) with $K(\alpha) = |\alpha|^n$ for every $\alpha \neq 0$, where n is the degree of homogeneity.*

The main result of this section now comes.

Theorem 3.3.5. *Let X and Y be Banach spaces, E be an invariant sequence space over X, $f \colon X \longrightarrow Y$ be a function and $\Gamma \subseteq (0, \infty]$. We have:*

(a) *If f is non-contractive, then $C(E, f, \Gamma)$ and $C(E, f, 0)$ are either empty or spaceable.*

(b) *If f is strongly non-contractive, then $C^w(E, f, \Gamma)$ is either empty or spaceable.*

Proof. (a) For $\alpha = (\alpha_n)_{n=1}^{\infty} \in \mathbb{K}^{\mathbb{N}}$ and $w \in X$ consider

$$w \otimes \alpha = \alpha \otimes w := (\alpha_n w)_{n=1}^{\infty} \in X^{\mathbb{N}}.$$

Assume that $C(E, f, \Gamma)$ is non-empty and choose $x \in C(E, f, \Gamma)$. Since E is an invariant sequence space, then $x^0 \in E$ and since $f(0) = 0$ we conclude that $x^0 \in C(E, f, \Gamma)$. Denoting $x^0 = (x_j)_{j=1}^{\infty}$ we obviously have $x_j \neq 0$ for every j. Now, as usual, we divide $\mathbb{N}$ into countably many infinite pairwise disjoint subsets $(\mathbb{N}_i)_{i=1}^{\infty}$. For every $i \in \mathbb{N}$ set $\mathbb{N}_i = \{i_1 < i_2 < \ldots\}$ and define

$$y_i = \sum_{j=1}^{\infty} x_j \otimes e_{i_j} \in X^{\mathbb{N}}.$$

Note that $y_i^0 = x^0$, so $0 \neq y_i^0 \in E$. We thus conclude that $y_i \in E$ for every i because E is an invariant sequence space. For $q \in \Gamma$, $q < \infty$, we have $\sum_{j=1}^{\infty} \|f(x_j)\|_Y^q = \infty$ because $x^0 \in C(E, f, \Gamma)$. If $\infty \in \Gamma$, by the same reason we have $\sup_i \|f(x_i)\|_Y = \infty$. It follows that each $y_i \in C(E, f, \Gamma)$. Let K be the constant of the definition of invariant sequence spaces and define $\tilde{s} = 1$ if E is a Banach space and $\tilde{s} = s$ if E is an s-Banach space, $0 < s < 1$. For $(a_i)_{i=1}^{\infty} \in \ell_{\tilde{s}}$,

$$\sum_{i=1}^{\infty} \|a_i y_i\|_E^{\tilde{s}} = \sum_{i=1}^{\infty} |a_i|^{\tilde{s}} \cdot \|y_i\|_E^{\tilde{s}} \leq K^{\tilde{s}} \cdot \sum_{i=1}^{\infty} |a_i|^{\tilde{s}} \cdot \|y_i^0\|_E^{\tilde{s}}$$

$$= K^{\tilde{s}} \cdot \|x^0\|_E^{\tilde{s}} \cdot \sum_{i=1}^{\infty} |a_i|^{\tilde{s}} = K^{\tilde{s}} \cdot \|x^0\|_E^{\tilde{s}} \cdot \|(a_i)_{i=1}^{\infty}\|_{\tilde{s}}^{\tilde{s}} < \infty.$$

Thus $\sum_{i=1}^{\infty} \|a_i y_i\|_E < \infty$ if E is a Banach space and $\sum_{i=1}^{\infty} \|a_i y_i\|_E^s < \infty$ if E is an

s-Banach space, $0 < s < 1$. From Lemma 3.2.5, the series $\sum_{i=1}^{\infty} a_i y_i$ converges in E, and the operator

$$T\colon \ell_{\tilde{s}} \longrightarrow E \ , \quad T\left((a_i)_{i=1}^{\infty}\right) = \sum_{i=1}^{\infty} a_i y_i,$$

is well defined. It is easy to see that T is linear and injective. Thus $\overline{T\left(\ell_{\tilde{s}}\right)}$ is a closed infinite-dimensional subspace of E. We just have to show that if $z = (z_n)_{n=1}^{\infty} \in \overline{T\left(\ell_{\tilde{s}}\right)}$, $z \neq 0$, then $(f(z_n))_{n=1}^{\infty} \notin \bigcup_{q \in \Gamma} \ell_q(Y)$. The rest of the proof follows the lines of the proofs of the previous sections and we have left the details for the reader.

(b) When $\infty \in \Gamma$ the proof is immediate because, in this case, $C^w(E, f, \Gamma) = C(E, f, \Gamma)$, and this situation is encompassed in (a) with a weaker assumption on f. So now we assume that $\infty \notin \Gamma$. Suppose that $C^w(E, f, \Gamma)$ is non-empty and choose $x \in C^w(E, f, \Gamma)$. We know that $x^0 \in E$ because E is an invariant sequence space; also $x^0 \in C^w(E, f, \Gamma)$ because $f(0) = 0$ and $x^0 = (x_j)_{j=1}^{\infty}$ with $x_j \neq 0$ for every j. The proof also follows the steps of the previous proofs. □

For the forthcoming consequences of the main result we recall the notion of summing operators and polynomials (these notions were commented *en passant* in the previous section).

Definition 3.3.6. Let X and Y be Banach spaces and $1 \leq p \leq q < \infty$. We say that

- a linear operator $u\colon X \longrightarrow Y$ is *absolutely (q, p)-summing* if $(u(x_j))_{j=1}^{\infty} \in \ell_q(Y)$ for each $(x_j)_{j=1}^{\infty} \in \ell_p^w(X)$.

- an n-homogeneous polynomial $P\colon X \longrightarrow Y$ is *p-dominated* if $(P(x_j))_{j=1}^{\infty} \in \ell_{p/n}(Y)$ for each $(x_j)_{j=1}^{\infty} \in \ell_p^w(X)$.

Corollary 3.3.7. *Let X and Y be Banach spaces, and $1 \leq p \leq q < \infty$.*

(a) *Let $u\colon X \longrightarrow Y$ be a non-absolutely (q, p)-summing linear operator. Then the set*

$$\left\{ (x_j)_{j=1}^{\infty} \in \ell_p^w(X) : (u(x_j))_{j=1}^{\infty} \notin \ell_q(Y) \right\}$$

is spaceable in $\ell_p^w(X)$.

(b) *Let $P\colon X \longrightarrow Y$ be a non-p-dominated n-homogeneous polynomial. Then the set*

$$\left\{ (x_j)_{j=1}^{\infty} \in \ell_p^w(X) : (P(x_j))_{j=1}^{\infty} \notin \ell_{p/n}(Y) \right\}$$

is spaceable in $\ell_p^w(X)$.

Proof. Use Theorem 3.3.5 and recall that linear operators and homogeneous polynomials are (strongly) non-contractive maps and that $\ell_p^w(X)$ is an invariant sequence space over X. □

3.4 Lineability and spaceability in $L_p[0,1]$

We shall now investigate lineability and spaceability in quite general function spaces $L_p(\Omega, \Sigma, \mu)$. As a starting point it seems interesting to deal with the case of classical spaces $L_p[0,1]$ (here the sigma-algebra is the Borel sigma-algebra of subsets of $[0,1]$ and μ is the Lebesgue measure). For $p > 0$ a variation of the argument of the proof of Proposition 3.2.4 yields that $L_p[0,1] \setminus \bigcup_{q>p} L_q[0,1]$ is non void. Thus, it is a natural environment to investigate lineability and spaceability. Next theorem shows that in fact $L_p[0,1] \setminus \bigcup_{q>p} L_q[0,1]$ is spaceable, and the proof uses again a variation of the technique used in the previous section.

Theorem 3.4.1. $L_p[0,1] \setminus \bigcup_{q>p} L_q[0,1]$ *is spaceable for every* $p > 0$.

Proof. Let us write $[0,1)$ as

$$[0,1) = \bigcup_{n=1}^{\infty} I_n,$$

where $I_n := [a_n, b_n) = \left[1 - \frac{1}{2^{n-1}}, 1 - \frac{1}{2^n}\right)$. Notice that, for every $n \in \mathbb{N}$ and every $x \in I_n$, there is a unique $x_n \in [0,1)$ such that

$$x = (1 - x_n)a_n + x_n b_n.$$

Fix a vector $f \in L_p[0,1] - \bigcup_{q>p} L_q[0,1]$, and define a sequence of functions $(f_n)_{n=1}^{\infty}$, with $f_n \colon [0,1] \longrightarrow \mathbb{R}$, as follows:

$$f_n(x) = \begin{cases} f(x_n) & \text{if } x \in I_n, \\ 0 & \text{if } x \notin I_n. \end{cases}$$

The maps f_n are a kind of small copies of f on the interval I_n. Note that $\|f_n\|_{L_p} \leq \|f\|_{L_p}$ for every $n \in \mathbb{N}$ and also that the functions f_n are linearly independent. Moreover,

$$\operatorname{span}\{f_n : n \in \mathbb{N}\} \subset L_p[0,1] \setminus \bigcup_{q>p} L_q[0,1]. \tag{3.2}$$

So, we conclude that $L_p[0,1] \setminus \bigcup_{q>p} L_q[0,1]$ is $\aleph_0$-lineable. In order to achieve the spaceability, let us begin by showing that there are a Banach space F and a continuous and injective linear operator $T \colon F \longrightarrow L_p[0,1]$ such that $T(F) \cap L_q[0,1] = \{0\}$ for every $q > p$.

If $(\alpha_j)_{j=1}^{\infty} \in \ell_s$, where $s = 1$ if $p \geq 1$, and $s = p$ if $0 < p < 1$, we have

$$\sum_{n=1}^{\infty} \|\alpha_n f_n\|_{L_p}^{s} = \sum_{n=1}^{\infty} |\alpha_n|^{s} \|f_n\|_{L_p}^{s} \leq \sum_{n=1}^{\infty} |\alpha_n|^{s} \|f\|_{L_p}^{s} = \|f\|_{L_p}^{s} \left\|(\alpha_j)_{j=1}^{\infty}\right\|_{s}^{s} < \infty.$$

Since $L_p[0,1]$ is a Banach space for $p > 1$ and a quasi-Banach space for $0 < p < 1$, it follows from Lemma 3.2.5 that $\sum_{n=1}^{\infty} \alpha_n f_n \in L_p[0,1]$ and, thus,

$$T \colon \ell_s \longrightarrow L_p[0,1] \ , \quad T((\alpha_j)_{j=1}^{\infty}) = \sum_{n=1}^{\infty} \alpha_n f_n$$

is a well defined linear operator. Note that T is injective; in fact, if $\sum_{n=1}^{\infty} \alpha_n f_n = 0$, then (given any $k \in \mathbb{N}$) we have that, for all $x \in I_k$,

$$\alpha_k f_k(x) = \sum_{n=1}^{\infty} \alpha_n f_n(x) = 0$$

and, since each $f_k \neq 0$, we conclude that $\alpha_k = 0$. Now we shall prove that the closure $\overline{T(\ell_s)}$ of $T(\ell_s)$ in $L_p[0,1]$ is such that $\overline{T(\ell_s)} \cap L_q[0,1] = \{0\}$ for every $q > p$. Let $g \in \overline{T(\ell_s)} \setminus \{0\}$; thus the set $A = \{x \in [0,1] : g(x) \neq 0\}$ has positive measure. Let $\left(a_i^{(k)}\right)_{i=1}^{\infty} \in \ell_s$ $(k \in \mathbb{N})$ be such that $g = \lim_{k\to\infty} T\left(\left(a_i^{(k)}\right)_{i=1}^{\infty}\right)$ in $L_p[0,1]$. In other words, we have $\lim_{k\to\infty} \left\| \sum_{n=1}^{\infty} a_n^{(k)} f_n - g \right\|_p = 0$ and, consequently, there is a subsequence $\left(\sum_{n=1}^{\infty} a_n^{(k_j)} f_n \right)_{j=1}^{\infty}$ such that

$$\lim_{j\to\infty} \sum_{n=1}^{\infty} a_n^{(k_j)} f_n(x) = g(x) \tag{3.3}$$

almost everywhere. Here we have used that convergence in L_p implies an almost everywhere convergence. Let $B = \{x \in [0,1] : \text{the limit above holds}\}$, and note that $B \cap A$ has positive measure (because $\mu\left((B \cap A)^{\complement}\right) = \mu\left(B^{\complement} \cup A^{\complement}\right) = \mu\left(A^{\complement}\right) = 1 - \mu(A) < 1$). Hence, there is $x_0 \in B \cap A$ with $x_0 \neq 1$, that is, $g(x_0) \neq 0$ and

$$\sum_{n=1}^{\infty} a_n^{(k_j)} f_n(x_0) \longrightarrow g(x_0) \text{ when } j \to \infty.$$

Since $x_0 \in [0,1)$, there is $r \in \mathbb{N}$ such that $x_0 \in I_r$. Thus

$$a_r^{(k_j)} f_r(x_0) = \sum_{n=1}^{\infty} a_n^{(k_j)} f_n(x_0) \longrightarrow g(x_0) \text{ when } j \to \infty.$$

Note that f can be chosen such that $f(x) \neq 0$, for all $x \in [0,1]$; consequently, $f_r(x) \neq 0$, for all $x \in I_r$. Thus

$$\lim_{j\to\infty} a_r^{(k_j)} = \frac{g(x_0)}{f_r(x_0)} = \eta \neq 0.$$

But from (3.3) we have

$$f_r(x)a_r^{(k_j)} \longrightarrow g(x) \ (j \to \infty) \text{ a.e. in } I_r,$$

and since the limit is unique we obtain

$$g(x) = \eta f_r(x) \text{ a.e. } x \in I_r$$

which, from (3.2), implies that $g \notin L_q[0,1]$ (regardless of the $q > p$), finishing the proof. ☐

3.5 Spaceability in Lebesgue spaces

Spaceability in L_p spaces has been investigated, with a different terminology (of course), since the 1950's. As we mentioned in the introductory part of this chapter, a celebrated theorem due to Grothendieck (see [352, Chapter 6]) asserts that if $0 < p < \infty$ and μ is finite then $L_\infty(X,\mu)$ is not spaceable in $L_p(X,\mu)$.

Let $(E, \|\cdot\|)$ be a Banach space, and (x_n) be a sequence in E. We recall that (x_n) is said to be a *basic sequence* whenever, for each vector $x \in \overline{\text{span}}\{x_n : n \geq 1\}$, there is a unique sequence (a_n) of scalars such that $\sum_{n=1}^{\infty} a_n x_n = x$. The last equality means that $\|\sum_{n=1}^{N} a_n x_n - x\| \longrightarrow 0$ as $N \longrightarrow \infty$. In other words, (x_n) is a Schauder basis of the subspace $\overline{\text{span}}\{x_n : n \geq 1\}$. The following criterium–which is known as Banach–Grumblum Criterion and also as Nikolskii Criterion–is very useful in Banach Space Theory; see for instance [65, pp. 81–83] or [7, Proposition 1.1.9].

Lemma 3.5.1. *Assume that E is a Banach space and that $(x_n) \subset E \setminus \{0\}$. The following properties are equivalent:*

(i) *(x_n) is a basic sequence.*

(ii) *There is a constant $C \in (0,\infty)$ such that, for every pair $r,s \in \mathbb{N}$ with $s \geq r$ and every finite sequence of scalars $a_1,\ldots,a_s$, one has $\|\sum_{n=1}^{r} a_n x_n\| \leq C \|\sum_{n=1}^{s} a_n x_n\|$.*

The constant C is sometimes called basic constant. The famous Haar system (h_n) given by $h_1 = 1$,

$$h_{2^k+l} = \chi_{[\frac{2l-2}{2^{k+1}}, \frac{2l-1}{2^{k+1}})} - \chi_{[\frac{2l-1}{2^{k+1}}, \frac{2l}{2^{k+1}}]} \ (k = 0,1,2,\ldots; \ l = 1,2,3,\ldots,2^k),$$

is a basic sequence for $L_p[0,1]$ for $p \in [1,\infty)$ (see, for instance, [65, Chapter 2]). As a matter of fact, the Haar system is, also, a Schauder basis.

We also need the following concepts:

- The *support* of a function $f : X \to \mathbb{R}$ is the set

$$\sigma(f) = \{x \in X : f(x) \neq 0\}.$$

- A *cone* in a vector space is a subset S with $cS \subset S$ for all $c \in \mathbb{K}$.

- A topological vector space E of $\mathbb{K}$-valued functions on a set X is a *PCS-space* if convergence in E implies pointwise convergence of a subsequence; this means that given a sequence (g_n) in E, if there is $g \in E$ with $g_n \longrightarrow g$ in E, then there is a subsequence $\{n_1 < n_2 < \cdots\} \subset \mathbb{N}$ depending on (g_n) such that, for every $x \in X$, $g_{n_k}(x) \longrightarrow g(x)$ as $k \to \infty$.

For instance, $L_p(X, \mu)$ are PCS-spaces (in the case of $L_\infty(X, \mu)$, we even choose $(n_k) = \mathbb{N}$). Note that when evaluation functionals are continuous in E, then E is a PCS-space.

The next technical result is crucial for the proof of the main theorem of this section (Theorem 3.5.5).

Theorem 3.5.2. *Let X be a nonempty set. Assume that $(E, \|\cdot\|)$ is a Banach space of $\mathbb{K}$-valued functions on X and that B is a nonempty subset of E satisfying the following properties:*

(1) *E is a PCS-space.*

(2) *There is a constant $C \in (0, \infty)$ such that $\|f+g\| \geq C\|f\|$ for all $f, g \in E$ with $\sigma(f) \cap \sigma(g) = \varnothing$.*

(3) *B is a cone.*

(4) *If $f, g \in E$ are such that $f + g \in B$ and $\sigma(f) \cap \sigma(g) = \varnothing$, then $f, g \in B$.*

(5) *There is a sequence of functions f_n $(n \in \mathbb{N})$ with pairwise disjoint supports such that, for all $n \in \mathbb{N}$, $f_n \in A$, where*

$$A := E \setminus B.$$

Then A is spaceable.

Proof. Note that the sequence (f_n) defined in (5) is a basic sequence. Indeed, by (3) we have $0 \in B$, and thus from (5) we conclude that $f_n \neq 0$ for all n. Besides, for every pair $r, s \in \mathbb{N}$ with $s \geq r$ and any scalars $a_1, \ldots, a_s$, from (2) and (5) we have

$$\left\| \sum_{n=1}^{s} a_n f_n \right\| = \left\| \sum_{n=1}^{r} a_n f_n + \sum_{n=r+1}^{s} a_n f_n \right\| \geq C \left\| \sum_{n=1}^{r} a_n x_n \right\|$$

since the supports of $\sum_{n=1}^{r} a_n f_n$ and $\sum_{n=r+1}^{s} a_n f_n$ have empty intersection. Invoking the Banach–Grumblum Criterium (Lemma 3.5.1), (f_n) is a basic sequence (with basic constant $1/C$). Also, the functions f_n ($n \geq 1$) are linearly independent. Defining the set

$$M := \overline{\operatorname{span}\{f_n : n \in \mathbb{N}\}},$$

we obviously note that M is a closed infinite dimensional vector subspace of E. It suffices to show that $M \setminus \{0\} \subset A$ and this is what we shall prove now. Let us fix a function $f \in M \setminus \{0\}$ and we shall prove that $f \in A$.

Since $f \in M \setminus \{0\}$, there is a uniquely determined sequence $(c_n) \subset \mathbb{K}$ such that

$$f = \sum_{n=1}^{\infty} c_n f_n = \|\cdot\|\text{-}\lim_{n \to \infty} \sum_{k=1}^{n} c_k f_k.$$

Defining $N = \min\{n \in \mathbb{N} : c_n \neq 0\}$, we can write

$$f = c_N f_N + h,$$

with $h = \|\cdot\|\text{-}\lim_{n \to \infty} T_n$ and $T_n := \sum_{k=N+1}^{n} c_k f_k$ ($n \geq N + 1$). If $x \in \sigma(c_N f_N)$ then $x \in \sigma(f_N)$ and, by (5), $x \notin \sigma(f_k)$ for all $k > N$. Therefore $T_n(x) = 0$ for all $n > N$. However, from (1) we know that there is a subsequence $(n_k) \subset \mathbb{N}$ such that T_{n_k} pointwise converges to h. Thus $h(x) = 0$ or, in other words, $x \notin \sigma(h)$. Therefore

$$\sigma(c_N f_N) \cap \sigma(h) = \varnothing.$$

If we had $f \notin A$ then we would have $f \in B$. But, since $f = c_N f_N + h$, we would also have $c_N f_N \in B$ due to (4). From (3), since $f_N = c_N^{-1} c_N f_N$, we would conclude that $f_N \in B$, which contradicts (5). Therefore we have $f \in A$. $\square$

From now on, (α) and (β) denote the following special conditions on the measure space $(X, \mathcal{M}, \mu)$:

(α) $\inf\{\mu(A) : A \in \mathcal{M}, \mu(A) > 0\} = 0$.

(β) $\sup\{\mu(A) : A \in \mathcal{M}, \mu(A) < \infty\} = \infty$.

The precise conditions under which the inclusions among the Lebesgue spaces $L_p(X, \mu)$ hold can be found in the papers by B. Subramanian [371] and J.L. Romero [347] (see also [327, Section 14.8]).

Theorem 3.5.3. *Let $(X, \mathcal{M}, \mu)$ be a measure space and p and q be extended real numbers satisfying $1 \leq p < q \leq \infty$. We have:*

(a) $L_p(X, \mu) \subset L_q(X, \mu) \iff \inf\{\mu(A) : A \in \mathcal{M}, \mu(A) > 0\} > 0$.

(b) $L_q(X, \mu) \subset L_p(X, \mu) \iff \sup\{\mu(A) : A \in \mathcal{M}, \mu(A) < \infty\} < \infty$.

Note that (a) and (b) of the last theorem can be, respectively, reformulated as follows:

(a) Let $1 \leq p < q \leq \infty$. Then $L_p(X, \mu) \setminus L_q(X, \mu) \neq \varnothing$ if and only if (α) holds.

(b) Let $1 \leq q < p \leq \infty$. Then $L_p(X, \mu) \setminus L_q(X, \mu) \neq \varnothing$ if and only if (β) holds.

Remark 3.5.4. *The following characterization of the property (α) (see [327, pp. 233–235]) will be important in the proof of our main theorem: (α) is true if and only if there exists a sequence (A_n) of pairwise disjoint measurable sets with $0 < \mu(A_n) < 1/2^n$ $(n \in \mathbb{N})$, while (β) holds if and only if there exists a sequence (A_n) of pairwise disjoint measurable sets with $1 < \mu(A_n) < \infty$ $(n \in \mathbb{N})$.*

The previous results remind us of well known inclusion (and non-inclusion) results (if $p < q$):

- If μ is a finite measure on $(X, \mathcal{M})$, then $L_q(X, \mu) \subset L_p(X, \mu)$;

- If ν is the counting measure on an infinite set Y then $L_p(Y, \nu) \subset L_q(Y, \nu)$;

- $L_q(I) \subset L_p(I)$, where I is a bounded interval of $\mathbb{R}$;

- $\ell_p \subset \ell_q$ as well as the non-inclusion relation $L_r(J) \not\subset L_s(J)$ $(r, s \in [1, \infty]$ with $r \neq s$ and J is an unbounded interval of $\mathbb{R}$).

Conditions (α) and (β) will be very important in the search for conditions for the spaceability in this section.

Let (X, M, μ) be a measure space and $p \in [1, \infty]$. From now on we define

$$L^p_{l\text{-}strict} := L_p(X, \mu) \setminus \bigcup_{q \in [1, p)} L_q(X, \mu).$$

The elements of $L^p_{l\text{-}strict}$ will be called left-strictly p-integrable functions. The members of the set

$$L^p_{r\text{-}strict} := L_p(X, \mu) \setminus \bigcup_{q \in (p, \infty]} L_q(X, \mu)$$

will be called right-strictly p-integrable functions. Finally, the members of the set

$$L^p_{strict} := L_p(X, \mu) \setminus \bigcup_{q \in [1, \infty] \setminus \{p\}} L_q(X, \mu)$$

are said to be strictly p-integrable functions.

Now, we present the main result of this section:

Theorem 3.5.5. *Assume that $p \in [1, \infty]$ and that $(X, \mathcal{M}, \mu)$ is a measure space. We have:*

(a) *If $p < \infty$, then set $L^p_{r\text{-}strict}$ is spaceable if and only if (α) holds.*

(b) *If $p > 1$, then $L^p_{l\text{-}strict}$ is spaceable if and only if (β) holds.*

(c) *If $1 < p < \infty$, then L^p_{strict} is spaceable if and only if both (α) and (β) hold.*

Proof. From Theorem 3.5.3 it follows that the conditions (α), (β) and (α)+(β) are, respectively, necessary in (a), (b) and (c). In order to prove that these conditions are also sufficient, we shall invoke Theorem 3.5.2 for the space $E = L_p$. Observe that it satisfies property (1) from Theorem 3.5.2. The property (2) is also satisfied with $C = 1$, because the norm $\|\cdot\|_p$ behaves monotonically for all p, i.e., $\|f\|_p \leq \|h\|_p$ if $|f| \leq |h|$ on X. In (a), (b) and (c), we define, respectively,

$$B = L_p \cap \bigcup_{q \in (p,\infty]} L_q,$$

$$B = L_p \cap \bigcup_{q \in [1,p)} L_q$$

and

$$B = L_p \cap \bigcup_{q \in [1,\infty] \setminus \{p\}} L_q.$$

So, in any case B is a subset of E and, since each L_q is a cone, it straightforwardly follows that B is a cone as well in all three cases above. Therefore, the property (3) in Theorem 3.5.2 is also fulfilled. But note that (4) also holds true, using one more time the monotonicity of the norms $\|\cdot\|_q$. Having in mind the notation of Theorem 3.5.2, we have $A = L^p_{r\text{-}strict}$, $A = L^p_{l\text{-}strict}$ and $A = L^p_{strict}$, respectively, for (a), (b) and (c). Now, according to Theorem 3.5.2, it just suffices to obtain (for each case) a sequence (f_n) in A whose members have mutually disjoint supports.

Let us prove the converse of (a). Suppose that (α) is valid and let $p \in [1,\infty)$. From Remark 3.5.4, we can find a sequence $(A_n) \subset \mathcal{M}$ of pairwise disjoint sets with $0 < \mu(A_n) < 1/2^n$ ($n \geq 1$). Choose a countable family $\{\{p(n,k)\}_{k\geq 1} : n \in \mathbb{N}\}$ of mutually disjoint strictly increasing sequences of positive integers. Since $p(n,k) \geq k$ for all n, k, we obtain $0 < \mu(A_{n,k}) < 1/2^k$ ($n,k \geq 1$), where $A_{n,k} := A_{p(n,k)}$. For every $n \in \mathbb{N}$, define the function $f_n : X \to [0,\infty)$ given by

$$f_n = \sum_{k=1}^{\infty} \frac{1}{k^{1/p}(\log(k+1))^{2/p}\mu(A_{n,k})^{1/p}} \cdot \chi_{A_{n,k}}. \tag{3.4}$$

It is straightforward that all f_n's are measurable and have pairwise disjoint supports. From the disjointness of the sets $A_{n,k}$, we have

$$f_n^p = \sum_{k=1}^{\infty} \frac{\chi_{A_{n,k}}}{k(\log(n+1))^2\mu(A_{n,k})}.$$

Moreover, we also have

$$\|f_n\|_p = \left(\int_X f_n^p \, d\mu\right)^{1/p} = \left(\sum_{k=1}^\infty \frac{1}{k(\log(k+1))^2}\right)^{1/p} < \infty.$$

Then it follows that each f_n belongs to L_p. Now, fix q finite with $q > p$. We have

$$\|f_n\|_q^q = \int_X f_n^q \, d\mu = \sum_{k=1}^\infty \frac{\mu(A_{n,k})}{k^{q/p}(\log(k+1))^{2q/p}\mu(A_{n,k})^{q/p}}$$

$$\geq \sum_{k=1}^\infty \frac{(2^{(q/p)-1})^k}{k^{q/p}(\log(k+1))^{2q/p}} = \infty,$$

since the general term of this series is unbounded (observe that $2^{(q/p)-1} > 1$). Thus $f_n \notin L_q$ and, moreover, we have also that $f_n \notin L_\infty$, as the measure of each $A_{n,k}$ is positive and on this set we have

$$|f_n| = \frac{1}{k^{1/p}(\log(k+1))^{2/p}\mu(A_{n,k})^{1/p}} > \frac{(2^{1/p})^k}{k^{1/p}(\log(k+1))^{2/p}} =: \gamma_k$$

for every $k \in \mathbb{N}$, from which it follows that $\|f_n\|_\infty \geq \gamma_k \to \infty$ as $k \to \infty$. So, in fact, $\|f_n\|_\infty = \infty$ and combining all these results we obtain $(f_n) \subset L^p_{r\text{-strict}}$ and the proof of (a) is done.

For the proof of the converse of (b) let us suppose that (β) holds. The proof is similar; we just need to have in mind that, this time, the mutually disjoint measurable sets $A_{n,k}$ can be chosen such that $1 < \mu(A_{n,k}) < \infty$ for all n (see Remark 3.5.4). Define f_n ($n \geq 1$) as in (3.4). The only thing that remains to do is to prove that $f_n \notin L_q$ whenever $q < p$. This holds because

$$\|f_n\|_q^q = \sum_{k=1}^\infty \frac{\mu(A_{n,k})^{1-q/p}}{k^{q/p}(\log(k+1))^{2q/p}} \geq \sum_{k=1}^\infty \frac{1}{k^{q/p}(\log(k+1))^{2q/p}},$$

and note that this series diverges (for $q/p < 1$).

In order to conclude the proof we need to prove the converse of (c). So, let us assume that (α) and (β) hold and let us consider $p \in (1, \infty)$. From (α), we know that there are infinitely many pairwise disjoint measurable sets B_n with $0 < \mu(B_n) < 1/2^n$ ($n \in \mathbb{N}$). We shall choose a countable family $\{\{p(n, k)\}_{k \geq 1} : n \in \mathbb{N}\}$ of mutually disjoint strictly increasing sequences in $\mathbb{N}$ satisfying $p(n, k) \geq n + 1$ for all n, k. Therefore $p(n, k) \geq n + k$ for all n, k. We then obtain

$$0 < \mu(B_{n,k}) < \frac{1}{2^{n+k}} \quad (n, k \geq 1),$$

where we have set $B_{n,k} := B_{p(n,k)}$. But, note that

$$Y := \bigcup_{n,k=1}^\infty B_{n,k}$$

has finite measure. So (β) is also satisfied by the measure subspace $(X \setminus Y, \mathcal{M}_{X \setminus Y}, \mu|_{X \setminus Y})$. From this we assure the existence of infinitely many mutually disjoint measurable sets $C_{n,k}$ with $1 < \mu(C_{n,k}) < \infty$ and $B_{n,k} \cap C_{m,j} = \emptyset$ $(n, k, m, j \in \mathbb{N})$. Defining

$$A_{n,1} := B_{n,1}, A_{n,2} := C_{n,1}, A_{n,3} := B_{n,2}, A_{n,4} := C_{n,2}, \ldots.$$

we consider the sequence (f_n) as in (3.4) and the proof can be concluded by following a suitable combination of the proofs of (a) and (b). $\qquad\square$

3.6 Lineability in sets of norm attaining operators in sequence spaces

In this section the set of all continuous linear operators from a Banach space E to a Banach space F will be represented by $\mathcal{L}(E; F)$. We recall that a continuous linear operator $u : E \to F$ *attains its norm* at $x_0 \in B_E$ if

$$\|u\| := \sup \|u(x)\| = \|u(x_0)\|.$$

In this case we say that u is *norm attaining*. Let us denote by $NA^{x_0}(E; F)$ the set of all operators from E to F that attain the norm at x_0. The classical theory of norm attaining functionals (i.e., operators with range in the scalar field) has deep and intriguing results. For instance, the Bishop–Phelps theorem, proved in 1961 by E. Bishop and R.R. Phelps, asserts that the set of norm attaining functionals in a Banach space X is dense in the set of all continuous linear functionals defined on X, the cornerstone of the theory is a penetrating result due to R.C. James (1957) asserting that a Banach space is reflexive if and only if every continuous linear functional defined on X is norm attaining.

Now, as an illustration of lineability in the context of norm attaining operators, we prove that $NA^{x_0}(E; \ell_q)$ is $\mathfrak{c}$-lineable. As we shall see, the proof uses again the separation of the set of natural numbers into infinitely many pairwise disjoint copies; a novelty here is the use of the Hahn–Banach theorem. We recall that one of the versions of this famous theorem of Functional Analysis asserts that if the Banach space G is a subspace of the Banach space E and $u : G \to \mathbb{K}$ is a continuous linear functional, then there is a (linear and continuous) extension $v : E \to \mathbb{K}$ of u such that $\|u\| = \|v\|$.

Proposition 3.6.1. *Let E be a Banach space and x_0 be a norm-one vector of E. Then $NA^{x_0}(E; \ell_q)$ is $\mathfrak{c}$-lineable.*

Proof. As in the proof of Theorem 3.2.6 we separate $\mathbb{N}$ into countably many infinite pairwise disjoint subsets $(A_k)_{k=1}^{\infty}$. For each positive integer k, let us denote

$$A_k = \left\{ a_1^{(k)} < a_2^{(k)} < \cdots \right\}$$

and consider

$$\ell_q^{(k)} = \{x \in \ell_q : x_j = 0 \text{ if } j \notin A_k\}.$$

Using the Hahn-Banach theorem we can easily find a nonzero operator $u \in NA^{x_0}(E; \ell_q)$. Now, for each fixed positive integer k, we define

$$u_k : E \to \ell_q^{(k)}$$

given by

$$(u_k(x))_{a_j^{(k)}} = (u(x))_j$$

for all positive integer j. Thus, for any fixed k, let $v_k : E \to \ell_q$ be given by

$$v_k = i_k \circ u_k,$$

where $i_k : \ell_q^{(k)} \to \ell_q$ is the canonical inclusion. It is plain that

$$\|v_k(x)\| = \|u_k(x)\| = \|u(x)\|$$

for every positive integer k and $x \in E$. Thus, each v_k attains its norm at x_0. Note also that the operators v_k have disjoint supports and thus the set $\{v_1, v_2, \dots\}$ is linearly independent. Consider the operator

$$T : \ell_1 \to \mathcal{L}(E; \ell_q)$$

given by

$$T((a_k)_{k=1}^{\infty}) = \sum_{k=1}^{\infty} a_k v_k.$$

Since

$$\sum_{k=1}^{\infty} \|a_k v_k\| = \sum_{k=1}^{\infty} |a_k| \, \|v_k\|$$

$$= \sum_{k=1}^{\infty} |a_k| \, \|v_k(x_0)\|$$

$$= \|u(x_0)\| \sum_{k=1}^{\infty} |a_k| < \infty$$

we conclude that T is well defined (i.e., the series converges in $\mathcal{L}(E; \ell_q)$); it is obvious that T is linear and injective. But, recalling that the supports of the operators v_k are pairwise disjoint, we conclude that

$$T(\ell_1) \subset NA^{x_0}(E; \ell_q)$$

and the proof is done, because $\dim(\ell_1) = \mathfrak{c}$. $\square$

In a different direction we can investigate the lineability of non norm attaining operators.

Proposition 3.6.2. *Let E and F be Banach spaces such that F contains an isometric copy of ℓ_q for some $1 \le q < \infty$. If $\mathcal{L}(E, \ell_q) \ne NA(E.\ell_q)$, then $\mathcal{L}(E, F) \setminus NA(E, F)$ is lineable in $\mathcal{L}(E, F)$. In particular, if $1 \le p \le q < \infty$, then $\mathcal{L}(\ell_p, \ell_q) \setminus NA(\ell_p, \ell_q)$ is lineable in $\mathcal{L}(\ell_p, \ell_q)$.*

Proof. It suffices to consider the case $F = \ell_q$. Let $T : E \to \ell_q$ be a non norm-attaining operator. Splitting $\mathbb{N}$ into infinitely many pairwise disjoint copies,

$$\mathbb{N} = \bigcup_{k=1}^{\infty} A_k,$$

and writing

$$A_k : \{a_1^{(k)} < a_2^{(k)} < \cdots \},$$

we define again

$$\ell_q^{(k)} = \{x \in \ell_q : x_j = 0 \text{ if } j \notin A_k\}.$$

For each fixed k, let $T_k : E \to \ell_q^{(k)}$ be given by

$$(T_k(x))_{a_j^{(k)}} = (T(x))_j$$

for all positive integers j, k. Now we consider, for all fixed k, the operators $V_k : E \to \ell_q$ given by

$$V_k = I_k \circ T_k,$$

where $I_k : \ell_q^{(k)} \to \ell_q$ is the natural inclusion. Since

$$\|V_k(x)\| = \|T_k(x)\| = \|T(x)\|$$

for every positive integer k and $x \in E$, we conclude that each V_k do not attain their norm. Since the operators V_k have disjoint supports it is routine to verify that $\{V_1, V_2, \dots \}$ is a linearly independent set. Now we shall show that the nontrivial vectors in span $\{V_1, V_2, \dots \}$ does not attain its norm. To simplify the notation let us restrict to the case of two vectors. Let us show that $aV_1 + bV_2$ does not attain its norm whenever $a \ne 0$ and $b \ne 0$. Note that since V_1 and V_2 are essentially the same operators (just with disjoint supports) we have

$$
\begin{aligned}
\|aV_1 + bV_2\|^q &= \sup_{\|x\| \le 1} \|aV_1(x) + bV_2(x)\|^q \\
&= \sup_{\|x\| \le 1} \left(\sum_k |a(V_1(x))_k|^q + \sum_k |b(V_2(x))_k|^q \right) \\
&= |a|^q \|V_1\|^q + |b|^q \|V_2\|^q .
\end{aligned}
$$

Finally, if x is a norm-one vector and since V_1 and V_2 do not attain their norms, we obtain

$$
\begin{aligned}
\|(aV_1 + bV_2)(x)\|^q &= |a|^q \, \|V_1(x)\|^q + |b|^q \, \|V_2(x)\|^q \\
&< |a|^q \, \|V_1\|^q + |b|^q \, \|V_2\|^q \\
&= \|(aV_1 + bV_2)\|^q \, .
\end{aligned}
$$

The general case is analogous. The final conclusion is a consequence of the fact that the operator $u : \ell_p \to \ell_q$ given by $u(x) = \left(\frac{1}{2}x_1, \frac{2}{3}x_2, \frac{3}{4}x_3, \frac{4}{5}x_4, ...\right)$ does not attain its norm. $\qquad\square$

3.7 Riemann and Lebesgue integrable functions and spaceability

The first main result of this section asserts that given any unbounded interval I, the set of all almost everywhere continuous bounded functions on I which are not Riemann integrable is spaceable. In fact, it contains an infinite closed algebra.

Lemma 3.7.1. *The set $(\ell_\infty \backslash c_0) \cup \{0\}$ contains a closed infinitely generated subalgebra in ℓ_∞.*

Proof. For every prime number p, consider the sequence x_p given by

$$
x_p(j) = \begin{cases} 1, & \text{if } j = p^k \text{ for some positive integer } k, \\ 0, & \text{otherwise.} \end{cases}
$$

Let $P = \{p_1, p_2, ...\}$ be the set of prime numbers with $p_i < p_j$ whenever $i < j$ and consider

$$
V = \left\{ \sum_{i=1}^{\infty} \lambda_i x_{p_i} : (\lambda_i)_{i \in \mathbb{N}} \in \ell_\infty \right\}.
$$

With a little bit of effort the reader can verify that $\overline{V} \cap c_0 = \{0\}$. Since ℓ_∞ with the pointwise product is a Banach algebra, note that V (and consequently $\overline{V}$) is a subalgebra with an infinite number of generators. In fact, this is a consequence of the simple fact that for any $p, q \in P$, the sequences x_p and x_q have disjoint supports. $\qquad\square$

If I is an interval of $\mathbb{R}$, by $\mathcal{B}(I)$ we will denote the set of all bounded functions on $I \to \mathbb{R}$. It is well known that $\mathcal{B}(I)$ becomes a Banach space under the supremum norm.

Theorem 3.7.2. *Given an arbitrary unbounded interval I, the set of all almost everywhere continuous bounded functions on I which are not Riemann integrable contains an infinitely generated closed subalgebra. In particular, this set is spaceable and algebrable.*

Proof. Let M be the set of all almost everywhere continuous functions on I. There is no loss of generality in assuming that I contains the interval $[1, \infty)$. Consider the function

$$\phi : \ell_\infty \to \mathcal{B}(I)$$

given by $\phi(x) = \sum\limits_{n=1}^{\infty} x(n) \chi_{(n,n+1)}$. Note that ϕ is a linear isometry and an algebra homomorphism, $\phi(x)$ is almost everywhere continuous for all $x \in \ell_\infty$. Besides, observe that $\phi(x)$ is Riemann integrable if and only if $\sum\limits_{n=1}^{\infty} x(n)$ converges and in this case

$$\int_I \phi(x)(t)\, dt = \sum_{n=1}^{\infty} x(n).$$

Now we invoke Lemma 3.7.1 to know that there is an infinitely generated closed subalgebra W of ℓ_∞ such that $c_0 \cap W = \{0\}$. But $\phi(W) \subset M \cup \{0\}$ and since ϕ is a linear isometry and an algebra homomorphism we know that $\phi(W)$ has the same properties of W and the proof is done. $\qquad\square$

The existence of functions which are not Lebesgue integrable is a simple consequence of the existence of nonmeasurable sets. Let us recall the classical example of G. Vitali, of a subset of $\mathbb{R}$ that is not Lebesgue measurable.

Let $x, y \in [0, 1]$, and consider the equivalence relation in $[0, 1]$ given by

$$x \sim y \iff y - x \in \mathbb{Q}_1 := \mathbb{Q} \cap [-1, 1].$$

So, we can choose $\alpha \in [0, 1]$ such that we have the disjoint union $[0, 1] = \bigcup E_\alpha$, and each E_α is the equivalence class of α. Note that $E_\alpha = \{x + \alpha; x \in \mathbb{Q}_1\} \cap [0, 1]$, and since $\mathbb{Q}_1$ is countable, we conclude that each E_α is countable. Now we use the Axiom of Choice to choose a representative of each equivalence class and define the set $V \subset [0, 1]$ composed by these choices. Let $\{r_1, r_2, ...\}$ be an enumeration of $\mathbb{Q}_1$ and, for all n, consider $V_n = V + r_n$. If $x \in [0, 1]$, then $x \in E_\alpha$ for some α. Thus $x \sim v$ for some $v \in V$. We thus conclude that $x - v \in \mathbb{Q}_1$ and therefore $x = v + r$, with $r \in \mathbb{Q}_1$. Thus $x \in \bigcup\limits_{n=1}^{\infty} V_n$. Note that of $x \in (V + r_n) \cap (V + r_m)$, then $x = v_1 + r_n = v_2 + r_m$ and thus

$$v_1 - v_2 = r_m - r_n. \tag{3.5}$$

Since $V \subset [0, 1]$, we have $v_1 - v_2 \in [-1, 1]$, and since $r_m - r_n \in \mathbb{Q}$, it follows that

$$v_1 - v_2 = r_m - r_n \in [-1, 1] \cap \mathbb{Q} = \mathbb{Q}_1.$$

So $v_1 \sim v_2$ and $v_1 = v_2$ (because $v_1 \in V$ and $v_2 \in V$). Hence, from (3.5) we have $r_m = r_n$ and $m = n$. Thus $V_n \cap V_m = \emptyset$, if $n \neq m$.

Finally, we have

$$[0,1] \subset \bigcup_{n=1}^{\infty} V_n = \bigcup_{n=1}^{\infty} (V + r_n) \subset \bigcup_{n=1}^{\infty} ([0,1] + [-1,1]) = [-1,2].$$

If V is measurable then each V_n is measurable (since $V_n = V + r_n$) and

$$\mu(V) = \mu(V_n).$$

Thus

$$1 = \mu([0,1]) \leq \sum_{j=1}^{\infty} \mu(V_n) \leq \mu([-1,2]) = 3.$$

However

$$\sum_{n=1}^{\infty} \mu(V_n) = \sum_{n=1}^{\infty} \mu(V) = 0 \text{ or } \infty.$$

We conclude this section by stating that given any unbounded interval I, the set of Riemann integrable functions f on I that are not Lebesgue integrable contains an infinite dimensional vector space. An illustration of a function belonging to this class is $f : \mathbb{R} \to \mathbb{R}$ given by $f(x) = \frac{\sin x}{x}$, because

$$\int_{\mathbb{R}} f(x)dx = \pi$$

but

$$\int_{\mathbb{R}} |f(x)| \, dx = \infty.$$

Recall that, if $f : I \to \mathbb{R}$ is measurable, then f is Lebesgue integrable if and only if $|f|$ is Lebesgue integrable.

Theorem 3.7.3. *Given any unbounded interval I, the set of Riemann integrable functions on I that are not Lebesgue integrable is lineable.*

The proof is left as an exercise for the interested reader (see Exercise 3.10).

3.8 Exercises

Exercise 3.1. Let E be an invariant sequence space over the Banach space X. Let $A \subset E$ be such that:

(i) For $x \in E$, $x \in A$ if and only if $x^0 \in A$.

(ii) If $x = (x_j)_{j=1}^\infty \in A$ and $y = (y_j)_{j=1}^\infty \in E$ is such that $(\|y_j\| x)_{j=1}^\infty$ is a multiple of a subsequence of $(\|x_j\| x)_{j=1}^\infty$, then $y \in A$.

(iii) There is $x \in E \setminus A$ with $x^0 \neq 0$.

Then $E \setminus A$ is spaceable. *Hint:* Rewrite the proof of Theorem 3.3.5.

Exercise 3.2. Let $I = [1, \infty)$. Prove that

$$\text{span} \left\{ \frac{1}{x^r} : x \in I \text{ and } r \in \left(\frac{1}{p}, \frac{1}{q} \right) \right\}$$

is a $\mathfrak{c}$-dimensional subspace of $(L_p(I) \setminus L_q(I)) \cup \{0\}$.

Exercise 3.3. Prove that

$$\text{span} \left\{ \left(\frac{1}{n^r} \right)_{n=1}^\infty : r \in \left(\frac{1}{p}, \frac{1}{q} \right) \right\}$$

is a $\mathfrak{c}$-dimensional subspace of $(\ell_p \setminus \ell_q) \cup \{0\}$.

Exercise 3.4. Let P denote the set of odd primes and for all $p \in P$, define

$$x_p = \left(\frac{1}{p}, \frac{1}{p^2}, \dots \right) \in \ell_\infty.$$

Prove that

$$\text{span} \{x_p : p \in P\}$$

is an infinite dimensional subspace of ℓ_∞ every non-zero element of which has a finite number of zero coordinates.

Exercise 3.5. Complete the proof of Theorem 3.3.5.

Exercise 3.6. Prove that if $1 \le p \le q < \infty$, then $\mathcal{L}(\ell_p; \ell_q) \setminus NA^{x_0}(\ell_p; \ell_q)$ is $\aleph_0$-lineable.

Exercise 3.7. Use the idea of the proof of Theorem 3.7.2 to prove that the set of bounded continuous functions which are not Riemann integrable is spaceable.

Exercise 3.8. Given any non-empty set A, there exists a measurable space (Ω, Σ) and a pairwise disjoint family $(V_a)_{a \in A}$ such that, for every $\varnothing \neq B \subset A$, $\bigcup_{b \in B} V_b$ is a non-measurable subset of Ω. In particular, V_a is non-measurable for every $a \in A$.

Exercise 3.9. Let A be a non-empty set. Let us consider a measurable space (Ω, Σ) and a pairwise disjoint family $(V_a)_{a \in A}$ verifying that, for every $\varnothing \neq B \subset A$, $\bigcup_{b \in B} V_b$ is a non-measurable subset of Ω. Then, the set $W_A := \{\sum_{a \in A} \alpha_a \chi_{V_a} : (\alpha_a)_{a \in A} \subset \mathbb{K}\}$ is a vector space contained in $N(\Omega, \mathbb{K}) \cup \{0\}$, and the dimension of W_A is bigger than $card(A)$.

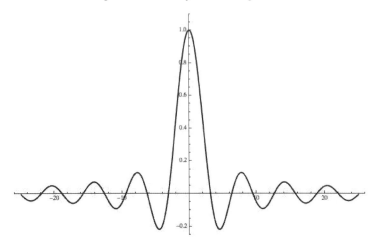

FIGURE 3.1: A function in $R(\mathbb{R}) \setminus L_1(\mathbb{R})$.

Exercise 3.10. Use the function (Figure 3.1)

$$f(x) = \begin{cases} \dfrac{\sin x}{x} & \text{if } x \neq 0 \\ 1 & \text{if } x = 0. \end{cases}$$

and a suitable modification of it in order to show that the set of (improper) Riemann integrable functions on an unbounded interval $I \subset \mathbb{R}$ that are not Lebesgue integrable on I is lineable.
Hint: Check the initial construction provided in the proof of [215, Theorem 3.2].

Exercise 3.11. Complete the proof of Theorem 3.3.5.

Exercise 3.12. Let X, Y be Banach spaces and $f \colon X \longrightarrow Y$ be a map. If f is discontinuous at the origin and non-contractive, then the set

$$\left\{ (x_j)_{j=1}^{\infty} \in c_0(X) : (f(x_j))_{j=1}^{\infty} \notin c_0(Y) \right\}$$

is spaceable.

Exercise 3.13. In the proof of Proposition 3.2.4 the Banach–Steinhaus theorem has been employed. In fact, going back to the classical proof of it (using the Open Mapping theorem, whose proof, in turn, uses Baire's theorem) it can be seen that, in fact, the set $\ell_p \setminus \bigcup_{0 < q < p} \ell_q$ is *residual* in ℓ_p. Prove directly Proposition 3.2.4 by exhibiting a sequence $(x_j)_{j \geq 1} \in \ell_p$ that is not in any ℓ_q $(0 < q < p)$. Prove also (this is easier) that $c_0 \setminus \bigcup_{p > 0} \ell_p \neq \varnothing$.
Hint: For the first question, try to find a sequence of the form $c_j = 1/\alpha_j$, with $\alpha_j > 0$ and $\alpha_j \to \infty$ (but not too rapidly) so that $(c_j) \in \ell_1$ but $(c_j) \notin \ell_r$ for all $r > 1$. Then define $x_j := c_j^{1/p}$ $(j \geq 1)$.

Exercise 3.14. Complete the proof of Theorem 3.5.5.

Exercise 3.15. Complete the proof of Lemma 3.7.1.

3.9 Notes and remarks

As we have already mentioned quite a few times, lineability and spaceability within the framework of Measure and Function spaces is a very fruitful topic of investigation and has been explored in quite different directions:

- Riemann integrable functions versus Lebesgue integrable functions ([215], see also Theorem 3.7.3);

- Sets of injective measures ([320]);

- Sets of composition of Riemann and Lebesgue integrable functions ([37]);

- L_p spaces ([83, 103]) among many others.

Section 3.2. In 1968, Rosenthal [348] showed that c_0 is quasi-complemented in ℓ_∞ (a closed subspace Y of a Banach space X is *quasi-complemented* if there is a closed subspace Z of X such that $Y \cap Z = \{0\}$ and $Y + Z$ is dense in X); this clearly implies that $\ell_\infty \setminus c_0$ is spaceable. In 2009, García-Pacheco, Martín and Seoane proved ([215]) that $\ell_\infty(\Gamma) \setminus c_0(\Gamma)$ is spaceable for every infinite set Γ. It is worth mentioning that Lindenstrauss (1968, [289]) proved that, if Γ is uncountable, then $c_0(\Gamma)$ is not quasi-complemented in $\ell_\infty(\Gamma)$. Proposition 3.2.4 was kindly shown to the authors by Professor M.C. Matos. Theorem 3.2.6 is due to Botelho, Diniz, Fávaro and Pellegrino [141]. Before the proof of Theorem 3.4.1, Muñoz, Palmberg, Puglisi and Seoane [320] proved that $\ell_p \setminus \ell_q$ is c-lineable. The following question was posed by Aron and Gurariy: Is there an infinite dimensional and closed subspace of ℓ_∞ every non-zero element of which has a finite number of zero coordinates? As remarked in [320], if P denotes the set of odd primes and $p \in P$, defining

$$x_p = \left(\frac{1}{p}, \frac{1}{p^2}, \ldots\right) \in \ell_\infty$$

we can verify that

$$\text{span}\,\{x_p : p \in P\}$$

is an infinite dimensional manifold enjoying the wished property (see Exercise 3.4). The original problem was finally solved in 2014 by Cariello and Seoane [153] (see Section 6.5).

Maximal spaceability in sequence spaces has been investigated by Barroso, Botelho, Fávaro and Pellegrino [44] and by Botelho, Cariello, Fávaro and Pellegrino [138]. It was proved, for instance, that

$$c_0(X) \smallsetminus \bigcup_{p>0} \ell_p^w(X),$$

$$\ell_p^w(X) \smallsetminus \bigcup_{0<q<p} \ell_q^w(X)$$

are maximal spaceable. In fact, the results of [138] are quite more general, and the proofs are based in a functorial general argument.

Section 3.3.   The results of this section are due to Botelho and Fávaro [143]. Their paper contains much more information than presented here. The following beautiful example is illustrative: Given a Banach space X and $p > 1$, defining $p_n := p - \frac{1}{\log(\log(n+4))}$ it is proved that there exists a Banach space of maximal dimension formed, up to the null vector, by X-valued p-summable sequences not belonging to the so-called Nakano space of X-valued sequences determined by the sequence $(p_n)_{n=1}^\infty$.

Section 3.4. Muñoz, Palmberg, Puglisi and Seoane [320] proved, by means of real analysis techniques, that if $1 \le p < q$ the set $L_p[0,1] \smallsetminus L_q[0,1]$ is c-lineable. The same kind of problem was investigated in 2014 by Ruiz and Sánchez [353] for the so-called Lorentz, Marcinkiewicz, Orlicz and Nakano spaces on $[0,1]$. These are generalizations of the classical Lebesgue spaces.

For unbounded intervals I, in 2008 Muñoz, Palmberg, Puglisi and Seoane [320] have shown that if $p > q \ge 1$, then $L_p(I) \smallsetminus L_q(I)$ is c-lineable. In 2009 Aron, García-Pacheco, Pérez-García and Seoane [26] extended the c-lineability to dense c-lineability. Part of the results of this section were extended to the framework of maximal dense lineability. In fact, the paper [104] is a source of criteria of lineability and presents some general procedures that may replace particular arguments regarding lineability and spaceability.

In the context of infinite dimensional Banach spaces, Bongiorno, Darji and Di Piazza [131] have recently proved that, if X is such a space and $[0,1]$ denotes the unit interval endowed with the Lebesgue measure, then the family $\mathcal{ND}$ of strongly measurable Pettis integrable functions $[0,1] \to X$ having nowhere weakly primitives is lineable. The reader is referred to [187] for a basic theory of vector-valued integration.

Section 3.5. The results of this section are due to Bernal and Ordóñez [103]. From Theorem 3.5.5 it can be inferred that, if $p \in [1,\infty)$ (if $p \in (1,\infty]$, if $p \in (1,\infty)$, resp.), then $L_{r\text{-}strict}^p$ ($L_{l\text{-}strict}^p$, L_{strict}^p, resp.) is spaceable if and only if it is nonempty.

If $p > 1$, then one obtains part (b) of Theorem 3.5.5 in the case $X = \mathbb{N}$, $\mu = $ the counting measure from Corollary 3.2.8. Also, part (a) of Theorem 3.5.5 can be obtained from Theorem 3.4.1 in the special case $\mu = m = $ the Lebesgue measure on $[0,1]$. The following technical and somewhat topological result can be inferred from Theorem 3.5.5 as a corollary:

Corollary. *Assume that $p \in [1, \infty)$ and that $(X, \mathcal{M}, \mu)$ is a measure space. Then $L^p_{r\text{-}strict}$ is spaceable if at least one of the following two properties is true:*

(1) *μ is semifinite, i.e., for each $A \in \mathcal{M}$, $\mu(A) = \sup\{\mu(B) : B \in \mathcal{M}, B \subset A$ and $\mu(B) < \infty\}$, and nonatomic, i.e., there is no set $A \in \mathcal{M}$ with $\mu(A) > 0$ such that, for every $B \in \mathcal{M}$ with $B \subset A$, one has $\mu(B) = 0$ or $\mu(A \setminus B) = 0$.*

(2) *X is a T_1 topological space, $\mathcal{M}$ contains the Borel sets of X and there is a non-isolated point $x_0 \in X$ satisfying:*

    (i) *for each closed set F with $x_0 \notin F$, there exist open sets A, B such that $x_0 \in A$, $F \subset B$ and $A \cap B = \varnothing$;*

    (ii) *x_0 possesses a countable fundamental system of neighborhoods, and*

    (iii) *there is an open neighborhood U of x_0 with $\mu(U) < \infty$ and $\mu(V) > 0$ for any nonempty open set $V \subset U$.*

Theorem 3.5.5 has been used by Akbarbaglu and Maghsoudi [4] in the investigation of lineability in certain subsets of the Orlicz spaces L^φ and Nakano spaces M^φ (depending on appropriate functions φ). In [4] several spaceability results of this section are extended to L^φ and M^φ, while the last corollary is extended to M^φ.

Although we have restricted our attention to spaceability, the maximal dense lineability of the diverse families of strictly Lebesgue integrable functions has been also investigated, yielding analogous results, as the following theorem shows (see [83]).

Theorem. *Assume that $p \in [1, \infty)$ and that $(X, \mathcal{M}, \mu)$ is a measure space such that $L_p(X, \mathcal{M}, \mu)$ is separable. We have:*

(a) *The set $L^p_{r\text{-}strict}$ is maximal dense-lineable if and only if (α) holds.*

(b) *If $p > 1$, then $L^p_{l\text{-}strict}$ is maximal dense-lineable if and only if (β) holds.*

(c) *If $p > 1$, then L^p_{strict} is maximal dense-lineable if and only if both (α) and (β) hold.*

Very recently, Bernal and Ordoñez [104] obtained a generalization of Theorem 3.5.5 encompassing the non-locally convex case; namely, the conclusion of this theorem holds for the whole range $0 < p \leq \infty$. For related results in this direction we also refer to [139].

Another interesting property to be studied from the point of view of lineability is that of (non-) locally integrability. If $0 < q \leq \infty$, X is a topological space and μ is a Borel measure on X, then a function $f : X \to \mathbb{K}$ is called nowhere q-integrable provided that, for each nonempty open subset U of X, the restriction $f|_U$ is not in $L_q(U)$. In [83] it is proved that if $1 \leq p < \infty$

then, under appropriate conditions, the family of functions in $L_p(X)$ that are nowhere q-integrable for each $q > p$ is dense in $L_p(X)$. Recently, Głąb, Kaufmann and Pellegrini [224] have been able to show –together with other properties and also under appropriate conditions– the maximal dense-lineability and the spaceability of the mentioned family, even for the range $0 < p < \infty$.

Concerning vector integration, García-Pacheco and Sofi [220] have shown the lineability of the family of scalarly measurable functions that are not strongly measurable, while J. Rodríguez [344] has proved the existence of infinite-dimensional linear spaces of Banach space-valued functions whose non-zero elements witness that two given notions of integrability are different: Bochner, Birkhoff, Mc-Shane, Pettis and Dunford integrability (see also [220]).

Section 3.6. The results in this Section are due to Pellegrino and Teixeira [332] and Botelho, Diniz, Fávaro and Pellegrino [141]. Besides the set of norm attaining operators, there is a classical lineability problem on whether the set of norm attaining functionals contains (or not) linear spaces. This problem was originally posed by Godefroy [226] and we will present and discuss it in Section 6.4.

Section 3.7. If f is Riemann integrable and g is continuous it is not always true that $f \circ g$ is Riemann integrable (examples of such f, g can be found in [221]). In this direction, in [37] it is constructed a $2^{\mathfrak{c}}$-dimensional linear space V formed by Riemann integrable functions on R and a $\mathfrak{c}$-dimensional linear space W of continuous functions on $[0, 1]$ such that for every $f \in V \smallsetminus \{0\}$ and every $g \in W \smallsetminus \{0\}$, the composition $f \circ g$ is not Riemann integrable on $[0, 1]$.

Chapter 4

Universality, Hypercyclicity and Chaos

In this chapter we deal with some classes of operators presenting chaotic dynamical behavior. The topic has been systematically studied during the last three decades. In contrast to the remaining chapters, lineability of families of universal vectors –which is the main concern of this one– has received a unified treatment in at least two major publications, namely, the 2009 book by Bayart and Matheron [61] and the 2011 book by Grosse-Erdmann and Peris [237]; see especially chapters 1 and 8 of [61] and chapters 2, 10 and 11 of [237]. Moreover, it should be said that even the mere enumeration of all known results about universality and close topics goes beyond the scope of this book. Consequently, a selection of the most relevant statements will be given in this chapter and, in turn, we will choose the proofs of a sufficient number of such statements, in order to illustrate the main techniques. Appropriate references will be provided for the remaining proofs, and some of these will be left as exercises. Of course, a reasonable updating of results has been tried.

4.1   What one needs to know

Denote by X, Y two topological vector spaces, and by $L(X, Y)$ the vector space of all continuous linear mappings $X \to Y$. Unless the contrary is explicitly established, in this section an *operator* T on X will be a member of $L(X) := L(X, X)$. The composition $T \circ S$ of two operators on X is usually denoted by TS, and the iterates of T are $T^1 = T$, $T^2 = TT$, $T^3 = TTT$, and so on. If $P(\lambda) = a_0 + a_1\lambda + a_2\lambda^2 + \cdots + a_N\lambda^N$ is a polynomial with coefficients a_j in the scalar field $\mathbb{K}$ of X, then $P(T)$ represents the operator $a_0 I + a_1 T + \cdots + a_N T^N$, where I is the identity operator.

If $(X, \|\cdot\|)$ is a normed space, then a Banach space $(Y, \|\cdot\|_0)$ is said to be a *completion* of $(X, \|\cdot\|)$ whenever there is a linear injective mapping $i : X \to Y$ such that $i(X)$ is dense in Y and $\|i(x)\|_0 = \|x\|$ for all $x \in X$. Hence we can assume that X is a dense vector subspace of Y. Such a completion Y always exists and, if Z is another completion of X, then Y and Z are topologically isomorphic.

For the content of this paragraph, see, e.g., Diestel's book [184]. A sequence (e_n) in a Banach space $(Z, \|\cdot\|)$ is said to be a *basic sequence* if it is a Schauder basis in its closed linear span, that is, each vector $x \in \overline{\text{span}}\{e_n : n \geq 1\}$ has a unique representation $x = \sum_{n=1}^{\infty} a_n e_n$ ($a_n \in \mathbb{K}$, $n \in \mathbb{N}$). The corresponding coefficient functionals $e_j^* : x \in X \mapsto a_j \in \mathbb{K}$ are well defined, linear and continuous, with norms $\|\|e_j^*\|\| := \sup\{|e_j(x)| : \|x\| = 1\}$ ($j \in \mathbb{N}$). *Mazur's theorem* states that every infinite dimensional Banach space contains a basic sequence. The *basis perturbation theorem* asserts that if (e_n) is a basic sequence in the Banach space Z and $(u_n) \subset X$ is such that $\sum_{n=1}^{\infty} \|e_n^*\| \|e_n - u_n\| =: \delta < 1$, then (u_n) is also a basic sequence. In addition, both basic sequences are *equivalent*, that is, a series $\sum_{n=1}^{\infty} a_n e_n$ converges if and only if $\sum_{n=1}^{\infty} a_n u_n$ does, and $\|u_n^*\| \leq \|e_n^*\|(1 - \delta)^{-1}$ for all $n \geq 1$.

Recall that an *F-space* is a completely metrizable topological vector space, while a *Fréchet space* is a locally convex F-space. If X is an F-space, its topology is induced by an *F-norm*, that is, a real functional $\|\cdot\| : X \to [0, \infty)$ satisfying the following conditions for all $x, y \in X$ and all scalars c: $\|x + y\| \leq \|x\| + \|y\|$, $\|cx\| \leq \|x\|$ if $|c| \leq 1$, $\lim_{c \to 0} \|cx\| = 0$ and, if $x \neq 0$, $\|x\| > 0$ (see [266, pp. 2–5]). Note that the first two axioms imply that $\|cx\| \leq (|c| + 1)\|x\|$ for all $(x, c) \in X \times \mathbb{K}$.

Recall that if G is a domain of the complex plane $\mathbb{C}$ then the space $H(G)$ of holomorphic functions on G becomes a Fréchet space under the topology generated by the collection of all sets $V(f, K, \varepsilon) := \{g \in H(G) : |g(z) - f(z)| < \varepsilon \text{ for all } z \in K\}$ ($f \in H(G)$, $\varepsilon > 0$, $K \subset G$ compact). Then we have that $f_n \to f$ in $H(G)$ for this topology if and only if $f_n \to f$ uniformly on each compact subset of G. For results about approximation of functions of $H(G)$ under appropriate conditions, it is convenient to take into account the theorems given in Section 2.1, or to consult some adequate book, such as Gaier [205].

Assume that X is a complex Banach space and that $T \in L(X)$. Let T^* be the adjoint of T, that is, T^* is the operator $T^* \in L(X^*)$ on the topological dual X^* of X defined as $T^*\varphi = \varphi \circ T$ if $\varphi \in X^*$. The *point spectrum* $\sigma_p(T)$ of T is the set of eigenvalues of T, that is, the set of $\lambda \in \mathbb{C}$ such that $T - \lambda I$ is not one-to-one, I being the identity operator. The *spectrum* $\sigma(T)$ of T is the set of scalars λ for which $T - \lambda I$ is not invertible. Then $\sigma_p(T) \subset \sigma(T)$ and $\sigma(T) = \sigma(T^*)$. We always have that $\sigma(T)$ is a nonempty compact subset of $\mathbb{C}$. The reader can find in e.g. [351] the concepts and properties given in this paragraph.

A useful result that will be used at least twice in this chapter is the so-called dominated convergence theorem for series, which can be stated as follows. Assume that $\{x_{n,k}\}$ is a double sequence of vectors in a Banach space X, such that $\lim_{n\to\infty} x_{n,k} = y_k \in X$ for every $k \in \mathbb{N}$. Suppose further that there is a sequence (d_k) of real positive numbers which dominates $\{x_{n,k}\}$, that is, (d_k) satisfies $\sum_{k=1}^{\infty} d_k < \infty$ and $\|x_{n,k}\| \leq d_k$ for all $n, k \in \mathbb{N}$. Then the series $\sum_{k=1}^{\infty} y_k$ converges in X and $\sum_{k=1}^{\infty} y_k = \lim_{n\to\infty} \sum_{k=1}^{\infty} x_{n,k}$.

4.2 Universal elements and hypercyclic vectors

Traditionally, chaotic processes had been associated to nonlinear settings. Surprisingly, in 1929 Birkhoff [121] showed the existence of an entire function $\mathbb{C} \to \mathbb{C}$ whose sequence of translates

$$\{f(\cdot + an) : n \geq 1\}$$

(with $a \in \mathbb{C} \setminus \{0\}$) approximates uniformly in compacta any prescribed entire function. This entails a rather wild dynamics for such a function f under the action of a continuous *linear* self-mapping of $H(\mathbb{C})$, namely, the translation operator $\tau_a g := g(\cdot + a)$. In 1952, MacLane [296] demonstrated the same denseness property for the orbit $\{f^{(n)} : n \geq 1\}$ of some entire function f under the action of the derivative operator $Dg := g'$. From these prominent examples, and others that do not necessarily come from iterates of one self-mapping, many analysts have invested much effort in studying these kinds of phenomena, mostly during the last thirty years. The adequate abstract framework for these results is given in the next definition. All topological spaces in this chapter are assumed to be *Hausdorff*, that is, two distinct points can be separated by disjoint open sets.

Definition 4.2.1. Let X and Y be two topological spaces and $T_n : X \to Y$ ($n \in \mathbb{N} := \{1, 2, \ldots\}$) be a sequence of continuous mappings. Then (T_n) is said to be *universal* provided that there exists an element $x_0 \in X$, called universal for (T_n), such that the orbit $\{T_n x_0 : n \in \mathbb{N}\}$ of x_0 under (T_n) is dense in Y.

We denote

$$\mathcal{U}((T_n)) := \{x \in X : x \text{ is universal for } (T_n)\}.$$

It is evident that the universality of some sequence (T_n) implies that Y is separable. If X and Y are topological vector spaces and $(T_n) \subset L(X,Y)$ then the words "universal" and "hypercyclic" are synonymous, although the term "hypercyclic" (coined by Beauzamy [66]) is mainly used to designate an operator $T \in L(X)$ such that the sequence (T^n) of its iterates is universal. We denote

$$HC(T) := \{\text{hypercyclic vectors for } T\} = \mathcal{U}((T^n)).$$

Excellent surveys for the theory of hypercyclicity and universality are the papers [128, 234, 235, 276] and the already cited books [61, 237].

Under the last terminology, the mentioned theorems by Birkhoff and MacLane can be reformulated as follows: both translation and derivation operators are hypercyclic on the space $H(\mathbb{C})$ endowed with the compact-open topology. In 1941, Seidel and Walsh gave a non-Euclidean version of Birkhoff's theorem by showing that if $H(\mathbb{D})$ is endowed with the compact-open topology and

$$C_\varphi : f \in H(\mathbb{D}) \mapsto f \circ \varphi \in H(\mathbb{D})$$

denotes the composition operator generated by a non-Euclidean translation $\varphi(z) = \frac{z+a}{1+\bar{a}z}$ $(a \in \mathbb{D} \setminus \{0\})$, then C_φ is hypercyclic. And in 1991, Godefroy and Shapiro [227] unified and strengthened the theorems by Birkhoff and MacLane in the following way: any $T \in L(H(\mathbb{C}))$ that is not a scalar multiple of the identity I and that commutes with the derivative operator D is hypercyclic. In particular, the differential operator $P(D)$ is hypercyclic on $H(\mathbb{C})$ for every nonconstant polynomial P with complex coefficients. In fact (see for instance [227]), an operator $T \in L(H(\mathbb{C}))$ commutes with D if and only if it commutes with translations, and if and only if $T = \Phi(D)$ for some entire function Φ with exponential type, meaning that there are positive constants A, B such that $|\Phi(z)| \leq Ae^{B|z|}$ for all $z \in \mathbb{C}$. Recall that if $\Phi(z) = \sum_{n=0}^{\infty} a_n z^n$ in $\mathbb{C}$ then $\Phi(D)f := \sum_{n=0}^{\infty} a_n f^{(n)}$ for all $f \in H(\mathbb{C})$; see Exercise 4.2.

In 1969, Rolewicz [346] provided the first example of a hypercyclic operator on a Banach space: if $X = c_0$ or ℓ_p $(1 \leq p < \infty)$ and

$$B : (x_1, x_2, x_3, \dots) \mapsto (x_2, x_3, x_4, \dots)$$

is the backward shift on X, then any scalar multiple λB $(|\lambda| > 1)$ is hypercyclic. In the same paper he proves that, in order that a topological vector space X supports a hypercyclic operator, X must be infinite dimensional. Rolewicz posed the problem of whether every separable infinite dimensional

Banach space X supports a hypercyclic operator. This was answered in the affirmative by Ansari [13] and Bernal [73]. In fact, a hypercyclic operator $T : X \to X$ can be constructed as a compact perturbation of the identity I. Specifically, if $\{e_n\}_{n=0}^{\infty} \subset X$, $\{\varphi_n\}_{n=0}^{\infty} \subset X^*$ (the topological dual of X) are sequences satisfying $\varphi_k(e_j) = \delta_{kj}$ $(k, j \geq 0)$, $\overline{\operatorname{span}}\{e_n\}_{n=0}^{\infty} = X$, $\|e_n\| = 1$ $(n \geq 0)$, $\sup_{n \geq 0} \|\varphi_n\| < \infty$ and $\bigcap_{n \geq 0} \ker \varphi_n = \{0\}$, then

$$T : x \in X \mapsto x + \sum_{n=0}^{\infty} a_{n+1} \varphi_{n+1}(x) e_n \in X$$

defines a hypercyclic operator on X (as usual, $\delta_{kj} = 0$ if $k \neq j$, and $\delta_{jj} = 1$). In [130] Bonet and Peris proved that this existence result was even valid for Fréchet spaces. Recently, Shkarin [367] has shown that every normed space of countable algebraic dimension supports a hypercyclic operator.

Another remarkable example in the setting of Banach spaces is the following. For $p \in [1, \infty)$ consider the Hardy space H^p on the open unit disc $\mathbb{D}$. The automorphisms (i.e. the bijective holomorphic self-mappings) of $\mathbb{D}$ are exactly the fractional linear mappings of the form $\varphi(z) = k\frac{z-a}{1-\bar{a}z}$. In Shapiro's book [362] it is proved that C_φ is hypercyclic on H^p if and only if φ lacks fixed points in $\mathbb{D}$; see [145] and [207] for extensions. Turning to sequences of operators, Bernal and Montes [101] showed that the sequence (C_{φ_n}) generated by a sequence (φ_n) of automorphisms of $\mathbb{D}$ is universal on $H(\mathbb{D})$ if and only if $\sup_{n \geq 1} |\varphi_n(0)| = 1$; moreover, if $\{\psi_n(z) = a_n z + b_n\}_{n \geq 1}$ is a sequence of automorphisms of $\mathbb{C}$, then (C_{ψ_n}) is universal on $H(\mathbb{C})$ if and only if the sequence $\{\min\{|b_n|, |b_n/a_n|\}\}_{n \geq 1}$ is unbounded. Furthermore, if $(\Phi_n(D))$ is a sequence of differential operators, such that each Φ_n is an entire function with exponential type and there exist subsets $A, B \subset \mathbb{C}$ each of them with at least one finite accumulation point, and satisfying $\lim_{n \to \infty} \Phi_n(z) = 0$ $(z \in A)$ and $\lim_{n \to \infty} \Phi_n(z) = \infty$ $(z \in B)$, then $(\Phi_n(D))$ is hypercyclic on $H(\mathbb{C})$ [75].

Let us now consider the topological size of the set of universal vectors. From now on, X and Y will stand for topological vector spaces on the same field $\mathbb{K}$ $(= \mathbb{R}$ or $\mathbb{C})$, and T_n, T will be linear and continuous. Firstly, for a single hypercyclic operator $T \in L(X)$, since each member $T^m x_0$ of the orbit of a hypercyclic vector $x_0 \in X$ is also hypercyclic (because T has dense range), one obtains that $HC(T)$ is dense in X. Now, since this property is not automatic for sequences (T_n), the following concept makes sense.

Definition 4.2.2. A sequence $(T_n) \subset L(X, Y)$ is said to be *densely universal* if $\mathcal{U}((T_n))$ is dense in X.

If Y is metrizable and separable (so second-countable) and (U_m) is a basis for the topology of Y then it is an easy exercise to see (see Figure 4.1) that

$$\mathcal{U}((T_n)) = \bigcap_{m \geq 1} \bigcup_{n \geq 1} T_n^{-1}(U_m), \qquad (U)$$

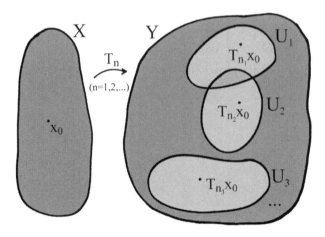

FIGURE 4.1: The vector x_0 is (T_n)-universal.

showing that $\mathcal{U}((T_n))$ is a G_δ subset. Thus, if in addition X is a Baire space, we get that (T_n) is densely universal if and only if $\mathcal{U}((T_n))$ is residual. Hence, in this case, $\mathcal{U}((T_n))$ has large size in a topological sense. In particular, the property of being a hypercyclic vector is topologically generic as soon as T is a hypercyclic operator on an F-space. The sequences of composition operators (C_{φ_n}), (C_{ψ_n}) and of differential operators $(\Phi_n(D))$ considered in the preceding paragraph –under the conditions specified there– are examples of densely universal sequences.

Several criteria guaranteeing large topological size for the family of universal vectors for (T_n) –and even for subsequences of (T_n)– are known; see [237]. By (n_k) it will be denoted a strictly increasing subsequence of $\mathbb{N}$.

Definition 4.2.3. A sequence $(T_n) \subset L(X, Y)$ is said to be *hereditarily universal* if (T_{n_k}) is universal for every $(n_k) \subset \mathbb{N}$, and *hereditarily densely universal* if (T_{n_k}) is densely universal for every $(n_k) \subset \mathbb{N}$.

Definition 4.2.4. Assume that X is an F-space and that Y is metrizable and separable. Then a sequence $(T_n) \subset L(X, Y)$ is said to satisfy the (hypotheses of the) *Universality Criterion* (UC) provided that there are respective dense sets $X_0 \subset X$, $Y_0 \subset Y$ and a sequence $(n_k) \subset \mathbb{N}$ such that $T_{n_k} x \to 0$ for all $x \in X_0$ and, for every $y \in Y_0$, there is a sequence $(u_k) \subset X$ with $u_k \to 0$ and $T_{n_k} u_k \to y$. In this case, (T_n) is said to satisfy the UC for (n_k). And if X is a separable F-space, then we say that an operator $T \in L(X)$ satisfy the *Hypercyclicity Criterion* (HCC) for a sequence $(n_k) \subset \mathbb{N}$ whenever the sequence of iterates (T^n) satisfies the UC for (n_k).

In [100] and [116] it is proved that, for separable F-spaces X and Y, the sequence (T_n) satisfies the UC if and only if some subsequence of it is hereditarily densely universal (in fact, the part "only if" is easy to see by using (U)). For instance, by applying this we get that, if one assumes that $(C_{\varphi_n}), (C_{\psi_n}), (\Phi_n(D))$ are the sequences of operators considered in the fifth paragraph, then each of the first two of them is hereditarily universal if and only if it is hereditarily densely universal, and this happens if and only if, respectively, $\lim_{n\to\infty} |\varphi_n(0)| = 1$ and $\lim_{n\to\infty} \min\{|b_n|, |b_n/a_n|\} = \infty$. And $(\Phi_n(D))$ is hereditarily densely universal if there are subsets A, B satisfying the conditions specified above.

Under conditions that are weaker than the UC, we can obtain results of topological genericity for the set $\mathcal{U}((T_n))$. Due to its large usefulness and the simplicity of its proof, we give a proof of one of such results. A direct application is, for instance, MacLane's theorem asserting the hypercyclicity of the derivative operator D on $H(\mathbb{C})$: take $X = H(\mathbb{C}) = Y$, $T_n = D^n$, $D_1 = \{\text{polynomials}\} = D_2$ in the following theorem, and use Runge's theorem and the fact $\frac{z^m}{(m+1)(m+2)\cdots(m+n)} \xrightarrow[n\to\infty]{} 0$ uniformly on compacta for every m.

Theorem 4.2.5. *Let X and Y be topological vector spaces such that X is a Baire space and Y is metrizable and separable. Assume that $(T_n) \subset L(X,Y)$ and that there are respective dense subsets $D_1 \subset X$, $D_2 \subset Y$ satisfying the following condition:*

> *For every $x \in D_1$ and every $y \in D_2$ there exist sequences $\{n_1 < n_2 < \cdots\} \subset \mathbb{N}$ and $(x_k) \subset X$ such that $T_{n_k} x \to 0$, $x_k \to 0$ and $T_{n_k}(x_k) \to y$ as $k \to \infty$.*

Then $\mathcal{U}((T_n))$ is residual in X.

Proof. We use (U) and its notation. Since $\mathcal{U}((T_n))$ is a G_δ subset and X is Baire, it suffices to show that, for each $m \in \mathbb{N}$, the set $H_m := \bigcup_{n\geq 1} T_n^{-1}(U_m)$ is dense. For this, fix a nonempty open subset $S \subset X$ and choose $a \in X$, $b \in Y$, $A \subset X$ and $B \subset Y$ such that $a + A + A \subset S$, $b + B + B + B \subset U_m$ and A (B, resp.) is a neighborhood of the origin of X (of Y, resp.). Due to the density of D_1 and D_2, we can find $x \in D_1$ and $y \in D_2$ with $x \in a + A$ and $y \in b + B$. Consider the corresponding sequences (n_k) and (x_k) given in the hypothesis. There must be $k \in \mathbb{N}$ such that $x_k \in A$, $T_{n_k} x \in B$ and $T_{n_k} x_k \in y + B$. Define $z \in X$ by $z := x_k + x$. Then $z \in x + A \subset a + A + A \subset S$ and

$$T_{n_k} z = T_{n_k} x_k + T_{n_p} x \in y + B + B \subset b + B + B + B \subset U_m,$$

so $z \in T_{n_k}^{-1}(U_m)$. Hence $z \in S \cap H_m$, showing that H_m is dense. $\square$

4.3 Lineability and dense-lineability of families of hypercyclic vectors

After topological genericity has been analyzed, we study under what conditions the family of universal vectors enjoys algebraic genericity. It is evident that the set of universal vectors is never a vector space.

Recall that a vector x_0 of a topological vector space X is called *cyclic* for an operator $T \in L(X)$ provided that the linear span of $\{x_0, Tx_0, T^2x_0, \dots\}$ is dense in X. If T has some cyclic vector then T is said to be cyclic. It is easy to see that every vector $x \in X \setminus \{0\}$ is cyclic for T if and only if there is not any nontrivial proper closed T-invariant vector subspace in X. Recall that the "invariant subspace problem" remains unsolved for Hilbert spaces H, that is, it is unknown whether or not there is an operator $T \in L(H)$ (with H separable and $\dim(H) = \infty$) lacking nontrivial proper closed subspaces M such that $T(M) \subset M$. The corresponding "invariant subset problem" is obtained by replacing "closed subspaces" by "closed subsets." Since hypercyclicity is a strong kind of cyclicity, this problem is still unsolved for Hilbert spaces.

An extreme case of lineability for $HC(T)$ is that in which $HC(T) = X \setminus \{0\}$. Analogously to the cyclic case, one has $HC(T) = X \setminus \{0\}$ if and only if X admits no nontrivial proper closed T-invariant *subset*. In fact, C. Read [340] solved the invariant subset problem for Banach spaces by exhibiting an operator T on the sequence space ℓ_1 for which any nonzero vector is hypercyclic.

But the last one is a very special operator. Let us go to a more general situation. In a chain of successive improvements, Herrero [248] (complex Hilbert case), Bourdon [144] (complex locally convex case), Bès [110] (real locally convex case) and, finally, Wengenroth [379] (general case) demonstrated that if T is a hypercyclic operator on an arbitrary topological vector space X then $HC(T)$ is dense-lineable. In fact, they proved much more, and their results are contained in the following theorem.

Theorem 4.3.1. *Let T be a hypercyclic operator on a topological vector space X. Then there is a dense T-invariant vector subspace M of X such that $M \setminus \{0\} \subset HC(T)$.*

Proof. Firstly, we are going to show that, for every scalar $\lambda \in \mathbb{K}$, the operator $T - \lambda I$ has dense range, where I denotes the identity operator. Suppose, by way of contradiction, that the range of $T \setminus \lambda I$ is not dense for some λ. Then $X \neq K := \overline{(T - \lambda I)(X)}$, and K is a closed vector subspace of X. Therefore, the quotient space X/K is a topological vector space of dimension at least 1. If

$$q : x \in X \mapsto x + K \in K$$

denotes the corresponding quotient mapping, then $q((T - \lambda I)x) = 0$ for all

$n \in \mathbb{N}$ and all $x \in X$. In particular, if we take a vector $x_0 \in HC(T)$, we have that the orbit $\{T^n x_0 : n \in \mathbb{N}\}$ is dense in X, so its image under q is dense in X/K because q is continuous and surjective. But this image is $\{\lambda^n q(x_0) : n \in \mathbb{N}\}$. Since the set $\{1, \lambda, \lambda^2, \dots\}$ is never dense in the field $\mathbb{K}$, one gets that $\{\lambda^n q(x_0) : n \in \mathbb{N}\}$ is a non-dense subset of a 1-dimensional subspace of X/K. This is the desired contradiction.

Denote by $\mathcal{P}$ the family of all polynomials with coefficients in $\mathbb{K}$. If $P(z) = a_0 + a_1 z + \cdots + a_m z^m \in \mathcal{P}$ and I is the identity on X, consider the operator $P(T) = a_0 I + a_1 T + \cdots + a_m T^m$. Our second step is to prove that, for every $P \in \mathcal{P} \setminus \{0\}$, the operator $P(T)$ has dense range. We distinguish two cases: $\mathbb{K} = \mathbb{C}$ and $\mathbb{K} = \mathbb{R}$. In the first case, factorize P as $P(z) = c(z - \lambda_1) \cdots (z - \lambda_m)$, with $c, \lambda_1, \dots, \lambda_m \in \mathbb{C}$ and $c \neq 0$. Then we can decompose P as a composition of operators

$$P(T) = cI(T - \lambda_1 I) \dots (T - \lambda_m I).$$

Since, by the first step, each of the factors $cI, T - \lambda_1 I, \dots, T - \lambda_m I$ has dense range, we get that $P(T)$ has also dense range. Assume now that $\mathbb{K} = \mathbb{R}$. Here a complexification technique will be used. We briefly recall the pertinent notions. The complexification $\widetilde{X}$ is nothing but the space $X \times X$, endowed with the product topology and with the structure of vector space over $\mathbb{C}$ defined by the coordinatewise sum and the external law

$$(a + bi, (x, y)) \in \mathbb{C} \times X \times X \mapsto (ax - by, ay + bx) \in X \times X.$$

The complexification of an operator $S \in L(X)$ is defined as the operator $\widetilde{S} \in L(\widetilde{X})$ given by $\widetilde{S}(x, y) = (Sx, Sy)$. If P is a nonzero polynomial on $\mathbb{R}$ then $\widetilde{p(T)} = p(\widetilde{T})$, because p has real coefficients. By the former case, $\widetilde{p(T)}$ has dense range, from where it is derived that, for every pair of nonempty sets $U, V \subset X$, the set $U \times V$ has some common point with $\{(p(T)x, p(T)y) : x, y \in X\}$. This entails the density of $p(T)(X)$ in X.

Let us construct the desired vector space M. Fix any $x_0 \in H(T)$ and let

$$M := \{P(T)x_0 : P \in \mathcal{P}\},$$

which is trivially a vector subspace. Since M contains the T-orbit of x_0, it is dense. Moreover, M is T-invariant because if $y_0 = P(T)x_0 \in M$ then $Ty_0 = Q(T)x_0$, where $Q(\lambda) := \lambda P(\lambda)$. Finally, each vector $y_0 = P(T)x_0 \in M \setminus \{0\}$ is hypercyclic, because $P(T)$ has dense range (as $P \neq 0$) and the T-orbit of y_0 is the image under $P(T)$ of the (dense) T-orbit of x_0. $\square$

The proof of Theorem 4.3.1 provides a dense vector subspace with dimension $\aleph_0$. If X is Banach then, as Theorem 4.3.2 will show, the maximal dense-lineability can be established, and the T-invariance is kept for the corresponding vector subspace; see [76].

Some background on general spectral theory is needed before establishing Theorem 4.3.2. We refer the reader to, for instance, [191, Chapter 1] or [352,

Chapter 10]. Assume that X is a complex Banach space and that $T \in L(X)$. Consider the adjoint T^* of T. Denote by $\mathcal{F}(T)$ the family of all functions which are analytic on some neighborhood of $\sigma(T)$. Hence $\mathcal{F}(T) = \mathcal{F}(T^*)$. Let $f \in \mathcal{F}(T)$ and Γ be a positively oriented Jordan cycle (i.e. Γ is an algebraic sum of finitely many, mutually disjoint Jordan closed curves) surrounding $\sigma(T)$, such that both γ and its geometric interior are contained in the domain of analyticity of f. Then the operator $f(T)$ is defined by the following equation, where the integral exists as a limit of Riemann sums in the norm of $L(X)$:

$$f(T) = \frac{1}{2\pi i} \cdot \oint_\Gamma f(\lambda)\,(\lambda I - T)^{-1}\,d\lambda.$$

This operator $f(T)$ is independent of Γ. Note that the factor $(\lambda I - T)^{-1}$ within the integral symbol makes sense because $\Gamma \cap \sigma(T) \neq \varnothing$. It happens that if $f(z)$ has power series expansion $f(z) = \sum_{n=0}^\infty a_n z^n$ valid in a neighborhood of $\sigma(T)$, then the series $\sum_{n=0}^\infty a_n T^n$ (with $T^0 := I$) converges to $f(T)$ in the norm of $L(X)$. Then, in this sense, the notion of $f(T)$ extends the definition $P(T) = \sum_0^m a_n T^n$ (with $T^0 = I$) when $P(z)$ is the polynomial $P(z) = \sum_0^m a_n z^n$. We have also that $f(T)^* = f(T^*)$. Note that this equality is especially easy to prove in the case in which f is a polynomial or an entire function. Moreover, observe also that the definition

$$f(T) = \sum_{n=0}^\infty a_n T^n \quad \text{if} \quad f(z) = \sum_{n=0}^\infty a_n x^n$$

also makes sense if X is a real Banach space, $T \in L(X)$ and $f : \mathbb{R} \to \mathbb{R}$ is a real entire function.

Theorem 4.3.2. *Assume that T is a hypercyclic operator on a Banach space X. Then there exists a dense T-invariant linear subspace M with maximal algebraic dimension (i.e. $\dim M = \mathfrak{c}$) such that $M \setminus \{0\} \subset HC(T)$.*

Proof. The Banach space X must be infinite dimensional and separable, so its dimension is $\mathfrak{c}$. Suppose first that the field of X is $\mathbb{K} = \mathbb{C}$. Let us consider a function $f \in \mathcal{F}(T)$, and call $D(f)$ its domain of analyticity. If f is nonconstant on every connected component of $D(f)$ then $\sigma_p(f(T^*)) = \varnothing$. Indeed, by a special version of the spectral mapping theorem [352, Theorem 10.33], $\sigma_p(f(T^*)) = f(\sigma_p(T^*))$. But, since T is hypercyclic, $\sigma_p(T^*) = \varnothing$ (see Exercise 4.6), so $\sigma_p(f(T^*)) = \varnothing$.

Assume now that f is not identically zero and that $D(f)$ is connected. The previous results yield $f(T)(HC(T)) \subset HC(T)$. Let us prove this. If f is nonconstant (the case $f = $ constant $\neq 0$ is trivial), then $\sigma_p(f(T^*)) = \varnothing$. But $f(T^*) = f(T)^*$, so $\sigma_p(f(T)^*) = \varnothing$. In particular, $0 \notin \sigma_p(f(T)^*)$, so $f(T)^*$ is one-to-one. From the Hahn-Banach theorem, we get that $f(T)$ has dense range, which implies that the image under $f(T)$ of a T-hypercyclic vector is also a T-hypercyclic vector. We have used that $T^n f(T) = f(T)T^n$ for all

$n \in \mathbb{N}_0$, which is true due to the commutative property $z^n f(z) = f(z)z^n$ and the definition of image of T under an analytic function.

Fix any vector $x_0 \in HC(T)$ and define

$$M := \{f(T)\,x_0 : f \text{ is entire}\}.$$

From the last paragraph, $M \setminus \{0\} \subset HC(T)$; for this note that if $u \in M \setminus \{0\}$, then $z = f(T)x_0$ for some non-identically zero entire function f. It is obvious that M is a vector subspace, and that M is T-invariant since $zf(z)$ is entire for each entire function $f(z)$. It is dense because M contains the orbit of T (take $f(z) = z, z^2, z^3, \dots$). Finally, the linear spaces M and $\{\text{entire functions}\}$ are evidently algebraically isomorphic, so $\dim(M) = \mathfrak{c}$. This completes the proof in the case $\mathbb{K} = \mathbb{C}$.

In the case $\mathbb{K} = \mathbb{R}$, we consider

$$M := \{f(T)\,x_0 : f \text{ is a real entire function}\}.$$

Following the previous steps, all that must be proved is that each operator $f(T)$ (with f a nonzero real entire function) has dense range. As in the proof of Theorem 4.3.1, this can be achieved by using complexifications. Once more, the case $f = \text{constant} \neq 0$ is obvious. Let f be nonconstant. It is not difficult to show that $\widetilde{f(T)} = f(\widetilde{T})$ and that, since T is hypercyclic, $\sigma_p((\widetilde{T})^*) = \varnothing$ (see [130]). Again by the spectral mapping theorem [352, Theorem 10.33], we derive

$$\sigma_p((\widetilde{f(T)})^*) = \sigma_p((f(\widetilde{T}))^*) = \sigma_p(f((\widetilde{T})^*)) = f(\sigma_p((\widetilde{T})^*)) = \varnothing.$$

In particular, $0 \notin \sigma_p((\widetilde{f(T)})^*)$, hence $(\widetilde{f(T)})^*$ is one-to-one, and so $\widetilde{f(T)}$ has dense range. Therefore $f(T)$ must have dense range, as required. $\square$

Still in the setting of Banach spaces, Sophie Grivaux [231] established the following important result about existence of common dense hypercyclic subspaces for a countable family of operators. The details of its long proof can be found in [231]; we just give a sketch of it.

Theorem 4.3.3. *If X is a Banach space and $(T_j)_{j \in \mathbb{N}}$ is a countable family of hypercyclic operators on X then $\bigcap_{j=1}^{\infty} HC(T_j)$ is dense-lineable.*

Sketch of the proof of Theorem 4.3.3. Consider the norm topology in the space $L(X)$ of operators on X. For every vector subspace M of X, denote $\Omega_M := \{T \in L(X) : M \setminus \{0\} \subset HC(T)\}$. Let M be a dense subspace of X of countable algebraic dimension. For each $j \in \mathbb{N}$, let

$$G_j := \{S \in L(X) : \|S - I\| < 1/2 \quad \text{and} \quad S^{-1}T_j S \in \Omega_M\}.$$

Then G_j is a dense G_δ subset of the open ball $B(I, 1/2)$ in $L(X)$. This can be seen as follows. Clearly, the map $S \in L(X) \mapsto S^{-1}T_j S \in L(X)$ is continuous.

Using the crucial fact that, for any $\varepsilon > 0$ and any pair of countable dense sets of linearly independent vectors of X, there is an isomorphism S of X with $\|I - S\| < \varepsilon$ such that $L(V) = W$, it is possible to deduce that Ω_M is a dense subset of $L(X)$. Therefore G_j is a G_δ subset of $L(X)$, so it is a G_δ subset of $B(I, 1/2)$ for the induced topology. In order to prove the density of Ω_M in $B(I, 1/2)$, consider an element $S_0 \in B(I, 1/2)$, and let $\varepsilon > 0$. The operator $S_0^{-1} T_j L_0$ is hypercyclic, since it is a conjugate of a hypercyclic one. If α is any positive number, one has that there exists $S_1 \in L(X)$ with $\|I - S_1\| < \alpha$ such that $S_1^{-1} S_0^{-1} T_j S_0 S_1 \in \Omega_M$. Set $S_2 = S_0 S_1$. If α is small enough, $\|I - S_2\| < 1/2$ and $\|S_2 - S_0\| < \varepsilon$. Since $S_2^{-1} T_j S_2 \in \Omega_M$, we have that S_2 is in G_j, which proves the density of Ω_M in $B(I, 1/2)$.

This open ball being a Baire space, the intersection $\bigcap_{j \geq 1} G_j$ is residual in $B(I, 1/2)$. In particular, $\bigcap_{j \geq 1} G_j \neq \varnothing$, and every operator S in this intersection satisfies that $S(M)$ is a vector subspace with $S(M) \setminus \{0\} \subset HC(T_j)$ for all $j \in \mathbb{N}$. $\qquad \square$

Grivaux also proved in [231] the following nice assertion (see Exercise 4.4).

Proposition 4.3.4. *If $(T_\lambda)_{\lambda \in \Lambda}$ is a family of operators on a topological vector space such that some T_{λ_0} commutes with each T_λ ($\lambda \in \Lambda$) then the set of common hypercyclic vectors $\bigcap_{\lambda \in \Lambda} HC(T_\lambda)$ is either empty or dense-lineable.*

It is worth mentioning that, as Bayart [54] showed, commutativity is not needed under adequate conditions. These conditions include the so-called "Common Hypercyclicity Criterion," which is also applicable to the universality of families of sequences of operators; see [54] for details.

As for sequences of linear mappings, it should be said that the mere residuality of the set of universal vectors does not entail lineability. For instance, let $\alpha = (a_k) \in \mathbb{C}^{\mathbb{N}_0}$ be a sequence with $\limsup_{k \to \infty} |a_k|^{1/k} < \infty$, and define the associated diagonal operator Δ_α as

$$\Delta_\alpha : \sum_{k=0}^{\infty} f_k z^k \in H(\mathbb{C}) \mapsto \sum_{k=0}^{\infty} a_k f_k z^k \in H(\mathbb{C}).$$

Consider a sequence $\{\Delta_{\alpha_n}\}_{n \geq 1}$ of diagonal operators on $H(\mathbb{C})$, where $\alpha_n = (a_{k,n})_{k \geq 0}$. Then (see [96]) (Δ_{α_n}) is universal if and only if $\{(a_{k,n})_{k \geq 0} : n \in \mathbb{N}\}$ is dense in $\mathbb{C}^{\mathbb{N}_0}$, in which case $\mathcal{U}((\Delta_{\alpha_n}))$ is residual. But no linear manifold contained in $\mathcal{U}((\Delta_{\alpha_n})) \cup \{0\}$ has dimension ≥ 2; see Exercise 4.5.

Fortunately, lineability properties hold for families of universal vectors of sequences of mappings under not too strong restrictions on the spaces and the mappings. In the following theorem a number of related results, starting from 1999, due to Bernal, Calderón and Prado-Tendero [74, 90, 106] are collected.

Theorem 4.3.5. *Assume that X, Y and Y_k ($k \in \mathbb{N}$) are topological vector spaces. We have:*

(a) *If Y is metrizable and $(T_n) \subset L(X,Y)$ is hereditarily universal then $\mathcal{U}((T_n))$ is lineable. If, in addition, X is metrizable and separable and (T_n) is hereditarily densely universal then $\mathcal{U}((T_n))$ is dense-lineable.*

(b) *Suppose that X and the Y_k's are metrizable and separable, X is Baire, $(T_{k,n})_{n\geq 1} \subset L(X,Y_k)$ for each $k \in \mathbb{N}$ and each sequence $(T_{k,n})_{n\geq 1}$ is hereditarily densely universal. Then the set $\bigcap_{k=1}^{\infty} \mathcal{U}((T_{k,n})_{n\geq 1})$ of common universal vectors is dense-lineable.*

Proof. (a) We assume that Y is metrizable. Pick a vector $x_1 \in \mathcal{U}((T_n))$. Then we can find a subsequence $\{p(1,j) : j \in \mathbb{N}\}$ of positive integers such that

$$T_{p(1,j)}x_1 \to 0 \quad (j \to \infty). \tag{4.1}$$

Now, since (T_n) is hereditarily universal, the sequence $(T_{p(1,j)})$ is universal, so we can choose a vector $x_2 \in HC((T_{p(1,j)}))$. By (4.1) it is clear that x_2 is linear independent of x_1, because $(T_{p(1,j)}x_2)$ cannot tend to zero, by density. Now choose a subsequence $\{p(2,j) : j \in \mathbb{N}\}$ of $(p(1,j))$ with

$$T_{p(2,j)}x_2 \to 0 \quad (j \to \infty). \tag{4.2}$$

Note that $T_{p(2,j)}x_1 \to 0$ $(j \to \infty)$ too. The new sequence $(T_{p(2,j)})$ is universal. Choose a vector $x_3 \in HC((T_{p(2,j)}))$. By (4.1) and (4.2) it is clear that x_3 does not belong to the linear span of $\{x_1,x_2\}$.

It is evident that this process can be continued by induction, getting a sequence $\{x_N : N \in \mathbb{N}\} \subset X$ and a family $\{\{p(n,j) : j \in \mathbb{N}\} : n \in \mathbb{N}\}$ of sequences of positive integers satisfying

$$x_N \in G_{N-1} \quad \text{for all } N \in \mathbb{N}, \tag{4.3}$$

$$x_N \in HC((T_{p(N-1,j)})) \quad \text{for all } N \in \mathbb{N} \text{ and} \tag{4.4}$$

$$T_{p(k,j)}x_N \to 0 \quad (j \to \infty) \quad \text{for all } k \geq N, \tag{4.5}$$

where $(p(0,j))$ is the whole sequence of positive integers, $G_0 = X$ and $G_N = X \setminus \text{span}(\{x_1,...,x_N\})$ for $N \in \mathbb{N}$. Define

$$M = \text{span}(\{x_N : N \in \mathbb{N}\}).$$

It is plain from (4.3) that M is an infinite dimensional linear subspace of X.

It remains to prove that each nonzero vector of M is universal for (T_n). Fix $x \in M \setminus \{0\}$. Then there are finitely many scalars $a_1, \dots, a_N$ with $a_N \neq 0$ such that $x = \sum_{k=1}^{N} a_k x_k$. We may assume that $a_N = 1$ because if λ is a nonzero scalar, then x is universal if and only if λx is universal. Let y be in Y. Let us exhibit a subsequence $\{T_{r(j)} : j \in \mathbb{N}\}$ of (T_n) such that

$$T_{r(j)}x \to y \quad (j \to \infty).$$

By (4.4), there is a subsequence $(r(j))$ of $(p(N-1,j))$ such that

$$T_{r(j)}x_N \to y \quad (j \to \infty). \tag{4.6}$$

But, since $(r(j))$ is a subsequence of $(p(N-1,j))$, we see from (4.5) that $T_{r(j)}x_k \to 0$ $(j \to \infty)$ for all $k \in \{1,\ldots,N-1\}$, so $\sum_{k=1}^{N-1} a_k T_{r(j)}x_k \to 0$ $(j \to \infty)$. Finally, by (4.6) and linearity,

$$T_{r(j)}x = T_{r(j)}x_N + \sum_{k=1}^{N-1} a_k T_{r(j)}x_k \to y + 0 = y \quad (j \to \infty),$$

as required.

The second part of (a) can be derived as a special case of the proof of (b) below, if we take $Y_k = Y$ and $T_{k,n} = T_n$ for all $k, n \geq 1$. Note that we do not assume that X is Baire in (a). This additional assumption is needed in (b) in order to assure that a certain countable intersection of dense open subsets is still dense.

(b) Each space Y_k is second-countable. Let us choose a dense sequence (z_n) in X and denote by d a distance on X compatible with its topology. We will consider later the open balls

$$G_N = \{x \in X : d(x, z_N) < \frac{1}{N}\} \quad (N \in \mathbb{N}).$$

Since X is a Baire space and every Y_k is second-countable, each of the sets $\mathcal{U}((T_{k,n})_{n\geq 1})$ $(k \in \mathbb{N})$ is residual in X by the equality (U) of the preceding section, because they are dense. Therefore their intersection $\bigcap_{k \in \mathbb{N}} \mathcal{U}((T_{k,n})_{n\geq 1})$ is also residual, so dense, whence we can pick a vector

$$x_1 \in G_1 \cap \bigcap_{k \in \mathbb{N}} \mathcal{U}((T_{k,n})_{n\geq 1}).$$

Then for every $k \in \mathbb{N}$ we can find a (strictly increasing) subsequence $\{p(1,k,j) : j \in \mathbb{N}\}$ of positive integers such that

$$T_{k,p(1,k,j)}x_1 \to 0 \quad (j \to \infty).$$

But, since each $(T_n^{(k)})$ $(k \in \mathbb{N})$ is hereditarily densely universal, every set $HC((T_{k,p(1,k,j)}))$ is again residual. Thus, as above, a vector x_2 can be selected in $G_2 \cap \bigcap_{k \in \mathbb{N}} HC((T_{k,p(1,k,j)}))$. Now choose for every k a subsequence $\{p(2,k,j) : j \in \mathbb{N}\}$ of $(p(1,k,j))$ with

$$T_{k,p(2,k,j)}x_2 \to 0 \quad (j \to \infty).$$

Note that also $T_{k,p(2,k,j)}x_1 \to 0$ $(j \to \infty)$ for each $k \in \mathbb{N}$. Since the new sequences $(T_{k,p(2,k,j)})$ $(k \in \mathbb{N})$ are again densely hypercyclic, one can choose a vector $x_3 \in G_3 \cap \bigcap_{k \in \mathbb{N}} HC((T_{k,p(2,k,j)}))$.

It is evident that this process can be continued by induction, getting a sequence $\{x_N : N \in \mathbb{N}\} \subset X$ and a family $\{\{p(n,k,j) : j \in \mathbb{N}\} : n,k \in \mathbb{N}\}$ of sequences of positive integers satisfying

$$x_N \in G_N \quad \text{for all } N \in \mathbb{N}, \tag{4.7}$$

$$x_N \in \bigcap_{k \in \mathbb{N}} HC((T_{k,p(N-1,k,j)})) \quad \text{for all } N \in \mathbb{N} \text{ and} \tag{4.8}$$

$$T_{k,p(n,k,j)} x_N \to 0 \quad (j \to \infty) \quad \text{for all } n \geq N \text{ and all } k \in \mathbb{N}, \tag{4.9}$$

where, in order to make the notation consistent, $(p(0,k,j))$ stands for the whole sequence of positive integers for every $k \in \mathbb{N}$. Define

$$M = \text{span}(\{x_N : N \in \mathbb{N}\}).$$

Since $\{z_n : n \in \mathbb{N}\}$ is dense in X and $d(x_n, z_n) < \frac{1}{n} \to 0$ $(n \to \infty)$ (by (4.7)), the set $\{x_n : n \in \mathbb{N}\}$ is also dense, hence M is a dense linear subspace of X.

It remains to prove that each nonzero vector of M is hypercyclic for each sequence $(T_{k,n})$ $(k \in \mathbb{N})$. Fix $x \in M \setminus \{0\}$. Then there are finitely many scalars $a_1, \ldots, a_N$ with $a_N \neq 0$ such that $x = \sum_{n=1}^{N} a_n x_n$. Since a nonzero multiple of a hypercyclic vector is still hypercyclic, we may assume that $a_N = 1$. Fix a positive integer k and a vector $y \in Y$. Let us show a subsequence $\{T_{k,r(j)} : j \in \mathbb{N}\}$ of $(T_{k,n})$ such that

$$T_{k,r(j)} x \to y \quad (j \to \infty).$$

By (4.8), there is a subsequence $(r(j))$ of $(p(N-1,k,j))$ such that

$$T_{k,r(j)} x_N \to y \quad (j \to \infty). \tag{4.10}$$

But, since $(r(j))$ is a subsequence of $(p(N-1,k,j))$ we see from (4.9) that $T_{k,r(j)} x_n \to 0$ $(j \to \infty)$ for all $n \in \{1, \ldots, N-1\}$, so $\sum_{n=1}^{N-1} a_n T_{k,r(j)} x_n \to 0$ $(j \to \infty)$. Finally, by (4.10) and linearity,

$$T_{k,r(j)} x = T_{k,r(j)} x_N + \sum_{n=1}^{N-1} a_n T_{k,r(j)} x_n \to y + 0 = y \quad (j \to \infty),$$

as required. $\qquad\qquad\qquad\qquad\qquad\qquad\qquad\qquad\qquad\qquad\qquad\qquad\qquad\qquad \Box$

Observe that the conclusions of (a) hold if there is a subsequence of (T_n) that is hereditarily (densely) universal. Analogously, the conclusion of (b) remains valid if each $(T_{k,n})_{n \geq 1}$ admits a subsequence (which may depend on k) being hereditarily densely universal.

For instance, the sets of functions which are, respectively, universal for the sequences (C_{φ_n}), (C_{ψ_n}), $(\Phi_n(D))$ considered in Section 4.2 –under the restrictions imposed there– are dense-lineable (see also [106] for combinations of composition operators with other kinds of operators). In particular, $\mathcal{U}((\tau_{a_n}))$ is dense-lineable provided that (a_n) is an unbounded sequence in $\mathbb{C}$.

4.4 Wild behavior near the boundary, universal series and lineability

Certain holomorphic functions exhibiting some kind of wild behavior near the boundary can be considered universal in some sense, hence susceptible to be analyzed from the point of view of lineability.

One of these kinds of functions is that of holomorphic monsters created by Luh [295] in 1985. If $G \subset \mathbb{C}$ is a simply connected domain, a function $f \in H(G)$ is said to be a *holomorphic monster* on G whenever, for each derivative or antiderivative F of f of any order, each $g \in H(\mathbb{D})$ and each $\xi \in \partial G$, there exists a sequence (τ_n) of affine linear transformations (i.e. of the form $\tau_n(z) = a_n z + b_n$, with $a_n \neq 0$) with $\tau_n(\mathbb{D}) \subset G$ ($n \in \mathbb{N}$) such that

$$\tau_n(z) \to \xi \quad \text{uniformly on } \mathbb{D} \quad \text{and} \quad F(\tau_n(z)) \to g(z) \quad \text{compactly in } \mathbb{D}.$$

These "monsters" were further investigated by Grosse-Erdmann, who proved its residuality in [233]. For more on this topic, including operators under the action of which many holomorphic functions become holomorphic monsters, the reader is referred to [17, 89–92, 150]. Let $G \subset \mathbb{C}$ be a domain with $G \neq \mathbb{C}$, and $T : H(G) \to H(G)$ be an operator. According to [90], T is said to be *totally omnipresent* provided that, for every $g \in H(\mathbb{D})$, every $\varepsilon > 0$, every $r \in (0, 1)$ and every sequence (τ_n) of nonconstant affine linear transformations with $\tau_n(\mathbb{D}) \subset G$ ($n \geq 1$) and $\lim_{n \to \infty} \sup_{z \in \mathbb{D}} \operatorname{dist}(\tau_n(z), \partial G) = 0$, the set

$$\{ f \in H(G) : \text{ there is } n \in \mathbb{N} \text{ with } \sup_{|z| \leq r} |(Tf)(\tau_n(z)) - g(z)| < \varepsilon \}$$

is dense in $H(G)$. Examples of totally omnipresent operators are $P(D)$ and $P(D_a^{-1})$ (the last one if G is simply connected), where P is any nonconstant polynomial, $a \in G$ and D_a^{-1} is the operator assigning to a function $f \in H(G)$ the unique $F \in H(G)$ such that $F(a) = 0$; see [90–92]. These ingredients are employed in [90] to prove the following theorem.

Theorem 4.4.1. *The set $\mathcal{M}$ of holomorphic monsters on a simply connected domain $G \subset \mathbb{C}$ is dense-lineable.*

Proof. An easy argument (see [233]) shows that if an antiderivative F of order N of a function $f \in H(G)$ satisfies the approximation property in the definition of holomorphic monster, then any other antiderivative of order N also satisfies it. Then we can fix any $a \in G$ and consider only the antiderivatives $D_a^{-j} f$ ($j = 1, 2, \dots$) for $f \in H(G)$ in order to check the condition of being a holomorphic monster. Moreover, one can easily realize (see [90, Proposition 2.2]) that an operator T on $H(G)$ is totally omnipresent if and only if, for every $\xi \in \partial G$ and every sequence (τ_n) of nonconstant affine linear transformations such that $\tau_n(\mathbb{D}) \subset G$ ($n \geq 1$) and $\tau_n(z) \to \xi$ uniformly on $\mathbb{D}$, the sequence

$$C_{\tau_n} \circ T : H(G) \to H(G) \quad (n \geq 1)$$

is hereditarily densely universal. Now, define the operators S_j $(j \in \mathbb{Z})$ as $S_j = D^j$ for $j \in \mathbb{N}_0$ and $S_j = D_a^j$ for $-j \in \mathbb{N}$. Fix a dense sequence $\{t_k : k \in \mathbb{N}\}$ in ∂G. Fix also, for each $k \in \mathbb{N}$, a sequence $\{\tau_n^{(k)}\}_{n \geq 1}$ of nonconstant affine linear transformations with $\tau_n^{(k)}(\mathbb{D}) \subset G$ for all n and $\tau_n^{(k)}(z) \to t_k$ uniformly on $\mathbb{D}$. Thus, for each k and j, the sequence $\{C_{\tau_n^{(k)}} S_j\}_{n \geq 1}$ is hereditarily densely universal on $H(G)$. According to Theorem 4.3.5, the set $\bigcap_{k,j} \mathcal{U}((C_{\tau_n^{(k)}} S_j))$ is dense-lineable. Finally, it is an easy exercise to check that each member of the last intersection must be a holomorphic monster. Hence $\mathcal{M}$ is dense-lineable. □

Another outstanding example to which methods close to Theorem 4.3.5 can be applied is that of universal series. During the seventies, Chui and Parnes [164] and Luh [294] provided a holomorphic function in the unit disc which is universal with respect to overconvergence. More precisely, they constructed a function $f(z) = \sum_{n=0}^{\infty} a_n z^n \in H(\mathbb{D})$ satisfying that, given a compact set K with connected complement and $K \cap \overline{\mathbb{D}} = \varnothing$, a function g continuous on K and holomorphic in its interior K^0, and $\varepsilon > 0$, there is $n \in \mathbb{N}$ such that

$$\left| \sum_{k=0}^{n} a_k z^k - g(z) \right| < \varepsilon \text{ for all } z \in K.$$

The topological generic nature of this property was shown in 1996 by Nestoridis [324] who, in fact, proved that K can be allowed to meet $\partial \mathbb{D}$ (i.e. $K \cap \mathbb{D} = \varnothing$). In 2005, Bayart [54] established the dense-lineability in $H(\mathbb{D})$ of the class of these functions f, which are known as *universal Taylor series;* see Figure 4.2. Since 1996, many extensions of these results, for other domains and more restrictive classes of functions, have been performed (see the recent papers [38,58,62,183,316] and the references therein). They can be put into a more general context, as we show with the next concept, given by Nestoridis and Papadimitropoulos in [325].

Let X be a metrizable topological vector space over the field $\mathbb{K} = \mathbb{R}$ or $\mathbb{C}$, $\{x_n\}_{n \geq 0} \subset X$ be a fixed sequence, $\{e_n\}_{n \geq 0}$ be the canonical basis of $\mathbb{K}^{\mathbb{N}_0}$ and A be a subspace of $\mathbb{K}^{\mathbb{N}_0}$ carrying a complete metrizable vector space topology. Assume that the coordinate projections

$$a = (a_n) \in A \mapsto a_m \in \mathbb{K}$$

are continuous for any m and that the set of "polynomials" $G = \{a = (a_n) \in \mathbb{K}^{\mathbb{N}_0} : \{n : a_n \neq 0\} \text{ is finite}\}$ is a dense subset of A. Then a sequence $a \in A$ is said to belong to the set $\mathcal{U}_A$ of *restricted universal series in A* with respect to (x_n) provided that, for every $x \in X$, there exists a sequence $(k_n) \subset \mathbb{N}_0$ such that

$$\sum_{j=0}^{k_n} a_j x_j \to x \quad \text{and} \quad \sum_{j=0}^{k_n} a_j e_j \to a \quad \text{as} \quad n \to \infty.$$

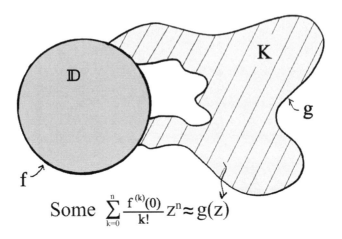

FIGURE 4.2: A universal Taylor series f.

In 2008, Bayart, Grosse-Erdmann, Nestoridis and Papadimitropoulos [58] established the following result, that characterizes the nonemptiness of $\mathcal{U}_A$, also proving that this is equivalent to its dense-lineability. If μ is an increasing sequence of positive integers, the symbol $\mathcal{U}_A^\mu$ will denote the set of $a \in A$ satisfying that, for every $x \in X$, there exists a subsequence (k_n) of μ satisfying the same double approximation property above.

Theorem 4.4.2. *Assume that X and A are as before, with respective translation-invariant metrics ρ and d defining their topologies. Then the following properties are equivalent:*

(a) $\mathcal{U}_A \neq \varnothing$.

(b) *For every $p \in \mathbb{N}_0$, $x \in \mathbb{N}$ and $\varepsilon > 0$, there exist $n \geq p$ and $a_p, a_{p+1}, \dots, a_n \in \mathbb{K}$ such that*

$$\rho\left(\sum_{j=p}^{n} a_j x_j, x\right) < \varepsilon \text{ and } d\left(\sum_{j=p}^{n} a_j e_j, 0\right) < \varepsilon.$$

(c) *For every $x \in \mathbb{N}$ and $\varepsilon > 0$, there exist $n \geq 0$ and $a_0, a_1, \dots, a_n \in \mathbb{K}$ such that*

$$\rho\left(\sum_{j=0}^{n} a_j x_j, x\right) < \varepsilon \text{ and } d\left(\sum_{j=0}^{n} a_j e_j, 0\right) < \varepsilon.$$

(d) *For every increasing sequence μ of positive integers, the set $\mathcal{U}_A^\mu$ is residual.*

(e) *For every increasing sequence μ of positive integers, the set $\mathcal{U}_A^\mu$ is dense-lineable.*

Proof. (a) $\Rightarrow$ (b): Let $a \in \mathcal{U}_A$ and $p \in \mathbb{N}_0$. Observe that for any polynomial $a' \in G$, we have $a + a' \in \mathcal{U}_A$. Thus, there exists some $q \geq p$ so that $b = (0, 0, \ldots, 0, a_q, a_{q+1}, \ldots)$ belongs to $\mathcal{U}_A$ and $d(b, 0) < \varepsilon/2$, where we have used the definition of restricted universal series. Since $b \in \mathcal{U}_A$, we may find $n > q$ so that

$$\rho\left(\sum_{j=q}^n a_j x_j, x\right) < \varepsilon \quad \text{and} \quad d\left(\sum_{j=q}^n a_j e_j, b\right) < \varepsilon/2.$$

By the triangle inequality we obtain that $d(\sum_{j=q}^n a_j e_j, 0) < \varepsilon/2$. This implies (b) upon setting $a_j = 0$ for $p \leq j < q$.

The implications "(b) $\Rightarrow$ (c)" and "(e) implies (a)" are obvious.

(c) $\Rightarrow$ (d): Fix an increasing sequence $\mu = (\mu_n) \subset \mathbb{N}$. Taking the a_j in $\mathbb{Q}$ or in $\mathbb{Q} + i\mathbb{Q}$, condition (c) implies that X is separable. Let (y_l) be a dense sequence in X. For $n \in \mathbb{N}_0$ and $l, s \geq 1$, we consider the sets

$$E^\mu(n, l, s) = \left\{a \in A : \rho\left(\sum_{j=0}^{\mu_n} a_j x_j, y_l\right) < 1/s\right\} \quad \text{and}$$

$$F^\mu(n, s) = \left\{a \in A : d\left(\sum_{j=0}^{\mu_n} a_j e_j, a\right) < 1/s\right\}.$$

Obviously, $E^\mu(n, l, s)$ and $F^\mu(n, s)$ are open subsets of A. Moreover, one has

$$\mathcal{U}_A^\mu = \bigcap_{l, s \in \mathbb{N}_0} \bigcup_{n \geq 0} [E^\mu(n, l, s) \cap F^\mu(n, s)].$$

Indeed, one inclusion is trivial, and the second one follows easily from the density of (y_l) and from the triangle inequality. Therefore, $\mathcal{U}_A^\mu$ is G_δ in A. Since A is a Baire space, we are reduced to proving that, for any $l, s \geq 1$, $\bigcup_{n \geq 0} [E^\mu(n, l, s) \cap F^\mu(n, s)]$ is dense in A. Therefore, let us consider $b \in G$ and $\varepsilon > 0$. By (c), there exists a $c \in G$ such that

$$d(c, 0) < \varepsilon \quad \text{and} \quad \rho\left(\sum_{j=0}^\infty a_j x_j, y_l - \sum_{j=0}^\infty b_j x_j\right) < 1/s.$$

We set $a = b + c$ and choose n such that μ_n is greater than the support of a (the maximum of those $k \in \mathbb{N}_0$ such that $a_k \neq 0$). Then the proof is complete, since $d(a, b) = d(c, 0) < \varepsilon$, $\rho(\sum_{j=0}^{\mu_n} a_j x_j, y_l) = \rho(\sum_{j=0}^\infty a_j x_j, y_l) = \rho(\sum_{j=0}^\infty c_j x_j, y_l - \sum_{j=0}^\infty b_j x_j) < 1/s$ and $d(\sum_{j=0}^{\mu_n} a_j e_j, a) = 0 < 1/s$.

(d) $\Rightarrow$ (e): Observe that A is separable, so we can fix a dense sequence (c^l) in A. Let $\mu^0 = \mu$. Choose first a^1 in A such that a^1 belongs to $\mathcal{U}_A^{\mu^0}$ and $d(a^1, c^1) < 1$. There exists a subsequence $\mu^1 = (\mu_n^1)$ of μ^0 such that

$\sum_{j=0}^{\mu_n^1} a_j^1 x_j \to 0$ and $\sum_{j=0}^{\mu_n^1} a_j^1 e_j \to 0$ as $n \to \infty$. Then consider a^2 in $\mathcal{U}_A^{\mu^1}$ with $d(a^2, c^2) < 1/2$. Proceeding inductively, one constructs a sequence (a^l) in A and sequences of integers μ^l so that, for any $l \geq 1$, μ^l is a subsequence of μ^{l-1}, $d(a^l, c^l) < 1/l$, $a^l \in \mathcal{U}_A^{\mu^{l-1}}$, $\sum_{j=0}^{\mu_n^l} a_j^l x_j \to 0$ and $\sum_{j=0}^{\mu_n^1} a_j^l e_j \to 0$ as $n \to \infty$. Now, we consider $M := \mathrm{span}\{a^l : l \in \mathbb{N}\}$. Clearly, B is a dense subspace of A. Our goal is to show that $M \setminus \{0\} \subset \mathcal{U}_A^\mu$.

To this end, take $a = \alpha_1 a^1 + \cdots + \alpha_m a^m \in M \setminus \{0\}$, with $\alpha_m \neq 0$, and choose any $x \in X$. There exists a subsequence (λ_n) of μ^{m-1} such that

$$\alpha_m \sum_{j=0}^{\lambda_n} a_j^m x_j \longrightarrow x \quad \text{and} \quad \sum_{j=0}^{\lambda_n} a_j^m e_j \longrightarrow a^m \quad \text{as } n \to \infty.$$

On the other hand, since (λ_n) is a subsequence of any μ^l with $l < m$, one has, for any $l < m$, that

$$\sum_{j=0}^{\lambda_n} a_j^l x_j \longrightarrow 0 \quad \text{and} \quad \sum_{j=0}^{\lambda_n} a_j^l e_j \longrightarrow a^l \quad \text{as } n \to \infty.$$

This implies that a belongs to $\mathcal{U}_A^\mu$, which finishes the proof. □

As shown in [58], the last theorem (or variants of it, see [58]) can be applied to a great deal of function spaces, so improving the corresponding results of existence and large topological size of families of special functions up to dense-lineability of such families: Fekete's series (Taylor series $\sum_{j \geq 1} a_j x^j$ whose partial sums approximate uniformly any given continuous function $f : [-1, 1] \to \mathbb{R}$ with $f(0) = 0$), universal series in the sense of Chui-Parnes-Luh, Menshov's functions (trigonometric series $\sum_{n \in \mathbb{Z}} a_n e^{int}$ such that every complex measurable 2π-periodic functions is the almost-everywhere limit of a subsequence of the partial sums $\sum_{j=-n}^{n} a_n e^{int}$), Seleznev's functions (power series having universal approximation properties in $\mathbb{C} \setminus \{0\}$), Dirichlet series $\sum_{n \geq 1} a_n n^{-z}$ whose partial sums approximate any entire function on a half-plane, universal expansions of harmonic functions, universal Laurent series, universal Faber series (a Faber series is an analog of a Taylor series, when a disc is replaced by a complex domain with certain properties), universal expansions of $\mathcal{C}^\infty$-functions and many more. In the same vein, Koumandos, Nestoridis, Smyrlis and Stefanopoulos [278] have recently investigated under what conditions dense-lineability holds when $A = \bigcap_{p>1} \ell_p$, ℓ_q $(1 < q < \infty)$, c_0 or $\mathbb{K}^{\aleph_0}$, endowed with their natural topologies. Then they apply their findings to trigonometric series in $\mathbb{R}^{\mathbb{N}}$ and Dirichlet series.

4.5 Hypercyclicity and spaceability

Regarding large closed subspaces, the following assertion seems to be the first result in the setting of universality (if one disregards the aforementioned "extreme" example by Read [340]). It was obtained by Bernal and Montes [102] in 1995. Recall that two domains Ω and G are said to be conformally equivalent whenever there is a bijective holomorphic function $h : \Omega \to G$. If this is the case, the inverse function h^{-1} is automatically holomorphic.

Theorem 4.5.1. *Assume that $G \subset \mathbb{C}$ is a domain that is not conformally equivalent to $\mathbb{C} \setminus \{0\}$. Suppose also that (φ_n) is a sequence of automorphisms of G that is runaway, in the sense that, given a compact subset $K \subset G$, there exists $n \in \mathbb{N}$ such that*

$$K \cap \varphi_n(K) = \varnothing.$$

Then, if the space of holomorphic functions $H(G)$ is endowed with the compact-open topology, the sequence (C_{φ_n}) of composition operators defined on $H(G)$ satisfies that $\mathcal{U}((C_{\varphi_n}))$ is spaceable.

In 1996, Montes [307] stated, in the setting of Banach spaces, the following criterium on existence of closed subspaces within the set of hypercyclic vectors. The proof is based on the construction of appropriate basic sequences. We provide here a version that is more general than the original one, in the setting of F-spaces; additional improvements are enumerated later. The proof given here is due to Bonilla and Grosse-Erdmann, and can be found in [237, Section 10.5]. For the Hypercyclicity Criterion (HCC), see Definition 4.2.4. We say that a topological vector space X supports a continuous norm provided there is a norm $\||\cdot\|| : X \to [0,\infty)$ that is continuous with the original topology of X.

Theorem 4.5.2. *If X is a separable F-space, T is an operator on X satisfying the Hypercyclicity Criterion for some $(n_k) \subset \mathbb{N}$ and there is an infinite dimensional closed vector subspace M_0 of X for which*

$$T^{n_k} x \to 0 \quad \text{for all} \ \ x \in M_0,$$

then $HC(T)$ is spaceable.

Proof. We can suppose that $(n_k) = \mathbb{N}$. Only small changes of notation are necessary to perform the proof in the general case. Since X is an F-space, its topology is induced by an F-norm $\|\cdot\| : X \to [0,\infty)$. Let $\||\cdot\||$ be a continuous norm on X, and let $(Z, \||\cdot\||)$ be the completion of $(X, \||\cdot\||)$. Then $(Z, \||\cdot\||)$ is a Banach space for which X is a dense subset. Let M_1 be the closure of M_0 in Z. By Mazur's theorem, $(M_1, \||\cdot\||)$ contains a basic sequence (u_n). The denseness of M_0 in M_1 together with the basis perturbation theorem yield the existence of a sequence $(e_n) \subset M_0$ that is a basic sequence in Z. The idea is perturbing (e_n) into a basic sequence (h_n) of hypercyclic vectors.

To do this, let $K_n = \max\{1, \|\|e_n^*\|\|\}$ $(n = 1, 2, \dots)$. By using induction with respect to the strict order $<$ on $\mathbb{N} \times \mathbb{N}$ defined by

$$(1,1) < (1,2) < (2,1) < (1,3) < (2,2) < (3,1) < (1,4) < \cdots ,$$

it is easily seen that from the HCC we can obtain a dense sequence (y_n) in X, vectors $x_{j,k} \in X$ and infinitely many strictly increasing sequences $(n(j,k)_{k \geq 1}) \subset \mathbb{N}$ $(j \in \mathbb{N})$ such that, for all $j, k, j', k' \geq 1$,

$$\max\{\|x_{j,k}\|, \|\|x_{j,k}\|\|\} \leq \frac{1}{2^{j+k+1} K_j}, \tag{4.11}$$

$$\|T^{n(j,k)} x_{j,k} - y_k\| \leq \frac{1}{2^k},$$

$$\|T^{n(j',k')} x_{j,k}\| \leq \frac{1}{2^{j+k+k'} K_j}, \text{ if } (j',k') \neq (j,k),$$

$$\|T^{n(j,k)} e_j\| \leq \frac{1}{2^k}.$$

In (4.11) the continuity of the inclusion $X \to Z$ has been used. Moreover, (4.11) implies that each series $\sum_{k=1}^{\infty} x_{j,k}$ converges. Then we can define $h_j := e_j + \sum_{k=1}^{\infty} x_{j,k} \in X$ $(j = 1, 2, \dots)$. From the above inequalities, these vectors h_j satisfy, for all $j, k, j' \in \mathbb{N}$,

$$\max\{\|e_j - h_j\|, \|\|e_j - h_j\|\|\} \leq \frac{1}{2^{j+1} K_j}, \tag{4.12}$$

$$\|T^{n(j,k)} h_j - y_k\| \leq \frac{3}{2^k}, \tag{4.13}$$

$$\|T^{n(j',k)} (e_j - h_j)\| \leq \frac{1}{2^{j+k} K_j}, \text{ if } j' \neq j. \tag{4.14}$$

By (4.12), $\sum_{j=1}^{\infty} \|\|e_j^*\|\| \, \|\|e_n - h_j\|\| \leq 1/2 < 1$. Hence the basis perturbation theorem yields that (h_n) is a basic sequence in Z with $\|\|h_j^*\|\| \leq 2K_j$ for all $j \geq 1$. In particular, the h_n's are linearly independent. Define M as

$$M := \overline{\text{span}} \{h_n : n \geq 1\},$$

where the closure is taken in $(X, \| \cdot \|)$.

For a finite linear combination $z = \sum_{j=1}^{N} a_j h_j$ with $a_m = 1$ for some m, let $w := \sum_{j=1}^{N} a_j e_j - e_m$. From (4.13) and (4.14) we get for every $k \geq 1$ that

$$\|T^{n(m,k)} z - y_k\| \leq \|T^{n(m,k)} h_m - y_k\|$$
$$+ \Big\| \sum_{j \leq N, j \neq m} a_j T^{n(m,k)}(h_j - e_j) \Big\| + \|T^{n(m,k)} w\|$$
$$\leq \frac{3}{2^k} + \sum_{j \leq N, j \neq m} (|a_j| + 1)\|T^{n(m,k)}(h_j - e_j)\| + \|T^{n(m,k)} w\|$$
$$\leq \frac{4}{2^k} + \sum_{j=1}^{\infty} \|e_j^*\| \, \|w\| \frac{1}{2^{j+k} K_j} + \|T^{n(m,k)} w\|$$
$$\leq \frac{4}{2^k} + \frac{1}{2^k} \|w\| + \|T^{n(m,k)} w\|. \tag{4.15}$$

Finally, let $z \in M \setminus \{0\}$. Then z has a representation $z = \sum_{j=1}^{\infty} a_j h_j$, with convergence in Z. Since $HC(T)$ is invariant under multiplication by scalars, we may assume $a_m = 1$ for some m. By definition of M, there is a sequence of vectors

$$z_\nu := \sum_{j=1}^{N_\nu} a_{\nu,j} h_j \to z \quad (\nu \to \infty)$$

in X. Since we also have convergence in Z and the convergence functionals are continuous, we have that $a_{\nu,j} \to a_j$ for each j. In particular, $a_{\nu,m} \to a_m$, and we may assume without loss of generality that $a_{\nu,m} = 1$ for all ν. Consider the vectors $w_\nu = \sum_{j=1}^{N_\nu} a_{\nu,j} e_j - a_m \in M_0$. Setting $a_{\nu,j} = 0$ for $j > N_\nu$, we find for $\nu, \mu \in \mathbb{N}$ that

$$\|(w_\nu - z_\nu) - (w_\mu - z_\mu)\| \leq \sum_{j \neq m} \|(a_{\nu,j} - a_{\mu,j})(e_j - h_j)\|.$$

Since $(a_{\nu,j})_\nu$ converges for all $j \geq 1$ and since, by (4.12) and the fact $\|h_j^*\| \leq 2K_j$,

$$\|(a_{\nu,j} - a_{\mu,j})(e_j - h_j)\| \leq (|a_{\nu,j} - a_{\mu,j}| + 1)\|e_j - h_j\|$$
$$\leq \|h_j^*\| \, \|z_\nu - z_\nu\| \, \|e_j - h_j\| + \frac{1}{2^{j+1}}$$
$$\leq \frac{\|z_\nu - z_\mu\|}{2^j} + \frac{1}{2^{j+1}} \leq \frac{C}{2^j}$$

with some constant $C > 0$, the dominated convergence theorem implies that $(w_\nu - z_\nu)$ is a Cauchy sequence in X. Since $z_\nu \to z$ and M_0 is closed, the sequence (w_ν) converges to some $w \in M_0$.

By the previous argument, z_ν and w_ν must satisfy (4.15) for any $\nu \in \mathbb{N}$. Letting $\nu \to \infty$ we have by continuity that, for all $k \geq 1$,

$$\|T^{n(m,k)} z - y_k\| \leq \frac{4}{2^k} + \frac{1}{2^k} \|w\| + \|T^{n(m,k)} w\| \longrightarrow 0 \quad (k \to \infty)$$

because $w \in M_0$. Then each y_k is approximated by a subsequence of $(T^n z)$, which shows that z is hypercyclic. In other words, $M \setminus \{0\} \subset HC(T)$, as required. $\qquad\qquad\qquad\qquad\qquad\qquad\qquad\qquad\qquad\qquad\qquad\qquad\qquad\quad\square$

For instance, if $1 \leq p < \infty$ and φ is an automorphism of $\mathbb{D}$ without fixed points then $H(C_\varphi)$ is spaceable in H^p [307]. Also, consider a weight, that is, a bounded sequence $w = (w_n) \subset \mathbb{K} \setminus \{0\}$, as well as its associated weighted backward shift

$$B_w : x = (x_n) \in \ell_p \mapsto (w_n x_{n+1}) \in \ell_p.$$

Then $HC(B_w)$ is spaceable provided that the set $\sup_{n \geq 1} \prod_{k=1}^n |w_k| = \infty$ and $\sup_{n \geq 1} \lim_{k \to \infty} \prod_{\nu=1}^n |w_{\nu+k}| < \infty$ [237, Chap. 10]. As a negative result, in [307] it is shown that the set $HC(T)$ for the Rolewicz operator $T = \lambda B$ ($|\lambda| > 1$) is not spaceable, although (T^n) satisfies the HCC. León and Montes [282] studied the spaceability of $HC(B_w)$ on ℓ_2. Recently, Menet [303] has obtained the following complete characterization.

Theorem 4.5.3. *Let $X = \ell_p$ ($p \geq 1$) or c_0. Consider a weight $w = (w_n)$ and the corresponding backward shift $B_w : X \to X$. Then the following conditions are equivalent:*

(a) *$HC(B_w)$ is spaceable.*

(b) *$\sup_{n \geq 1} \inf_{k \geq 1} \prod_{\nu=1}^n |w_{\nu+k}| < \infty$.*

In 2000, González, León and Montes [229] characterized the spaceability of $HC(T)$ for Banach space operators satisfying the HCC. Their findings can be summarized as follows. Recall that an operator S is called Fredholm provided that S has finite dimensional kernel $S^{-1}(\{0\})$ and cofinite dimensional closed range $S(X)$.

Theorem 4.5.4. *Let X be a complex separable Banach space and T be an operator on X such that (T^n) satisfies the Hypercyclicity Criterion. Then the following are equivalent:*

(a) *$HC(T)$ is spaceable.*

(b) *There exists an increasing sequence $(n_k) \subset \mathbb{N}$ and an infinite dimensional closed subspace M_0 of X such that $T^{n_k} x \to 0$ for all $x \in M_0$.*

(c) *There exists an increasing sequence $(m_k) \subset \mathbb{N}$ and an infinite dimensional closed subspace M_1 of X such that $\sup_{k \geq 1} \|T^{m_k}|_{M_1}\| < \infty$.*

(d) *The essential spectrum $\sigma_e(T) := \{\lambda \in \mathbb{C} : \lambda I - T$ is not Fredholm$\}$ meets the closed unit disc $\mathbb{D}$.*

Complementary criteria of spaceability and non-spaceability of $HC(T)$ and $HC((T_n))$ were provided by León and Müller [284] in 2006 for Banach spaces, and recently by Grosse-Erdmann and Peris [237, Chap. 10] and Ménet [303] in the setting of Fréchet spaces. Among them, we select the following *non-spaceability* result from [303], which extends results from [284] and [237].

Theorem 4.5.5. *Let X be an infinite dimensional Fréchet space, Y be a separable Fréchet space and $(T_n) \subset L(X,Y)$. Let (p_n) be an increasing sequence of seminorms defining the topology of X. If there exists a continuous seminorm p on X such that $p(x) > 0$ for every hypercyclic vector $x \in X$ and if there exist a sequence of subspaces $\{E_n\}_{n \geq 1}$ of finite codimension, a sequence $\{C_n\}_{n \geq 1} \subset (0, \infty)$ with $C_n \longrightarrow +\infty$ and a continuous seminorm q on Y such that*

$$q(T_n x) \geq C_n \, p_n(x) \quad \text{for all } n \in \mathbb{N} \text{ and all } x \in E_n$$

then $\mathcal{U}((T_n))$ is not spaceable.

Theorem 4.5.5 is based on the following lemma due to Grosse-Erdmann and Peris; see [237, Lemma 10.39]. Recall that the *codimension* of a vector subspace B of a vector space A is the dimension of any vector subspace C such that $B \cap C = \varnothing$ and $B + C = A$.

Lemma 4.5.6. *Let X be a Fréchet space, E a finite dimensional subspace of X, p a continuous seminorm on X and $\varepsilon > 0$. Then there exists a closed subspace L of finite codimension such that for any $x \in L$ and $y \in E$,*

$$p(x + y) \geq \max \left(\frac{p(x)}{2 + \varepsilon}, \frac{p(y)}{2 + \varepsilon} \right).$$

Proof. Since $\operatorname{Ker} p := p^{-1}(\{0\})$ is clearly a vector subspace of X, there is a subspace F of E such that $E = (\operatorname{Ker} p \cap E) \oplus F$ algebraically. Plainly, p is a norm on F. Since F is finite dimensional, all norms define the same topology on it. Hence the unit sphere $\{x \in F : p(x) = 1\}$ with respect to p in F, being bounded and closed, is compact. Consequently, there are finitely many points $y_k \in F$ with $p(y_k) = 1$ $(k = 1, \ldots, N)$ such that, for any $y \in F$ with $p(y) = 1$, there is some k with $p(y_k - y) \leq \frac{\varepsilon}{1+ve}$. More generally, if $y \in E$ with $p(y) = 1$ then $y = u + v$ with $p(u) = 0$ and $v \in F$, $p(v) = 1$. Hence there is some $k \in \mathbb{N}$ such that

$$p(y_k - y) = p(y_k - v) \leq \frac{\varepsilon}{1 + \varepsilon}.$$

By the Hahn–Banach theorem and the fact that p is continuous, there are continuous linear functionals $y_k^* \in X^*$ $(k = 1, \ldots, N)$ such that $y_k^*(y_k) = p(y_k) = 1$ and $|y_k^*(y)| \leq p(y)$ for all $y \in X$. Then

$$L := \bigcap_{k=1}^{N} \operatorname{Ker} y_k^*$$

is a closed vector subspace of X and, as an easy exercise shows, it has finite codimension.

Now let $x \in L$ and $y \in E$. If $p(y) = 0$ then $p(x + y) = 0$ and the claim is trivially true. Hence let $p(y) \neq 0$. Then $y' := y/p(y)$ satisfies $p(y') = 1$, so that there is some k with $p(y_k - y) \leq \frac{\varepsilon}{1+\varepsilon}$. Setting $x' := x/p(y) \in L$ we have that

$$p(x' + y') \geq p(y_k + x') - p(y_k - y') \geq |y_k^*(y_k + x)| - \frac{\varepsilon}{1+\varepsilon} = 1 - \frac{\varepsilon}{1+\varepsilon} = \frac{1}{1+\varepsilon},$$

and therefore $p(x+y) \geq \frac{p(y)}{1+\varepsilon}$. In addition,

$$(2+\varepsilon)p(x+y) = p(x+y) + (1+\varepsilon)p(x+y) \geq p(x+y) + p(y) \geq p(x),$$

which proves the lemma. □

Proof of Theorem 4.5.5. If p is a continuous seminorm on X, there exists an integer $N \geq 1$ and a positive number $K > 0$ such that for every $x \in X$, we have $p(x) \leq Kp_N(x)$. Without loss of generality, we can thus suppose that we have $p_1(x) > 0$ for every hypercyclic vector x in X. Assume that M is an infinite dimensional closed subspace of X with $M \setminus \{0\} \subset \mathcal{U}((T_n))$. We show that there exists a vector $x \in M$ such that $\lim_{n\to\infty} q(T_j x) = \infty$, which is a contradiction.

Since $C_n \to \infty$, there exists a sequence $\{k_1 < k_2 < \cdots\} \subset \mathbb{N}$ such that

$$n^3 \leq C_j \quad \text{for any } n \geq 2 \text{ and any } j \in (k_{n-1}, k_n].$$

We then construct a sequence $\{e_n\}_{n\geq 1} \subset M$ recursively such that

- $p_n(e_n) = 1/n^2$,

- $e_n \in \bigcap_{k_{n-1} < j \leq k_n} E_j$ and

- for any $j \leq k_n$, $T_j e_n \in \bigcap_{k \leq n-1} L_{k,j}$, where $L_{k,j}$ is the closed subspace of finite codimension given by Lemma 4.5.6 for $E_{k,j} = \text{span}\{T_j e_1, \dots, T_j e_k\}$, the seminorm q and $\varepsilon > 0$.

Such a construction is possible because M is an infinite dimensional space and each of the intersections $\bigcap_{k_{n-1} < j \leq k_n} E_j$ and $\bigcap_{k \leq n-1} L_{k,j}$ are spaces of finite codimension. Moreover we can choose e_n such that $p_n(e_n) = 1/n^2$ because we have $p_1(x) > 0$ for every hypercyclic vector x and each nonzero vector of M is universal. Letting $x := \sum_{\nu=1}^{\infty} e_\nu$, we have $x \in M$ and if $k_{n-1} < j \leq k_n$, we then have

$$q(T_j x) = q\Big(\sum_{\nu=1}^{\infty} T_j e_\nu\Big) \geq \frac{1}{1+\varepsilon} q\Big(\sum_{\nu=1}^{n} T_j e_\nu\Big) \geq \frac{1}{(1+\varepsilon)(2+\varepsilon)} q(T_j e_n)$$

$$\geq \frac{C_j p_j(e_n)}{(1+\varepsilon)(2+\varepsilon)} \geq \frac{C_j p_n(e_n)}{(1+\varepsilon)(2+\varepsilon)} \geq \frac{n}{(1+\varepsilon)(2+\varepsilon)},$$

where we have used the facts $\sum_{\nu=n+1}^{\infty} T_j e_\nu \in L_{n,j}$, $T_j e_n \in L_{n-1,j}$, $e_n \in E_j$, $n-1 \leq k_{n-1} < j$ and $n^3 \leq C_j$. The result follows. □

A special case of Theorem 4.5.5 is the following corollary, which of course can be proved with an easier, independent proof.

Corollary 4.5.7. *Let X be a separable infinite dimensional Banach space and $T \in L(X)$. If there exists $C > 1$, a closed subspace E of finite codimension in X and $n \in \mathbb{N}$ such that $\|T^n x\| \geq C \|x\|$ for any $x \in E$, then $HC(T)$ is not spaceable.*

Once it has been proved that a given topological vector space supports hypercyclic operators, the question of whether $HC(T)$ is spaceable for some T among them arises naturally. In 1997, León and Montes [282] gave a positive answer for every (separable, infinite dimensional) Banach space. In 2006, Petersson [333] and Bernal [78] independently solved the question in the affirmative for Fréchet spaces admitting a continuous norm. But nonexistence of continuous norms may be allowed: Bès and Conejero [112] constructed an operator T on $\omega := \mathbb{K}^{\mathbb{N}}$ for which $HC(T)$ is spaceable. Finally, Menet [302] has been able to prove that the assertion holds for every separable infinite dimensional Fréchet space.

The original Banach version of Theorem 4.5.2 has been improved and extended in several directions, for instance:

- The sequence of powers (T^n) is replaced by a sequence of continuous linear mappings (T_n).

- The arrival space is allowed to be different from X, that is, $(T_n) \subset L(X, Y)$.

- The space X may be just an F-space, as seen in the statement of Theorem 4.5.2.

- It is possible to obtain spaceability for the set of common universal vectors of a countable family $\{(T_{j,n})_{n \geq 1} : j \in \mathbb{N}\}$ of sequences of continuous linear mappings.

- The condition $T_{n_k} \to 0$ pointwise on M can be replaced by the mere convergence.

These extensions can be found in the papers by Bonet, Martínez-Giménez and Peris [129], Aron, Bès, León and Peris [20], León and Müller [284], Petersson [333], Bernal [78] and Bonilla and Grosse-Erdmann [137]. Specifically, by combining the results and the approaches of the proofs in these papers (for this action, Sections 10.1, 10.2, 10.5 and 11.4 of the book [237], as well of the mentioned equivalence for (T_n) of satisfying the UC and of being densely hereditarily hypercyclic for some $(n_k) \subset \mathbb{N}$, will show to be very useful), we obtain the following assertion. Note that for sequences (T_n) there is no need to introduce a subsequence (n_k) since one may always pass to subsequences.

Theorem 4.5.8. *Let X be a separable F-space with a continuous norm, Y_j $(j \in \mathbb{N})$ be separable metrizable topological vector spaces and let $(T_{j,n})_{n \geq 1} \subset L(X, Y_j)$ $(j \in \mathbb{N})$. Suppose that the following holds:*

(i) *For every $j \in \mathbb{N}$, $(T_{j,n})_{n \geq 1}$ is hereditarily densely universal.*

(ii) *There exists an infinite dimensional closed vector subspace M of X such that the sequence $(T_{j,n}x)_{n \geq 1}$ converges in Y_j for every $x \in M$ and every $j \in \mathbb{N}$.*

Then $\bigcap_{j=1}^{\infty} \mathcal{U}((T_{j,n})_{n \geq 1})$ is spaceable.

We furnish some examples, which may be interesting even in the case of a single operator or of one sequence of operators. In 2010, Shkarin [366] proved for the derivative operator D that $HC(D)$ is spaceable in $H(\mathbb{D})$ (he also notes that the essential spectrum $\sigma_e(D) = \varnothing$, so Theorem 4.5.4 breaks down for Fréchet spaces). His proof does not rely on Theorem 4.5.8, but it can be extracted from this theorem: see Exercise 4.14, which exhibits the approach given in [237, Example 10.13]. By the first result given in this section, the Birkhoff operator τ_a ($a \neq 0$) also enjoys spaceability for its family of hypercyclic functions; see [134] for a corresponding assertion in the space of harmonic functions on $\mathbb{R}^N$. More generally, Petersson [333] showed in 2006 the spaceability of $HC(\Phi(D))$ in $H(\mathbb{C})$, where Φ is any entire function of exponential type that is not a polynomial (this again carries as a consequence that $HC(\tau_a)$ is spaceable in $H(\mathbb{C})$ for any $a \neq 0$, because $\tau_a = \exp(aD)$). In fact, this can be extracted (see Exercise 4.13) from the following corollary of Theorem 4.5.2 due to Petersson [333].

Proposition 4.5.9. *Let X be a separable Fréchet space supporting a continuous norm, and let T be an operator on X that satisfies the HCC. If $\mathrm{Ker}(\lambda I - T)$ is infinite dimensional for some λ with $|\lambda| < 1$, then $HC(T)$ is spaceable.*

Proof. By continuity of T and the hypothesis, the set $M_0 := \mathrm{Ker}(\lambda I - T)$ is a closed infinite dimensional vector subspace, and for any $x \in M_0$ we have that $T^n x \to \lambda^n x \to 0$ as $n \to \infty$. Finally, apply Theorem 4.5.2. $\square$

Recently, Menet [303] has completed the Shkarin–Petersson results by proving that $HC(P(D))$ is also spaceable if P is a nonconstant polynomial. For this, he used appropriate backward shifts B_w on a certain sequence space associated to $H(\mathbb{C})$. Godefroy and Shapiro [227] proved that an operator $T \in L(H(\mathbb{C}))$ is of the form $\Phi(D)$, with Φ an entire function of exponential type, if and only if T commutes with all translations τ_a ($a \in \mathbb{C}$). Then we can establish the following assertion.

Theorem 4.5.10. *Let $T : H(\mathbb{C}) \to H(\mathbb{C})$ be an operator that commutes with all translations and that is not a scalar multiple of the identity. Then $HC(T)$ is spaceable.*

As another example, if Ω is a domain in $\mathbb{C}$ and φ, ψ are two automorphisms of Ω such that C_φ, C_ψ are hypercyclic on $H(\Omega)$ then $HC(C_\varphi) \cap HC(C_\psi)$ is spaceable [237, Chap. 11].

By using Theorem 4.5.8, it is demonstrated in [84] the following result, which complements the corresponding "generic" one stated by Müller [315] in 2009. Recall that the partial sums $S_n f$ of the Fourier series of an integrable function $f : \mathbb{T} \to \mathbb{C}$ are defined as $(S_n f)(e^{it}) = \sum_{k=-n}^{n} a_k \, e^{ikt}$, where a_k is the kth-Fourier coefficient, given by $a_k = \int_0^{2\pi} f(e^{it}) \, e^{-ikt} \, \frac{dt}{2\pi}$.

Theorem 4.5.11. *Let E be a countable subset of the unit circle $\mathbb{T} = \partial \mathbb{D}$, and consider the set $\mathbb{C}^E$ endowed with the product topology. Then the set of continuous functions $f : \mathbb{T} \to \mathbb{C}$ whose sequence $\{S_n f|_E\}_{n \geq 1}$ of partial Fourier sums restricted to E is dense in $\mathbb{C}^E$ is spaceable in the space $C(\mathbb{T})$ of complex continuous functions on the unit circle.*

The maximal dense-lineability of the set described in Theorem 4.5.11 is also shown in [84]; see Exercise 4.19.

There exist pairs of operators with spaceable sets of hypercyclic vectors, such that the set of their common hypercyclic vectors is not spaceable. For instance, consider the weights $w = \left(\frac{n+1}{n} \right)$ and $v = (2, 2, \ldots)$, the associated weighted backward shifts B_w, B_v, and the product maps $T_1 := B_w \oplus B_v$: $(x, y) \in \ell_2 \oplus \ell_2 \mapsto (B_w x, B_v y) \in \ell_2 \oplus \ell_2$, $T_2 := B_v \oplus B_w$ (for operators T and S, $T \oplus S(x, y) := (Tx, Sy)$). In [20] it is shown that $HC(T_1)$ and $HC(T_2)$ are spaceable, but $HC(T_1) \cap HC(T_2)$ is not. In the opposite side, Bayart [55] furnished in 2005 some criteria guaranteeing the spaceability of the set of common hypercyclic vectors of an uncountable family of operators on a Fréchet space (see improvements in [237, Chap. 11]). A corresponding result for dense-lineability of an uncountable family of sequences of operators on a Banach space, under rather strong assumptions, was obtained by the same author in [54]. Among the collection of existing results, let us select the following one dealing with multiples of operators. Its proof can be found in [237, Chap. 11].

Theorem 4.5.12. *Let X be a separable Fréchet space with a continuous norm, let $T \in L(X)$ with dense generalized kernel $X_0 = \bigcup_{n \geq 1} \operatorname{Ker}(T^n)$ and $\lambda_0 \geq 0$. Suppose that*

(i) *there is a map $S : X_0 \to X$ such that $TSx = x$ and $\lambda^{-n} S^n x \to 0$ for all $x \in X_0$ and all $\lambda > \lambda_0$, and*

(ii) *there exists an infinite dimensional closed subspace M_0 of X such that $\lambda^n T^n x \to 0$ for all $x \in M_0$, $\lambda > \lambda_0$.*

Then the family $\bigcup_{|\lambda| > \lambda_0} HC(\lambda T)$ is spaceable.

Recently, a number of criteria for spaceability of the set of common universal vectors for a family $\{\{T_{n,i}\}_{n \geq 1} : i \in I\}$ of continuous linear mappings between two Fréchet spaces have been obtained by Bès and Menet [115].

Turning to the universal Taylor series in the sense of Nestoridis described in the preceding section, Bayart [53] established in 2005 that the set of such functions admits large closed vector subspaces, that is, the following statement holds. The reader can observe that, as in the proof of Theorem 4.5.2, the use of basic sequences is crucial.

Theorem 4.5.13. *The set of universal Taylor series is spaceable in* $H(\mathbb{D})$.

Proof. For the sake of convenience, we denote by $\mathcal{M}$ the family of all compact subsets of $C \setminus \mathbb{D}$ with connected complement. Recall that, for every compact set K of $\mathbb{C}$, $A(K)$ stands for the space of all continuous functions $f : K \to \mathbb{C}$ such that $f \in H(K^0)$.

Step 1. We begin with the following refinement of Mergelyan's approximation theorem (see Section 2.1 or [351]). Let $K \in \mathcal{M}$, L be a compact subset of $\mathbb{D}$, and $g \in A(K)$. For any $\varepsilon > 0$ and $N \in \mathbb{N}$, there exists a polynomial $P(z) = \sum_{n=N}^{q} a_n z^n$ such that $|P(z)| < \varepsilon$ for all $z \in L$ and $|P(z) - g(z)| < \varepsilon$ for all $z \in K$. To prove it, set $R > 1$ and $r \in (0,1)$ such that $K \subset B(0,R)$ and $L \in B(0,r)$. By Mergelyan's theorem, there exists a polynomial $Q(z) = \sum_{n=0}^{q} b_n z^n$ such that

$$|Q(z)| < \frac{\varepsilon \, r^N}{2N \, R^N} \quad \text{for all } z \in \overline{B}(0,r) \quad \text{and} \quad |Q(z) - g(z)| < \frac{\varepsilon}{2} \quad \text{for all } z \in K.$$

By Cauchy's estimates for Taylor coefficients of an analytic function, for $k \leq N$ one has $|b_k| \leq \varepsilon/2NR^k$. We set $P(z) = \sum_{n=N}^{q} b_n z^n$. Observe that, for $z \in B(0,R)$,

$$|P(z) - Q(z)| \leq \sum_{k=0}^{N-1} |b_k| R^k \leq \frac{\varepsilon}{2} < \varepsilon.$$

Therefore P satisfies the desired properties.

Step 2. Now, fix a sequence (K_m) of compact subsets of $\mathbb{C} \setminus \mathbb{D}$ with connected complement such that (K_m) is "fundamental" for these compacta, that is, such that for every $K \in \mathcal{M}$, there exists an $m \in \mathbb{N}$ with $K \subset K_m$: this is an easy exercise; see, e.g., [324]. Let (Q_l) be an enumeration of all polynomials with coefficients in $\mathbb{Q} + i\mathbb{Q}$ and let $\varphi, \psi : \mathbb{N} \to \mathbb{N}$ be two functions such that, given any couple $(m, l) \in \mathbb{N} \times \mathbb{N}$, there exist infinitely many j with $(\varphi(j), \psi(j)) = (m, l)$. Let us finally fix (r_j) an increasing sequence in $(1/2, 1)$ that is converging to 1. We use induction to build sequences of polynomials $(f_{j,k})$ for $k \leq j$.

For this, define $g_{1,1}(z) = 2z + P(z)$, where P is given in Step 1 with $K = K_{\varphi(1)}$, $L = \overline{B}(0, r_1)$, $g = Q_{\psi(1)}$, $N = 2$ and $\varepsilon = 1/2^3$. The Taylor series of $g_{1,1}$ approaches $Q_{\psi(1)}$ on $K_{\varphi(1)}$. We now correct this value on $K_{\varphi(2)}$ for further expansions by setting $f_{1,1}(z) = g_{1,1}(z) + Q(z)$, where Q is given by Step 1 with $K = K_{\varphi(2)}$, $L = \overline{B}(0, r_1)$, $g = -g_{1,1}$, $N = \deg(P) + 1$ and $\varepsilon = 1/2^3$. Following this procedure, assume that polynomials $(f_{j-1,k})$ have been built for $k \leq j - 1$. Let $N_{j,1} = \max_{1 \leq k \leq j-1} \deg(f_{j-1,k}) + 1$. We define an intermediate polynomial $g_{j,1}$ by setting $g_{j,1}(z) = f_{j-1,1}(z) + P(z)$, where P is given by Step 1 for $K = K_{\varphi(j)}$, $g = Q_{\psi(j)} - f_{j-1,1}$, $L = \overline{B}(0, r_j)$,

$N = N_{j,1}$ and $\varepsilon = 1/2^{j+2}$. We now correct the value of the Taylor series by setting $f_{j,1}(z) = g_{j,1}(z) + Q(z)$, where Q is given by Step 1 for $K = K_{\varphi(j+1)}$, $g = -g_{j,1}$, $L = \overline{B}(0, r_j)$, $N = \deg(P) + 1$ and $\varepsilon = 1/2^{j+2}$. Therefore, we have the inequalities

$$|f_{j,1}(z) - f_{j-1,1}(z)| \leq \frac{1}{2^{j+1}} \ \forall z \in \overline{B}(0, r_j) \quad \text{and} \quad |f_{j,1}(z)| \leq \frac{1}{2^{j+1}} \ \forall z \in K_{\varphi(j+1)}. \tag{1}$$

Moreover, $S_n(f_{j,1}) = S_n(f_{j-1,1})$ for $n < N_{j,1}$, where $S_n(h)$ denotes the nth partial sum of the Taylor series of h at the origin. Taking $N_{j,2} = \deg(f_{j,1}) + 1$, we apply the same construction to deduce $f_{j,2}$ from $f_{j-1,2}$, and inductively we build polynomials $f_{j,k}$ ($1 \leq k \leq j-1$) satisfying inequalities similar to (1). Finally, if $N_{j,j}$ is an integer greater than the degree of all polynomials $f_{j,k}$ ($1 \leq k \leq j-1$), then $f_{j,j}$ is deduced from $2^{N_{j,j}} z^{N_{j,j}}$ by following the same process.

Step 3. The first inequality of (1) ensures that the sequence $(f_{j,k})_{j \geq k}$ converges uniformly on any compact set of $\mathbb{D}$ to a function $f_k \in H(\mathbb{D})$. Let E be the vector space consisting of all series $\sum_{k=1}^{\infty} \alpha_k f_k$ that converge uniformly on compacta of $\mathbb{D}$, and let F be the closure of E in $H(\mathbb{D})$. Clearly F is a closed infinite dimensional subspace of $H(\mathbb{D})$, and we need only prove that it consists –except for the null function– of universal Taylor series.

We begin with the case of a series

$$h = \sum_{k=1}^{\infty} \alpha_k f_k \in E,$$

where $\sum_{k=1}^{\infty} |\alpha_k|^2 < \infty$. Take $\varepsilon > 0$, K_m a compact set in the family described at the beginning of Step 2, and Q_l a polynomial with coefficients in $\mathbb{Q} + i\mathbb{Q}$. Without loss of generality, assume that $\alpha_1 = 1$ (if $\alpha_1 = 0$, we take the least integer k such that $\alpha_k \neq 0$; the proof is exactly the same). Let j be such that $(\varphi(j), \psi(j)) = (m, l)$, and let N be the degree of the polynomial $g_{j,1}$ built at step j of the above induction procedure. Then, for $z \in K_m$, we obtain from the triangle and the Cauchy–Schwarz inequalities that

$$|S_N(h)(z) - Q_l(z)| \leq |g_{j,1}(z) - Q_l(z)| + \left| \sum_{k=1}^{j-1} \alpha_k f_{j-1,k}(z) \right|$$

$$\leq \frac{1}{2^j} + \left(\sum_{k=1}^{\infty} |\alpha_k|^2 \right)^{1/2} \left(\sum_{k=1}^{j-1} |f_{j-1,k}|^2 \right)^{1/2} \leq \frac{1}{2^j} + \frac{\|\alpha\|_2 \, j^{1/2}}{2^j},$$

where the last inequality follows from the second one in (1). Letting $j \to \infty$, one obtains

$$|S_n(h)(z) - Q_l(z)| \leq \varepsilon \quad \text{for all } z \in K_m.$$

This shows that $\sum_{k=1}^{\infty} \alpha_k f_k$ is universal in the sense of Nestoridis. Note that the integer N does not depend specifically on h; it depends only on $\|\alpha\|_2$ and the condition $\alpha_1 = 1$.

Now, we are going to transfer the universal behavior to all elements of F. For this, basic sequences will be used. First of all, consider $H = L^2(\frac{1}{2}\mathbb{T}) = L^2(|z| = \frac{1}{2})$, endowed with its natural norm

$$\|g\|_H = \left(\int_0^{2\pi} |g(e^{i\theta}/2)|^2 \, \frac{d\theta}{2\pi} \right)^{1/2}.$$

If $g \in H(\mathbb{D})$ then $g \in H$ and $\|g\|_H \le \max_{|z| \le 1/2} |g(z)|$. Hence, by construction,

$$\|f_k - 2^{N_{k,k}} z^{N_{k,k}}\|_H^2 \le \sum_{j=k+1}^{\infty} \frac{1}{4^j} \le \frac{1}{3 \cdot 4^k}.$$

By the basis perturbation theorem (see Section 4.1), (f_j) is a basic sequence of $L^2(\frac{1}{2}\mathbb{T})$ that is equivalent to the canonical basis of ℓ_2. Now, if $f \in F$ then there is a sequence $(h^r) \subset E$ converging to h. By continuity of $\| \cdot \|_H$ with respect to the maximum norm on $\overline{B}(0, 1/2)$, this sequence of series converges also to h in $L^2(\frac{1}{2}\mathbb{T})$. Therefore h has a representation as a series $\sum_{k=1}^{\infty} \alpha_k f_k$ in $L^2(\frac{1}{2}\mathbb{T})$, perhaps not convergent in $H(\mathbb{D})$. Let us write

$$h^r = \sum_{k=1}^{\infty} \alpha_{k,r} f_k,$$

with $\alpha^r := (\alpha_{k,r})_{k \ge 1} \in \ell_2$ and $\alpha_r \to \alpha := (\alpha_k)$ $(r \to \infty)$ in $L^2(\frac{1}{2}\mathbb{T})$. Take $\varepsilon > 0$, K_m and Q_l as before. We suppose again that $\alpha_1 = 1$, and it is obvious that we may consider that $\alpha_{1,r} = 1$ for all $r \in \mathbb{N}$. Since (α^r) is convergent in ℓ_2, the sequence of its norms is bounded by M. If j is such that $(\varphi(j), \psi(j)) = (m, l)$, then our previous calculation yields the existence of $N_j \in \mathbb{N}$ such that, for every pair $(r, z) \in \mathbb{N} \times K_m$,

$$|S_{N_j}(h^r)(z) - Q_l(z)| \le \frac{1}{2^j} + \frac{M \cdot j^{1/2}}{2^j}.$$

Finally, fix j such that $\frac{1}{2^j} + \frac{M \cdot j^{1/2}}{2^j} \le \frac{\varepsilon}{2}$. Now, by taking the limit in the (finite) sum of the Taylor series, one obtains $r \in \mathbb{N}$ satisfying

$$|S_{N_j}(h)(z) - S_{N_j}(h^r)(z)| \le \frac{\varepsilon}{2}.$$

An application of the triangle inequality completes the proof that h is universal. $\qquad\square$

4.6 Algebras of hypercyclic vectors

In contrast with dense-lineability or spaceability, not much is known about algebrability of the family of hypercyclic vectors for concrete operators. The derivative operator belongs to the short list of lucky operators. The following theorem was proved in 2007 by Aron, Conejero, Peris and Seoane [22,23].

Theorem 4.6.1. *Consider the derivative operator* $D : H(\mathbb{C}) \to H(\mathbb{C})$. *Then there is a residual subset* M *of* $H(\mathbb{C})$ *such that, for each* $f \in M$, *the algebra generated by* f *is contained in* $HC(D) \cup \{0\}$.

Proof. Firstly, we prove the following result:

(R) Let (U, V) be a pair of nonempty open subsets of $H(\mathbb{C})$, and let $m \in \mathbb{N}$. Then there are $P \in U$ and $q \in \mathbb{N}_0$ such that $D^q(P^j) = 0$ for $j < m$ and $D^q(P^m) \in V$.

To do this, it is enough to show that, for any pair of polynomials (A, B), one can find a sequence of polynomials (R_n) and a sequence $(q_n) \subset \mathbb{N}$ such that $D^{q_n}((A + R_n)^j) = 0$ for all $j < m$, $D^{q_n}((A + R_n)^m) = B$ and $R_n \to 0$ uniformly on compacta. Indeed, having chosen A, B with $A \in U$ and $B \in V$, the polynomial $P := A + R_n$ and the integer $q := q_n$ will have the required properties if n is large enough. So let us fix A and B, with degree$(B) = p$. We need the next assertion:

(S) Let $n \in \mathbb{N}$ and set $q := mn + (m - 1)p$. Then there exists a polynomial R of the form $R(z) = z^n \sum_{i=0}^{p} c_i z^i$ such that $D^q(R^m) = B$. Moreover, writing $c_i = c_i(n)$ one has $c_i(n) = O(n^{p-i}/[(m(n+p))!]^{1/m})$ as $n \to \infty$, for each $i \in \{0, \ldots, p\}$.

Let us prove (S). If R is as in the assertion then degree$(R^m) \leq m(n+p) = q + p$. For each $i \in \{0, \ldots, p\}$, let us denote by $\delta_i(c_0, \ldots, c_p)$ the coefficient of z^{q+i} in $R^m(z)$ or, that is the same, the coefficient of $z^{mp-(p-i)}$ in $(\sum_{i=0}^{p} c_i z^i)^m$. Explicitly,

$$\delta_i(c_0, \ldots, c_p) = \sum_{\gamma \in \Gamma_i} u_\gamma c_0^{\gamma_0} \cdots c_p^{\gamma_p},$$

where $\Gamma_i := \{\gamma = (\gamma_0, \ldots, \gamma_p) \in \mathbb{N}_0^{p+1} : \sum_{j=0}^{p} j\gamma_j = mp - (p - i)$ and $\sum_{j=0}^{p} \gamma_j = m\}$ and u_γ is the multinomial coefficient $u_\gamma = \frac{m!}{\gamma_0! \cdots \gamma_p!}$. Note that u_γ does not depend on n.

By the definition of the coefficients $\delta_i(c_0, \ldots, c_p)$, the polynomial $D^q(R^m)$ is given by

$$D^q(R^m) = \sum_{i=0}^{p} \frac{(q + i)!}{i!} \delta_i(c_0, \ldots, c_p) z^i.$$

If now we write $B(z) = \sum_{i=0}^{p} b_i z^i$, then we have to solve the system of equations

$$\delta_i(c_0, \ldots, c_p) = \frac{i! b_i}{(q+i)!} \quad (i = 0, \ldots, p). \tag{4.16}$$

This system is upper-triangular in the unknowns $c_0, \ldots, c_p$. Therefore it can indeed be solved. Let us check by (finite) induction on j that any solution of the system satisfies

$$c_{p-j}(n) = O\left(\frac{n^j}{[(m(n+p))!]^{1/m}}\right) \quad \text{as } n \to \infty \tag{4.17}$$

for each $j \in \{0, \ldots, p\}$. For $i = p$, (4.15) reduces to $c_p^m = \frac{p! b_p}{(m(n+p))!}$, so (4.17) is true for $j = 0$.

Assume that it has been proved for all $j \leq l - 1$, where $1 \leq l \leq p$. From (4.16) for $i = p - l$, we get

$$\frac{(m(n+p) - l)!}{(p-l)!}\left(m c_{p-l} c_p^{m-1} + \sum_{\gamma \in \Gamma_{p-l}^*} u_\gamma c_{p-l+1}^{\gamma_{p-l+1}} \cdots c_p^{\gamma_p}\right) = b_{p-l}, \tag{4.18}$$

where $\Gamma_{p-l}^* = \Gamma_{p-l} \setminus \{(0, \ldots, 0, \underbrace{1}_{[p-l]}, 0, \ldots, 0, m-1)\}$. By the induction hypothesis, we have the asymptotic estimate (4.17) for all $j \leq l - 1$. Moreover, if $\gamma \in \Gamma_{p-l}^*$ then $\sum_{j=0}^{l-1} j\gamma_{p-j} = \sum_{j=0}^{p}(p-j)\gamma_j = mp - (mp - l) = l$ and $\sum_{j=0}^{l-1} \gamma_{p-j} = \sum_{j=0}^{p} \gamma_j = m$. Taking into account that the coefficients u_γ do not depend on n, we get

$$\frac{(m(n+p) - l)!}{(p-l)!} \sum_{\gamma \in \Gamma_{p-l}^*} u_\gamma c_{p-l+1}^{\gamma_{p-l+1}} \cdots c_p^{\gamma_p}$$

$$= O\left(\frac{(m(n+p) - l)!}{(m(n+p))!} n^l\right) = O(1) \quad \text{as } n \to \infty.$$

This together with (4.18) and the equality $c_p^m = \frac{p! b_p}{(m(n+p))!}$ yields $c_{p-l} = O\left(\frac{[(m(n+p))!]^{(m-1)/m}}{[(m(n+p))!]^{1/m}}\right)$, which proves (S).

Now, we conclude the proof of (R) as follows. For each $n \in \mathbb{N}$, set $q_n := mn + (m-1)p$ and let R_n be the polynomial given by (S). For any $k < m$, the polynomial R_n^k has degree at most $(m-1)(n+p) = q_n - n$. By the binomial theorem, it follows that if n is large enough then $D^{q_n}((A + R_n)^j) = 0$ for all $j < m$ and $D^{q_n}((A + R_n)^m) = D^{q_n}(R_n^m) = 0 = B$. Finally, $R_n \to 0$ uniformly on compact sets because $M^n c_i(n) \to 0$ for any $M > 0$. Thus, the sequences (R_n) and (q_n) have the required properties, so proving (R).

In order to demonstrate the theorem, fix $f \in H(\mathbb{C})$, $m \in \mathbb{N}$ and $\alpha = (\alpha_1, \ldots, \alpha_m) \in \mathbb{C}^m$, and put

$$f_\alpha := \alpha_1 f + \cdots + \alpha_m f^m.$$

Let $\{V_k\}_{k\geq 1}$ be a countable basis of open sets for $H(\mathbb{C})$, and define the set $M := \bigcap_{k,s,m\in\mathbb{N}} A(k,s,m)$, where

$$A(k,s,m) := \{f \in H(\mathbb{C}) : \text{for every } \alpha \in \mathbb{C}^m \text{ with } \alpha_m = 1 \text{ and } \sup_i |\alpha_i| \leq s,$$

$$\text{exists } q \in \mathbb{N}_0 \text{ such that } D^q f_\alpha \in V_k\}.$$

The complement of each set $A(k,s,m)$, being the projection of the closed set $\{(\alpha,f) : D^q f_\alpha \notin V_k \text{ for all } q \in \mathbb{N}\} \subset \mathbb{C}^m \times H(\mathbb{C}^m)$ along the compact set $\{\alpha \in \mathbb{C}^m : \alpha_m = 1 \text{ and } \sup_i |\alpha_i| \leq s\}$, is closed. Therefore $A(k,s,m)$ is open. But it is also dense. Indeed, let U be a nonempty open subset of $H(\mathbb{C})$. By the result (R), one can find an entire function P and $q \in \mathbb{N}_0$ such that $P \in U$, $D^q(P^m) \in V_k$ and $D^q(P^j) = 0$ for all $j < m$. By linearity, $P \in A(k,s,m)$, so

$$U \cap A(k,s,m) \neq \varnothing.$$

Thus, each $A(k,s,m)$ is open and dense, so M is residual in $H(\mathbb{C})$.

Finally, let $f \in M$ and g be a nonzero member of the algebra generated by f, so that $g = f_\alpha$ for some $\alpha \in \mathbb{C}^m$ with $\alpha_m \neq 0$. By the definition of M, $\alpha_m^{-1} g = \alpha_m^{-1} f_\alpha \in HC(D)$. Hence $g \in HC(D)$ and the proof is concluded. □

Nevertheless, the algebrability of $HC(D)$ is still unknown. For composition operators, the situation is even worse: the translation operator τ_a does *not* admit algebras contained, except for zero, in $HC(\tau_a)$; see [23]. In fact, the same argument given in [23] shows that, for any domain $G \subset \mathbb{C}$ and any sequence (φ_n) of holomorphic self-mappings of G, the set $HC((C_{\varphi_n})) \cup \{0\}$ does not contain any algebra: see Exercise 4.16.

On the positive side, Bayart and Matheron [61, Chap. 8] made the following observation, which is extracted from the approach of [23]. Recall that an F-algebra is a completely metrizable topological linear algebra.

Proposition 4.6.2. *Let X be an F-algebra and $T \in L(X)$. Assume that, for every pair (U,V) of nonempty open sets in X, any open neighborhood W of 0 in X and any $m \in \mathbb{N}$, one can find $u \in U$ and $q \in \mathbb{N}$ such that $T^q(u^j) \in W$ for all $j < m$ and $T^q(u^m) \in V$. Then there is a residual subset M of X such that, for each $f \in M$, the algebra generated by f is contained in $HC(T)\cup\{0\}$.*

Proposition 4.6.2 can be applied, for instance, to the Rolewicz operator $T := 2B$ acting on $X := \ell_1$ if X is endowed with the convolution product $(a_n) * (b_n) = \{\sum_{k=0}^n a_k b_{n-k}\}_{n\geq 1}$; see [61, Exercise 8.5].

4.7 Supercyclicity and lineability

In this section, we briefly pay attention to the phenomenon of supercyclicity, a notion that is halfway between cyclicity and hypercyclicity.

In 1974, Hilden and Wallen [250] introduced the notion of supercyclicity. Let X be a Hausdorff topological vector space over $\mathbb{K} = \mathbb{R}$ or $\mathbb{C}$. An operator $T \in L(X)$ is said to be *supercyclic* provided that there is a vector $x_0 \in X$, called supercyclic for T, such that the projective orbit

$$\{\lambda T^n x_0 : n \geq 0, \, \lambda \in \mathbb{K}\}$$

of x_0 under T is dense in X. Again, the separability of X is necessary for supercyclicity. It is plain that every hypercyclic operator is also supercyclic. The backward shift B on ℓ_2 is an example of a supercyclic operator that is not hypercyclic. As in the hypercyclic case, it is immediate that the set $SC(T)$ of supercyclic vectors for T is dense as soon as T is supercyclic. If X is a separable F-space and (U_k) is an open basis for X then

$$SC(T) = \bigcap_{k \geq 1} \bigcup_{\substack{n \geq 1 \\ \lambda \in \mathbb{K} \setminus \{0\}}} (T^n)^{-1}(\lambda U_k),$$

showing that if T is supercyclic then $SC(T)$ is a dense G_δ subset. Hence we have topological genericity in this case and the question of the algebraic size of $SC(T)$ arises naturally. The Hilbert version of the following theorem was proved by Herrero [248] in 1991, but it holds in fact in the context of locally convex spaces.

Theorem 4.7.1. *Let X be a complex locally convex space, and let $T \in L(X)$ such that $\sigma_p(T^*) = \varnothing$. Then $SC(T)$ is dense-lineable.*

For the proof, simply follow the approach of Bourdon [144] for the dense-lineability of $HC(T)$; see Exercise 4.6. Note that, in contrast to the hypercyclic case, dense invariant supercyclic linear manifolds are not always available because, as shown in [248], there are supercyclic operators T whose adjoints T^* possess eigenvalues; recall that this is not possible for hypercyclic operators.

As for large closed subspaces within $SC(T)$, there have been remarkable contributions due to Salas [355] and Montes and Salas [310] (see also [311]). The next two results keep some similarity to Theorem 4.5.2, and were, respectively, stated by Salas in 1999 [355] and by Montes and Salas in 2001 [310]. Some refinements of techniques of the proof of Theorem 4.5.2 (in the Banach case) are used in [310, 355] to prove them.

Theorem 4.7.2. *Let T be an operator on a complex separable Banach space X satisfying the following:*

(i) *There exist an increasing sequence $(n_k) \subset \mathbb{N}$, dense subsets Y, Z of X and a mapping $S : Z \to Z$ such that $TS =$ identity on Z in such a way that $\|T^{n_k} y\| \, \|S^{n_k} z\| \to 0$ for each $(y, z) \in Y \times Z$.*

(ii) $0 \in \sigma_e(T)$.

Then $SC(T)$ is spaceable.

Theorem 4.7.3. *Let T be an operator on a complex separable Banach space X for which there is a strictly increasing sequence $(n_k) \subset \mathbb{N}$, a sequence $(\lambda_k) \subset \mathbb{C} \setminus \{0\}$, dense subsets Y, Z of X, a mapping $S : Z \to Z$ and an infinite dimensional closed subspace X_0 of X satisfying:*

(i) *$TS = $ identity on Z, $\lambda_k T^{n_k} y \to 0$ for every $y \in Y$ and $\lambda_k^{-1} S^{n_k} z \to 0$ for every $z \in Z$.*

(ii) *$\lambda_k T^{n_k} x \to 0$ for every $x \in X_0$.*

Then $SC(T)$ is spaceable.

Using Chan's techniques [156], Montes and Romero [309] gave a simpler proof of Theorem 4.7.3. Moreover, this theorem is used in [310] to prove the following spectral sufficient condition for the existence of a large closed supercyclic subspace.

Theorem 4.7.4. *Let T be an operator on a complex separable Banach space X for which there is a strictly increasing sequence $(n_k) \subset \mathbb{N}$, a sequence $(\lambda_k) \subset \mathbb{C} \setminus \{0\}$, dense subsets Y, Z of X and a mapping $S : Z \to Z$ satisfying condition* (i) *of Theorem 4.7.3. Assume, in addition, that the sequence $(\lambda_k \alpha^{n_k})$ is bounded for some $\alpha \in \sigma_e(T)$. Then $SC(T)$ is spaceable.*

Another remarkable consequence of Theorem 4.7.3 –also due to Montes and Salas [310]– is the following, that deals with basic sequences. Its proof illustrates the use of these kind of sequences in this setting, so we provide it as given in [310].

Theorem 4.7.5. *Let T be an operator on a complex separable Banach space X for which there is a strictly increasing sequence $(n_k) \subset \mathbb{N}$, a sequence $(\lambda_k) \subset \mathbb{C} \setminus \{0\}$, dense subsets Y, Z of X and a mapping $S : Z \to Z$ satisfying condition* (i) *of Theorem 4.7.3. Assume, in addition, that there exists a normalized basic sequence (u_m) such that $Tu_m \to 0$ as $m \to \infty$. Then $SC(T)$ is spaceable.*

Proof. Consider the sequences (λ_k) and (u_m) provided in the hypotheses. Note that $\|u_m\| = 1$ for all $m \in \mathbb{N}$. Let (u_m^*) be the sequence of functional coefficients corresponding to the basic sequence (u_m). We now use a certain background on basic sequences; see for instance Diestel's book [184] or Section 3.5 of this book. By Nikolskii's theorem, there is a basis constant $K \in [1, \infty)$, with the property that

$$\Big\| \sum_{j=1}^{n} c_j u_j \Big\| \leq K \Big\| \sum_{j=1}^{m} c_j u_j \Big\|$$

for all m, n with $m \geq n$ and all scalars $c_1, \ldots, c_m$. Then $\|u_m^*\| \leq 2K$ for all m. Since Y is dense, we can choose a sequence $(u_m') \subset Y$ such that $\|u_m - u_m'\| < \frac{1}{2^{m+1} K}$. Since

$$\|Tu_m'\| \leq \|Tu_m\| + \|T(u_m - u_m')\| \leq \|Tu_m\| + \|T\| \, \|u_m - u_m'\|,$$

we find that $Tu'_m \to 0$ as $m \to \infty$. On the other hand, we have

$$\sum_{m=1}^{\infty} \|u_m^*\| \, \|u'_m - u_m\| < 1.$$

Then the basis perturbation theorem yields that (u'_m) is a basic sequence equivalent to (u_m). It is clear that we can also assume that Y contains $x/\|x\|$ for every $x \in X$. Therefore, we may suppose, from the beginning, that the normalized sequence (u_m) is contained in Y. Now we will see that condition (ii) of Theorem 4.7.3 is satisfied with respect to a subsequence of (λ_k) for an appropriate subspace X_0. Set $v_1 = u_1$. Since $v_1 \in Y$, we can find $k_1 \in \mathbb{N}$ such that $\|\lambda_{k_1} T^{n_{k_1}} v_1\| < 1/2$. Since $Tu_m \to 0$, we may suppose, by extracting a subsequence, that

$$\|Tu_m\| < \frac{1}{2^m \lambda_{k_1} \|T^{n_{k_1}-1}\|} \quad \text{for} \quad m > 1.$$

We set $v_2 = u_2$. Since $v_1, v_2 \in Y$, we can find $k_2 > k_1$ such that $\|\lambda_{k_2} T^{n_{k_1}} v_i\| \leq 1/2^2$ for $i = 1, 2$. By extracting a subsequence from (u_m) we may suppose that

$$\|Tu_m\| < \frac{1}{2^m \lambda_{k_2} \|T^{n_{k_2}-1}\|} \quad \text{for} \quad m > 2.$$

Proceeding in this way, we obtain a sequence $\{\lambda_{k_l}\}_{l\geq 1}$, a sequence (u_m) and vectors $v_1, \ldots, v_l$ such that

$$\|\lambda_{k_l} T^{n_{k_l}} v_i\| \leq \frac{1}{2^l} \quad \text{for} \quad i = 1, \ldots, l \tag{4.19}$$

and $\|Tu_m\| \leq \frac{1}{2^m \lambda_{k_l} \|T^{n_{k_l}-1}\|}$ for $m > l$. Therefore, we have

$$\|\lambda_{k_l} T^{n_{k_l}} u_m\| = \|\lambda_{k_l} T^{n_{k_l}-1} T u_m\| \leq \frac{1}{2^m} \quad \text{for} \quad m > l. \tag{4.20}$$

We set $v_{l+1} = u_{l+1}$, define $X_0 = \overline{\operatorname{span}}\{v_m : m \in \mathbb{N}\}$ and let L be the basis constant of $\{v_i\}_{i\geq 1}$. If $v = \sum_{m=1}^{\infty} \alpha_m v_m \in X_0$, we know that $|\alpha_m| \leq 2L\|v\|$. Using (4.19) and (4.20) in the second inequality below we have

$$\|\lambda_{k_l} T^{n_{k_l}} v\| \leq \sum_{m=1}^{l} |\alpha_m| \, \|\lambda_{k_l} T^{n_{k_l}} v_m\| + \sum_{m=l+1}^{\infty} |\alpha_m| \, \|\lambda_{k_l} T^{n_{k_l}} v_m\|$$

$$\leq \frac{2Ll\|v\|}{2^l} + 2L\|v\| \sum_{m=l+1}^{\infty} \frac{1}{2^m} = \frac{2L(l+1)\|v\|}{2^l} \longrightarrow 0 \text{ as } l \to \infty.$$

Thus we can apply Theorem 4.7.3 to get the desired result. $\qquad\square$

As a prominent example, a characterization of spaceability of $SC(B_w)$ in ℓ_p ($1 \leq p < \infty$) in terms of the weights w_n is provided in [310]: $SC(B_w)$ is spaceable if and only if there exist a strictly increasing sequence $(m_j) \subset \mathbb{N}$ and a sequence $(q(n)) \subset \mathbb{N}$ such that

$$\liminf_{n \to \infty} \Big(\limsup_{m_j > n} \frac{\prod_{i=1}^{n} w_{m_j - n + i}}{\min_{q \leq q(n)} \prod_{i=1}^{n} w_{q+i}} \Big) = 0.$$

In particular, for the unweighted backward shift B, the set $SC(B)$ is not spaceable. The last property is also true on c_0: see the survey [311], in which one can find further examples.

4.8 Frequent hypercyclicity and lineability

In 2006, Bayart and Grivaux [56] introduced the following notion as a quantified, stronger form of hypercyclicity, connected to ergodic theory.

Definition 4.8.1. An operator T on a topological vector space X is said to be *frequently hypercyclic* provided there exists a vector $x_0 \in X$ such that

$$\underline{\text{dens}} \{ n \in \mathbb{N} : T^n x_0 \in U \} > 0 \quad \text{for every nonempty open subset } U \text{ of } X.$$

In this case, x_0 is called a *frequently hypercyclic vector* for T, and the set of these vectors will be denoted by $FHC(T)$.

Recall that if $A \subset \mathbb{N}$ then the "lower density" and the "upper density" of A are respectively defined as

$$\underline{\text{dens}} \, (A) = \liminf_{n \to \infty} (1/n) \, \text{card}(A \cap \{1, \dots, n\}) \quad \text{and}$$

$$\overline{\text{dens}} \, (A) = \limsup_{n \to \infty} (1/n) \, \text{card}(A \cap \{1, \dots, n\}),$$

where card (B) denotes, as usual, the cardinality of the set B. It turns out that wide families of operators which are known to be hypercyclic (including, as seen in Section 4.2, composition operators C_φ on $H(\mathbb{D})$ with φ an automorphism without fixed points, nonscalar differential operators $\Phi(D)$ on $H(\mathbb{C})$ and the Rolewicz operator λB with $|\lambda| > 1$ on c_0 or ℓ_p, $1 \leq p < \infty$, among others) are also frequently hypercyclic; see [56], [57], [135] and [136]. But there are hypercyclic operators that are not frequently hypercyclic. An example of this is the weighted backward shift B_w on ℓ_2 with $w = ((1 + n^{-1})^{1/2})$; see [56].

Let us consider now the topological size of $FHC(T)$. Even on F-spaces, there is not a result analogous to the hypercyclic case. In fact, the more popular operators (derivative operators, weighted backward shifts, composition operators) satisfy that their respective sets of frequent hypercyclic vectors are *not* residual; see [56] and [136]. In fact, several authors (see [64, 232, 312]) have proved that if T is an operator on a Banach space, then $FHC(T)$ is always *meager*. Nevertheless, by following the approach of Bourdon [144] and Wengenroth [379], Bayart and Grivaux [56] showed in 2006 that dense-lineability is kept for $FHC(T)$:

Theorem 4.8.2. *Let T be a frequently hypercyclic operator on a separable F-space X. There is a dense T-invariant linear manifold M of X, every nonzero vector of which is frequently hypercyclic for T.*

Proof. As in the proof of Theorem 4.3.1, the operator $P(T)$ has dense range in X for every nonzero polynomial P, because T is, in particular, hypercyclic. Fix any vector $x_0 \in FHC(T)$. As in Theorem 4.3.1, the set

$$M := \operatorname{span}\{P(T)x_0 : P \text{ is a polynomial}\}$$

is a dense T-invariant linear manifold of X. Let us show that each vector $x \in M \setminus \{0\}$ is frequently hypercyclic for T. To this end, select a (necessarily nonzero) polynomial P such that $x = P(T)x_0$ and fix a nonempty open set U in X. Since $P(T)$ and T commute, the T-orbit $\{T^n x : n \in \mathbb{N}\}$ of x is the image under $P(T)$ of the T-orbit $\{T^n x_0 : n \in \mathbb{N}\}$ of x_0. By continuity, the preimage $W := P(T)^{-1}(U)$ is a nonempty open set of X. Then $\overline{\operatorname{dens}}\{n \in \mathbb{N} : T^n x_0 \in W\} > 0$. But

$$\{n \in \mathbb{N} : T^n x_0 \in W\} = \{n \in \mathbb{N} : P(T)T^n x_0 \in U\} = \{n \in \mathbb{N} : T^n x \in U\}.$$

Consequently, $\overline{\operatorname{dens}}\{n \in \mathbb{N} : T^n x \in U\} > 0$, hence $x \in FHC(T)$, as desired. $\qquad\square$

In [56] it is raised the question of whether –analogously to the case of simple hypercyclicity– the set $\bigcap_k FHC(T_k)$ is dense-lineable, where all T_k ($k \in \mathbb{N}$) are frequently hypercyclic operators on the same Banach space X.

Similarly to the mere hypercyclicity, the existence of large closed subspaces of frequently hypercyclic vectors calls for additional assumptions. The following concept was original from Bayart and Grivaux [56]. We introduce a modified version due to Bonilla and Grosse-Erdmann [136].

Definition 4.8.3. Let T be an operator on a separable F-space X. Then T is said to satisfy the *frequent hypercyclicity criterion* (FHCC) if there is a dense subset X_0 of X and a map $S : X_0 \to X_0$ such that, for any $x \in X_0$, $TSx = x$ and both series $\sum_{n=1}^{\infty} T^n x$, $\sum_{n=1}^{\infty} S^n x$ converge unconditionally.

Recall that a series $\sum_{n=1}^{\infty} x_n$ in an F-space X converges unconditionally if for every $\varepsilon > 0$ there is some $N \in \mathbb{N}$ such that $\left\| \sum_{n \in F} x_n \right\| < \varepsilon$ for every finite subset $F \subset \{N+1, N+2, \dots\}$. Here $\|\cdot\|$ is an F-norm generating the topology of X. If T satisfies the FHCC then T is frequent hypercyclic, and in fact the FHCC is a powerful tool to check the property for the main kinds of operators. In 2012, Bonilla and Grosse-Erdmann [137] established the following theorem.

Theorem 4.8.4. *Let $T \in L(X)$, where X is a separable F-space with a continuous norm. Suppose that*

(i) *T satisfies the FHCC, and*

(ii) *there exists an infinite dimensional closed subspace M_0 of X such that $T^n x \to 0$ for all $x \in M_0$.*

Then the set $FHC(T)$ is spaceable.

Proof. We need to invoke the following independent fact about distribution of natural numbers, which can be found in [137, Lemma 2.5]: there are pairwise disjoint sets $A(l, \nu) \subset \mathbb{N}$ ($l, \nu \geq 1$) of positive lower density such that, for all $n \in A(l, \nu)$, $k \in A(\lambda, \mu)$ we have that $n \geq \nu$ and $|n - k| \geq \nu + \mu$ if $n \neq k$.

Denote by $\| \cdot \|$ an F-norm generating the topology of X. The proof will be divided into four steps.

Step 1. The following result will be demonstrated: Let $(e_j) \subset X$ be a sequence with

$$T^n e_j \to 0 \quad \text{as } n \to \infty \quad \text{for every } j \in \mathbb{N} \tag{4.21}$$

and $(\varepsilon_n) \subset (0, \infty)$. Then there exists a dense sequence (y_n) in X, a sequence $(f_n) \subset X$ and sets $E(l, j) \subset \mathbb{N}$ ($l \geq j \geq 1$) of positive lower density such that

$$\|f_j - e_j\| \leq \varepsilon_j, \tag{4.22}$$

$$\|T^n f_j - y_l\| \leq 2^{-l} \text{ for } n \in E(l, j) \text{ with } l \geq j \text{ and} \tag{4.23}$$

$$\|T^n (f_j - e_j)\| \leq \varepsilon_j 2^{-l} \text{ for } n \in E(l, j') \text{ with } l \geq j' \text{ and } j' \neq j. \tag{4.24}$$

To this end, let $S : X_0 \to X_0$ be a map provided by the FHCC. Since X is separable, X_0 contains a sequence (y_n) that is dense in X. Then the FHCC and (4.21) imply that there is a strictly increasing sequence $(N_l) \subset \mathbb{N}$ such that, for any $j \leq l$ and any finite set $F \subset \{N_l, N_l + 1, N_l + 2, \dots\}$, we have that

$$\left\| \sum_{n \in F} T^n y_j \right\| \leq \frac{\varepsilon_l}{l 2^{l+3}}, \tag{4.25}$$

$$\left\| \sum_{n \in F} S^n y_j \right\| \leq \frac{\varepsilon_l}{l 2^{l+3}}, \tag{4.26}$$

$$\|T^n e_j\| \leq \frac{1}{2^{l+1}} \text{ for } n \geq N_l. \tag{4.27}$$

We may assume that (ε_j) is decreasing with $\varepsilon_1 \leq 1$. Let us define $E(l, j) := A(l, N_{l+j})$ ($l \geq j \geq 1$) and $E_j := \bigcup_{l=j}^{\infty} E(l, j)$. Since (N_j) is strictly increasing, the sets E_j ($j \in \mathbb{N}$) are pairwise disjoint. In addition we have that

$$n \geq N_{l+j} \geq \max(N_l, N_j) \text{ if } n \in E(l, j), \text{ and} \tag{4.28}$$

$$|n - k| \geq \max(N_{l+j}, N_{\lambda+\tau}) \geq \max(N_l, N_j, N_\lambda, N_\tau) \tag{4.29}$$

if $n \in E(l, j)$, $k \in E(\lambda, \tau)$, $n \neq k$.

Setting

$$z_{j,n} = y_l \quad \text{if} \quad n \in E(l, j) \ (l \geq j), \tag{4.30}$$

we define

$$f_j = e_j + \sum_{n \in E_j} S^n z_{j,n} \quad \text{for } j \geq 1. \tag{4.31}$$

In order to see that the series in (4.31) converges in X, let $l \geq j$ and F a finite subset of $\{N_l, N_l + 1, N_l + 2, \dots\}$. Then

$$\sum_{n \in E_j \cap F} S^n z_{j,n} = \sum_{\lambda=j}^{l} \sum_{n \in E(\lambda,j) \cap F} S^n y_\lambda + \sum_{\lambda=l+1}^{\infty} \sum_{n \in E(\lambda,j) \cap F} S^n y_\lambda.$$

By (4.26) we have for $\lambda \leq l$ that $\| \sum_{E(\lambda,j) \cap F} S^n y_\lambda \| \leq \varepsilon_l / (l3^{l+3})$. Moreover, (4.28) entails $\| \sum_{E(\lambda,j) \cap F} S^n y_\lambda \| \leq \varepsilon_\lambda / 2^{l+3}$. It follows that $\| \sum_{E_j \cap F} S^n z_{j,n} \| \leq \varepsilon_l / 2^{l+2}$, which implies that the series in (4.31) converges unconditionally.

By the same argument, we have

$$\Big\| \sum_{n \in E_j, \, n \leq m} S^n z_{j,n} \Big\| \leq \sum_{\lambda=j}^{\infty} \Big\| \sum_{n \in E(\lambda,j), \, n \leq m} S^n y_\lambda \Big\| \leq \sum_{\lambda=j}^{\infty} \frac{\varepsilon_\lambda}{2^\lambda} \leq \varepsilon_j,$$

from which (4.22) is obtained just by letting $m \to \infty$.

Now let $n \in E(l, j)$ with $l \geq j \geq 1$. Then the first condition in the FHCC and (4.30) imply that

$$T^n f_j - y_l = T^n e_j + \sum_{k \in Ej, \, k<n} T^{n-k} z_{j,k} + \sum_{k \in Ej, \, k>n} S^{k-n} z_{j,k}. \tag{4.32}$$

We consider the three terms in turn. First, (4.28) and (4.27) imply that

$$\| T^n e_j \| \leq \frac{1}{2^{l+1}}. \tag{4.33}$$

From (4.30) we get

$$\sum_{k \in Ej, \, k<n} T^{n-k} z_{j,n} = \sum_{\lambda=j}^{l} \sum_{k \in E(\lambda,j), \, k<n} T^{n-k} y_\lambda + \sum_{\lambda=l+1}^{\infty} \sum_{k \in E(\lambda,j), \, k<n} T^{n-k} y_\lambda. \tag{4.34}$$

From (4.29) and (4.25) it follows that

$$\Big\| \sum_{k \in E(\lambda,j), \, k<n} T^{n-k} y_\lambda \Big\| \leq \frac{\varepsilon_l}{l2^{l+3}} \leq \frac{1}{l2^{l+3}} \quad \text{for } \lambda \leq l, \text{ and} \tag{4.35}$$

$$\Big\| \sum_{k \in E(\lambda,j),\, k<n} T^{n-k} y_\lambda \Big\| \le \frac{\varepsilon_\lambda}{\lambda 2^{\lambda+3}} \le \frac{1}{2^{\lambda+3}} \quad \text{for } \lambda \ge 1. \tag{4.36}$$

Hence, (4.34), (4.35) and (4.36) imply that

$$\Big\| \sum_{k \in E_j,\, k<n} T^{n-k} z_{j,k} \Big\| \le \frac{1}{2^{l+2}}. \tag{4.37}$$

In the same way, using (4.26) instead of (4.25), we see that $\big\| \sum_{k \in E_j,\, n<k\le m} S^{k-n} z_{j,k} \big\| \le \frac{1}{2^{l+2}}$ for all $m > k$, and thus, after letting $m \to \infty$,

$$\Big\| \sum_{k \in E_j,\, k>n} S^{k-n} z_{j,k} \Big\| \le \frac{1}{2^{l+2}}. \tag{4.38}$$

Altogether, (4.32), (4.33), (4.37) and (4.38) yield (4.23).

Next, let $n \in E(l,j')$ with $l \ge j$ and $j' \ne j$. Then $n \notin E_j$ and, thanks to the FHCC, we obtain that

$$T^n(f_j - e_j) = \sum_{k \in E_j,\, k<n} T^{n-k} z_{j,k} + \sum_{k \in E_j,\, k>n} S^{k-n} z_{j,k}. \tag{4.39}$$

For the first term in (4.39) we have with (4.30) that

$$\begin{aligned}
\sum_{k \in E_j,\, k<n} T^{n-k} z_{j,k} &= \sum_{\lambda \ge j,\, \lambda \le l} \ \sum_{k \in E(\lambda,j),\, k<n} T^{n-k} y_\lambda \\
&\quad + \sum_{\lambda \ge j,\, \lambda \ge l+1} \ \sum_{k \in E(\lambda,j),\, k<n} T^{n-k} y_\lambda.
\end{aligned} \tag{4.40}$$

From (4.29) and (4.26) it follows that if $\lambda \le l$ and $j \le l$, then

$$\Big\| \sum_{k \in E(\lambda,j),\, k<n} T^{n-k} y_\lambda \Big\| \le \frac{\varepsilon_l}{l 2^{l+3}} \le \frac{\varepsilon_j}{l 2^{l+2}}, \text{ and} \tag{4.41}$$

$$\Big\| \sum_{k \in E(\lambda,j),\, k<n} T^{n-k} y_\lambda \Big\| \le \frac{\varepsilon_{\lambda+j}}{(\lambda+j)2^{\lambda+j+3}} \le \frac{\varepsilon_j}{2^{\lambda+2}} \quad \text{for } \lambda \ge 1. \tag{4.42}$$

Hence, (4.40), (4.41) and (4.42) imply that

$$\Big\| \sum_{k \in E_j,\, k<n} T^{n-k} z_{j,k} \Big\| \le \frac{\varepsilon_j}{2^{l+1}}. \tag{4.43}$$

Once again, we use (4.26) instead of (4.25) to obtain $\big\| \sum_{k \in E_j,\, n<k\le m} S^{k-n} z_{j,k} \big\| \le \frac{\varepsilon_j}{2^{l+1}}$, hence

$$\Big\| \sum_{k \in E_j,\, k>n} S^{k-n} z_{j,k} \Big\| \le \frac{\varepsilon_j}{2^{l+1}}. \tag{4.44}$$

Altogether, (4.39), (4.43) and (4.44) yield (4.24), which completes the proof of the first step.

Step 2. Recall that X supports a continuous norm $\||\cdot\||$. Then $(X, \||\cdot\||)$ is a normed space and hence a dense subspace of a Banach space $(\widetilde{X}, \||\cdot\||)$ (its completion). The closure $\overline{M_0}$ of M_0 in X is an infinite dimensional Banach space. By Mazur's theorem there exists a basic sequence $(\widetilde{e}_n)$ in $\overline{M_0}$. Approximating these vectors from within M_0 we can find a sequence $\{e_n\}_{n\geq 1} \subset M_0$ forming a basic sequence in $\widetilde{X}$; see [184]. In particular, $T^n e_j \to 0$ as $n \to \infty$, for all $j \geq 1$.

Step 3. Let e_n^* $(n \geq 1)$ be continuous linear functionals on $(X, \||\cdot\||)$ that extend the coefficient functionals of the basis (e_n) in $\overline{\text{span}}\{e_n : n \in \mathbb{N}\}$, where the closure is taken in $\widetilde{X}$. Let $K_n = \max\{1, \||e_n^*\||\}$. Since X is continuously embedded in $\widetilde{X}$, from Step 1 there exist a dense sequence (y_n) in X, a sequence $(f_n) \subset X$ and sets $E(l,j)$ $(l \geq j \geq 1)$ of positive lower density such that

$$\max(\|f_j - e_j\|, \||f_j - e_j\||) \leq \frac{1}{2^{j+1}K_j}, \tag{4.45}$$

$$\|T^n f_j - y_l\| \leq \frac{1}{2^l} \quad \text{for } n \in E(l,j),\, l \geq j, \tag{4.46}$$

$$\|T^n(f_j - e_j)\| \leq \frac{1}{2^{j+l}K_j} \quad \text{for } n \in E(l,j'),\, l \geq j',\, j' \neq j. \tag{4.47}$$

By (4.45), $\sum_{j=1}^{\infty} \||e_j^*\|| \, \||f_j - e_j\|| \leq 1/2 < 1$. Hence (f_n) is a basic sequence in $\widetilde{X}$ whose coefficient functionals satisfy (see [184, Theorem 2.9] and [237, Lemma 10.6])

$$\||f_j^*\|| \leq 2K_j \quad \text{for all } j \geq 1. \tag{4.48}$$

Step 4. We denote by M the closed linear span (in X) of the vectors f_n, $n \geq 1$. Since these vectors are linearly independent, M is an infinite dimensional closed subspace of X. We claim that M is a frequently hypercyclic subspace for T. To this end we first consider a finite linear combination $z := \sum_{j=1}^{N} a_j f_j$ with $a_m = 1$ for some m. Define $w := \sum_{j \leq N,\, j \neq m} a_j e_j$. Let $l \geq m$ and $n \in E(l,m)$. We obtain from (4.46), (4.47) and (4.48) that

$$\|T^n z - y_l\| \leq \|T^n f_m - y_l\| + \Big\| \sum_{j \leq N,\, j \neq m} a_j T^n(f_j - e_j)\Big\| + \|T^n w\|$$

$$\leq \frac{1}{2^l} + \sum_{j \leq N,\, j \neq m} (|a_j| + 1)\|T^n(f_j - e_j)\| + \|T^n w\|$$

$$\leq \frac{2}{2^l} + \sum_{j=1}^{\infty} \||f_j^*\|| \, \||z\|| \frac{1}{2^{j+l}K_j} + \|T^n w\| \leq \frac{2}{2^l} + \frac{2}{2^l}\||z\|| + \|T^n w\|. \tag{4.49}$$

Now let $z \in M \setminus \{0\}$ be arbitrary. We want to show that $z \in FHC(T)$. As z belongs to the closed linear span of the f_n in $\widetilde{X}$, there are scalars a_j such that

$z = \sum_{j=1}^{\infty} a_j f_j$, with convergence in $\widetilde{X}$. At least one a_m must be non-zero, so that we can assume that $a_m = 1$. By the definition of M there are vectors $z_\nu = \sum_{j=1}^{N_\nu} a_{\nu,j} f_j$ ($\nu \geq 1$) with $z_\nu \to z$ in X. Since we also have convergence in $\widetilde{X}$ we have that $a_{\nu,m} \to a_m$, so that we can assume that $a_{\nu,m} = 1$ for all ν. Define the vectors $w_\nu = \sum_{j \leq N_\nu,\, j \neq m} a_{\nu,j} e_j \in M_0$. Setting $a_{\nu,j} = 0$ for $j > N_\nu$, we find for $\nu, \mu \geq 1$ that

$$\|(w_\nu - z_\nu) - (w_\mu - z_\mu)\| \leq \sum_{j \neq m} \|(a_{\nu,j} - a_{\mu,j})(e_j - f_j)\|.$$

Since $(a_{\nu,j})$ converges for all $j \geq 1$ and since, by (4.45) and (4.48),

$$\|(a_{\nu,j} - a_{\mu,j})(e_j - f_j)\| \leq (|a_{\nu,j} - a_{\mu,j}| + 1)\|e_j - f_j\|$$

$$\leq \|f_j^*\| \, \|z_\nu - z_\mu\| \|e_j - f_j\| + \frac{1}{2^{j+1}} \leq \frac{\|z_\nu - z_\mu\|}{2^j} + \frac{1}{2^{j+1}} \leq \frac{C}{2^j},$$

the dominated convergence theorem implies that $(w_\nu - z_\nu)$ is a Cauchy sequence in X. Hence (w_ν) converges in X to some vector X, which necessarily belongs to M_0.

Now, applying (4.49) to z_ν and w_ν and letting $\nu \to \infty$ we have by continuity that

$$\|T^n z - y_l\| \leq \frac{2}{2^l} + \frac{2}{2^l}\|z\| + \|T^n w\|.$$

Since $\{y_l\}_{l \geq m}$ is dense in X, $T^n w \to 0$ as $n \to \infty$ and the sets $E(l,m)$ $(l \geq m)$ are of positive lower density, we obtain that $z \in FHC(T)$, as desired. $\square$

As a nice application of Theorem 4.8.4, Bès [111] has proved that if $G \subset \mathbb{C}$ is a simply connected domain and $\varphi : G \to G$ is a univalent holomorphic function, then the set $FHC(C_\varphi)$ is spaceable. For negative results, see the Notes and Remarks section.

4.9 Distributional chaos and lineability

In this section, we deal briefly with a related notion of chaos presenting some similarities with that of hypercyclicity, but revealing deep differences at the same time. The notion was introduced in 1994 by Schweizer and Smítal [364]. According to [364], if X is a metric space with distance d, a continuous map $T : X \to X$ is said to be *distributionally chaotic* if there exist an uncountable set $\Gamma \subset X$ and $\varepsilon > 0$ such that for every $\tau > 0$ and each pair of distinct points $x, y \in \Gamma$, we have that

$$\overline{\mathrm{dens}}\,\{n \in \mathbb{N} : d(T^n x, T^n y) < \tau\} = 1 \text{ and } \underline{\mathrm{dens}}\,\{n \in \mathbb{N} : d(T^n x, T^n y) < \varepsilon\} = 0.$$

In 1988, Beauzamy [67] coined the following notion in the setting of Banach spaces which, of course, can be given within Fréchet spaces.

Definition 4.9.1. If X is a Banach space and $T \in L(X)$, then a vector $x_0 \in X$ is called *irregular* for T provided that the sequence $(T^n x_0)$ is unbounded but it has a subsequence tending to 0.

The set of such vectors x_0 will be denoted by $Irr(T)$. Inspired by both notions of distributionally chaotic maps and irregular vectors, Bermúdez, Bonilla, Martínez-Giménez and Peris [69] presented in 2011 the following concept.

Definition 4.9.2. If X is a Banach space and $T \in L(X)$ then a vector $x_0 \in X$ is said to be *distributionally irregular* if there are increasing sequences $A = (n_k)$, $B = (m_k) \subset \mathbb{N}$ such that

$$\overline{\operatorname{dens}}(A) = \overline{\operatorname{dens}}(B) = 1, \quad \lim_{k \to \infty} \|T^{n_k} x_0\| = 0 \quad \text{and} \quad \lim_{k \to \infty} \|T^{m_k} x_0\| = \infty.$$

We denote by $DI(T)$ the set of distributionally irregular vectors for T. If $DI(T) \neq \varnothing$ then T is distributionally chaotic [69], and the converse is also true [107]. A number of lineability results have been obtained in [69]. The next theorem gathers a selection of them.

Theorem 4.9.3.   (a) *Assume that X is a Banach space, that $T \in L(X)$ and that there exists a dense subset $X_0 \subset X$ such that $\lim_{n \to \infty} T^n x = 0$. Then $DI(T)$ is dense-lineable if at least one of the following conditions holds:*

(i) *$DI(T) \neq \varnothing$.*

(ii) *There exist an increasing sequence $A = (n_k) \subset \mathbb{N}$ and a vector $y \in X$ satisfying $\overline{\operatorname{dens}}(A) = 1$ and $\lim_{k \to \infty} \|T^{n_k} y\| = \infty$.*

(iii) *There exists an increasing sequence $A = (n_k) \subset \mathbb{N}$ with $\overline{\operatorname{dens}}(A) = 1$ and $\sum_{n=1}^{\infty} \frac{1}{\|T_n\|} < \infty$.*

(b) *If X is infinite dimensional and separable, there exists a hypercyclic and distributionally chaotic operator T such that $DI(T)$ is dense-lineable.*

Recently, the authors of [298] have constructed a hypercyclic operator T and a non-hypercyclic operator S such that *every* nonzero vector is distributionally irregular for each of them. Concerning Beauzamy's irregular vectors, in [108] Bernardes *et al.* have proved, among other results, the following criteria for dense-lineability of $Irr(T)$. The first of them extends the corresponding one given in [69] for Banach spaces.

Theorem 4.9.4.   (a) *Suppose that X is a Fréchet space and that $T \in L(X)$ is an operator satisfying the following conditions:*

(i) *There is a dense subset X_0 of X such that $\lim_{n \to \infty} T_n x = 0$ for all $x \in X_0$.*

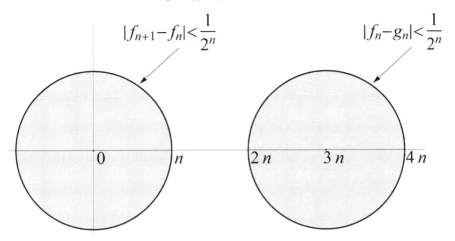

$$|f_{n+1}-f_n|<\frac{1}{2^n} \qquad |f_n-g_n|<\frac{1}{2^n}$$

$0 \qquad n \qquad 2\,n \qquad 3\,n \qquad 4\,n$

FIGURE 4.3: Construction of a Birkhoff-universal entire function.

(ii) *There is a bounded sequence $(a_n) \subset X$ such that the sequence $(T_n a_n)$ is unbounded.*

Then $Irr(T)$ is dense-lineable.

(b) *Suppose that X is a Banach space and that $T \in L(X)$ is an operator satisfying the following conditions:*

(i) *There are a dense subset X_0 of X and a sequence $(n_k) \subset \mathbb{N}$ with $\lim_{k\to\infty} T_{n_k} x = 0$ for all $x \in X_0$.*

(ii) $\sup_{n\geq 1} \|T^n\| = \infty.$

Then $Irr(T)$ is dense-lineable.

4.10 Exercises

Exercise 4.1. Construct a universal entire function in the sense of Birkhoff, that is, a τ_1-hypercyclic function $f \in H(\mathbb{C})$.

Hint: Select any dense sequence (h_n) in $H(\mathbb{C})$ and define $g_n(z) := h_n(z - 3n)$. Then choose $f_1 := g_1$ and use induction and Runge's approximation theorem to find entire functions f_n $(n \geq 1)$ satisfying $|f_{n+1} - f_n| < 2^{-n}$ on the ball $\overline{B}(0, n)$ and $|f_n - g_n| < 2^{-n}$ on the ball $\overline{B}(3n, n)$, which is disjoint with the first ball (see Figure 4.3). Finally, define $f := f_1 + \sum_{n=1}^{\infty}(f_{n+1} - f_n) = \lim_{n\to\infty} f_n$.

Exercise 4.2. Assume that $\Phi(z) := \sum_{n=0}^{\infty} a_n z^n$ is an entire function. Show that $\Phi(D) : f \in H(\mathbb{C}) \mapsto \sum_{n=0}^{\infty} a_n f^{(n)} \in H(\mathbb{C})$ is a well-defined operator if Φ is of exponential type. Suppose now that $G \subset \mathbb{C}$ is a domain and that Φ is of subexponential type, that is, given $\varepsilon > 0$, there is a constant positive $K = K(\varepsilon)$ such that $|\Phi(z)| \leq Ke^{\varepsilon|z|}$ for all $z \in \mathbb{C}$. Show that $\Phi(D) : H(G) \to H(G)$ is a well-defined operator.
Hint: By applying Cauchy's inequalities on Taylor coefficients, prove that Φ is of exponential type if and only if the sequence $\{(n!|a_n|)^{1/n}\}_{n \geq 1}$ is bounded, and that Φ is of subexponential type if and only if $\lim_{n \to \infty} (n!|a_n|)^{1/n} = 0$.

Exercise 4.3. Prove that if X is a real Banach space, f is a real entire function and $T \in L(X)$, then $\widetilde{f(T)} = f(\widetilde{T})$.

Exercise 4.4. Prove Proposition 4.3.4.
Hint: Assume that $\bigcap_{\lambda \in \Lambda} HC(T_\lambda) \neq \varnothing$ and that $T_\lambda T_{\lambda_0} = T_{\lambda_0} T_\lambda$ for all $\lambda \in \Lambda$. Pick a vector x_0 in the last intersection. By the proof of Theorem 4.3.1, $P(T_{\lambda_0})$ has dense range for any nonzero polynomial P. Show that $\{P(T_{\lambda_0})x_0 : P \text{ polynomial}\}$ is a dense subspace every nonzero member of which is T_λ-hypercyclic for every $\lambda \in \Lambda$.

Exercise 4.5. If $\{\Delta_{\alpha_n}\}_{n \geq 1}$ is a sequence of diagonal operators on $H(\mathbb{C})$ and M is a nontrivial linear manifold contained in $\mathcal{U} := \mathcal{U}((\Delta_{\alpha_n})) \cup \{0\}$, show that $\dim(M) = 1$.
Hint: Prove first that if $f(z) := \sum_{k=0}^{\infty} f_k z^k \in \mathcal{U}((\Delta_{\alpha_n}))$ then $f_k \neq 0$ for all $k \geq 0$; in particular $f_0 \neq 0$. Proceed by contradiction by assuming the existence of two functions $f, g \in \mathcal{U}$ such that $f + \lambda g \in \mathcal{U}$ for every $\lambda \in \mathbb{C}$.

Exercise 4.6. The key point in the original proof of Bourdon [144] is that, if E is a locally convex space and T is a hypercyclic operator on it, then $\sigma_p(T^*) = \varnothing$. Moreover, for any $T \in L(E)$, we have that $T - \lambda I$ has dense range for all $\lambda \in \mathbb{K}$ if and only if $\sigma_p(T^*) = \varnothing$. Prove these facts.
Hint: For the second fact, use the Hahn–Banach theorem.

Exercise 4.7. Prove that the sets of functions which are respectively universal for the sequences (C_{φ_n}), (C_{ψ_n}), $(\Phi_n(D))$ considered in Section 4.2 are dense-lineable if the restrictions imposed there hold.
Hint: Use Theorem 4.3.5.

Exercise 4.8. Demonstrate that the set of universal Taylor series is residual in $H(\mathbb{D})$.
Hint: Try to express this set of an appropriate countable intersection of subsets, and prove that each of them is dense by using Mergelyan's approximation theorem together with the fact that the nth Taylor partial sum of a polynomial P equals P as soon as $n \geq \text{degree}(P)$.

Exercise 4.9. Give a rapid proof of Bayart's theorem [54] establishing the dense-lineability in $H(\mathbb{D})$ of the class of universal Taylor series.
Hint: Apply appropriately Theorem 4.4.2.

Exercise 4.10. Assume that X is a separable F-space and that T is an operator on X satisfying the hypotheses of Theorem 4.5.2. Suppose, in addition, that there is a dense subset $D \subset X$ such that $(T^n x)$ converges for every $x \in D$. Prove that $HC(T)$ is maximal dense-lineable in X.

Hint: By Theorem 4.5.2, $HC(T)$ is spaceable, so there is an infinite dimensional F-space $A \subset HC(T) \cup \{0\}$. Since A is complete, we have $\dim(A) = \mathfrak{c}$. Also, $B := \mathrm{span}(D)$ is a dense vector subspace of X satisfying that $(T^n x)$ converges for every $x \in B$. Now, try to add an appropriate multiple of each x of a Hamel basis of A to a vector of B, and consider the corresponding span.

Exercise 4.11. Prove that, in Theorem 4.5.2, we can replace the condition $T^{n_k} x \to 0$ for all $x \in M_0$ by the mere pointwise convergence on M_0.

Exercise 4.12. This is a more intricate exercise, addressed to lovers of complex variables. Demonstrate Theorem 4.5.1 as a consequence of Theorem 4.5.8 in the case in which the domain G is simply connected.

Hint: Of course, only one sequence $(T_n) = (C_{\varphi_n})$ is enough when applying Theorem 4.5.8. Firstly, show that (C_{φ_n}) satisfies the UC for some sequence (m_n), which can be supposed to be the whole $\mathbb{N}$. Suppose, without loss of generality, that $G \subset \mathbb{D}$. Choose any sequence of compact sets (K_n) with $K_1 = \overline{\mathbb{D}} \subset K_2^0 \subset K_2 \subset K_3^0 \subset \cdots \subset G$. From the runaway condition, select compact sets L_k with connected complement and a sequence $\{n_1 < n_1 < \cdots < n_k < \cdots\} \subset \mathbb{N}$ such that $\varphi_{n_k}(L_k) \cap L_k = \varnothing$, $\varphi_{n_k}(L_k) \cup L_k$ has connected complement and $K_k \cup L_{k-1} \cup \varphi_{n_{k-1}}(L_{k-1}) \subset L_k$, where $L_0 := \varnothing$ and $n_0 := 0$. Consider the functions $e_n(z) := z^n$ $(n \geq 1)$, which form an orthonormal system (so a basic sequence) of the Hilbert space $L^2(\mathbb{T}) = \{f : \mathbb{T} \to \mathbb{C} : f$ is measurable and $\int_0^{2\pi} |f(e^{i\theta})| \, d\theta < \infty\}$, with norm $\|f\|_2 = (\int_0^{2\pi} |f(e^{i\theta})|^2 \frac{d\theta}{2\pi})^{1/2}$. Now, apply Arakelian's approximation theorem (see Section 2.1 or [205]) to obtain functions f_j $(j \geq 1)$ such that $\sup\{|f_j(z)| : z \in \varphi_{n_m}(L_m)\} < 1/2^{j+m}$ $(m \geq 1)$ and $\sup\{|e_j(z) - f_j(z)| : z \in L_1\} < 1/2^{j+1}$. Deduce that $\sum_{j=1}^{\infty} \|e_j - f_j\|_2 \leq 1/2$ and, from the basis perturbation theorem, derive that (f_n) is a basic sequence in $L^2(\mathbb{T})$ with $\|f_n^*\| \leq 2$. Define $M_0 := \overline{\mathrm{span}}\{f_n : n \in \mathbb{N}\}$, where the closure is taken in $H(G)$. Show that $C_{\varphi_{n_k}} g \to 0$ in $H(G)$ for any $g \in M_0$.

Exercise 4.13. From Theorem 4.5.9, derive the spaceability of the set $HC(\Phi(D))$ in $H(\mathbb{C})$, where Φ is an entire function with exponential type.

Hint: Use Picard's theorem asserting that for any $\lambda \in \mathbb{C}$ with at most one exception the equation $\Phi(z) = \lambda$ has infinitely many solutions; see, e.g., Ahlfors' book [2].

Exercise 4.14. Show the spaceability of $HC(D)$ for the differentiation operator D on $H(\mathbb{C})$.

Hint: By using the density of the set of polynomials in $H(\mathbb{C})$, prove that D satisfies the Hypercyclicity Criterion for the full sequence $\mathbb{N}$. After this, choose a strictly increasing sequence $(n_k) \subset \mathbb{N}$ such that $n_{k+1} \geq C_{n_k}$ for all $k \geq 1$, where the constants C_n have been selected so that $x^n \leq 2^x$ for all $x \geq C_n$.

Then define M_0 as the closed subspace of $H(\mathbb{C})$ of all entire functions of the form $f(z) = \sum_{k=1}^{\infty} a_k z^{n_k - 1}$, and apply Theorem 4.5.2.

Exercise 4.15. Prove Montes' result asserting that $HC(2B)$ is not spaceable in ℓ_2, where B is the unweighted backward shift.
Hint: Use Corollary 4.5.7.

Exercise 4.16. Let $G \subset \mathbb{C}$ be a domain and (φ_n) a sequence of holomorphic self-mappings of G. Show that if $f \in HC((C_{\varphi_n}))$ then $f^2 \in HC((C_{\varphi_n}))$.
Hint: Fix a point $z_0 \in G$ as well as a closed disc $K = B(z_0, r) \subset G$, and consider the function $g(z) := z - z_0 \in H(G)$. Proceed by way of contradiction and suppose that there is a sequence $\{n_1 < n_2 < \cdots\} \subset \mathbb{N}$ with $(f(\varphi_{n_j}(z)))^2 \to g(z)$ uniformly on K. Now, invoke Hurwitz's theorem (see, e.g., [2]) asserting that if $(f_j) \subset H(G)$, $h \in H(G)$ and $f_j \to h$ uniformly on K then there must exist j such that f_j and h have the same number of zeros (counted with their multiplicity) in K.

Exercise 4.17. Adapt the proof of Theorem 4.8.4 so as to weaken the condition "there exists an infinite dimensional closed subspace M_0 with $T^n x \to 0$ for all $x \in M_0$" to "there exist a set $A \subset \mathbb{N}$ with $\underline{\mathrm{dens}}\,(A) > 0$ and an infinite dimensional closed subspace M_0 with $T^n x \xrightarrow[n \to \infty]{n \in A} 0$ for all $x \in M_0$."

Exercise 4.18. Let X be a topological vector space and $T, S \in L(X)$. The operators T, S are called *conjugate* if there is $R \in L(X)$ invertible (that is, R is bijective and $R^{-1} \in L(X)$) such that $S = RTR^{-1}$. Prove that $HC(T)$ ($FHC(T)$, resp.) is spaceable if and only if $HC(S)$ ($FHC(S)$, resp.) is spaceable.

Exercise 4.19. Show that the set $\mathcal{F}$ described in Theorem 4.5.11 is maximal dense-lineable in $\mathcal{C}(\mathbb{T})$.
Hint: Trigonometrical polynomials (that is, polynomials in e^{it}, e^{-it}) with coefficients in $\mathbb{Q} + i\mathbb{Q}$ form a countable dense subset of $\mathcal{C}(\mathbb{T})$, say $\{h_n\}_{n \geq 1}$. Since $\mathcal{F}$ is spaceable, Baire's theorem yields the existence of a vector subspace $L \subset \mathcal{C}(\mathbb{T})$ with $L \subset \mathcal{F} \cup \{0\}$ and $\dim(L) = \mathfrak{c}$. Select a linearly independent set $\{g_\alpha\}_{\alpha \in I} \subset L$ such that $\mathrm{card}(I) = \mathfrak{c}$. There is a surjective mapping $\sigma : I \to \mathbb{N}$. Given $\alpha \in I$, define $\varphi_\alpha := \frac{g_\alpha}{n \|g_\alpha\|_\infty}$, where $n = \sigma(\alpha)$. Define $f_\alpha := \varphi_\alpha + h_n$. Prove that $M_0 := \mathrm{span}\,\{f_\alpha\}_{\alpha \in I}$ satisfies $\dim(M_0) = \mathfrak{c}$ and $M_0 \subset \mathcal{F} \cup \{0\}$.

4.11   Notes and remarks

Section 4.2. Hypercyclicity of operators similar to differential operators $\Phi(D)$ but on spaces of holomorphic functions in infinite dimensional complex Banach spaces has been investigated in [19], [109] and [152]. Moreover, additional hypercyclicity results for C_φ and (C_{φ_n}) on other domains in $\mathbb{C}$ and for other kinds of self-mappings can be found in [101], [308] and [236].

Shkarin [367] has characterized those inductive limits of sequences of separable Banach spaces which support a hypercyclic operator. See, e.g., Horváth's book [254] for the concept and properties of inductive limits.

If X is an F-space, an operator $T \in (X)$ is called *weakly mixing* whenever the operator $T \oplus T : (x, y) \in X \times X \mapsto (Tx, Ty) \in X \times X$ is hypercyclic. It is elementary that T is hypercyclic if T is weakly mixing. In 1992, Herrero [249] posed the problem of whether the reciprocal is true, and some years later León and Montes [283] raised the question of whether every hypercyclic operator satisfies the HCC. In 1999, Bès and Peris [116] proved that both problems are in fact equivalent. This question has been the "great open problem in hypercyclicity" for a long time, and has served as a primary motivation for a decade-long development of the theory. Finally, De la Rosa and Read [181] settled the problem in the negative. Several examples of hypercyclic operators defined on classical spaces and not satisfying the HCC have been provided by Bayart and Matheron in [60].

Section 4.3. Concerning the invariant subspace problem, in the recent paper [228] Goliński has given examples of operators S without nontrivial proper invariant subspaces on classical non-Banach spaces X. Hence the set of S-cyclic vectors is $X \setminus \{0\}$.

Exercise 4.10 is taken from a result of [104]. It is an open problem the maximal dense-lineability of $HC(T)$ for any operator T on an arbitrary F-space X. Recall that, by Theorem 4.3.2, the answer is affirmative if X is Banach.

Concerning dense-lineability of sets of Birkhoff-universal functions, it is worth mentioning that with more sophisticated techniques –invoking, for instance, Arakelian's approximation theorem– Bernal, Calderón and Luh established the following theorem about entire functions presenting rapid decay along a fixed large set $A \subset \mathbb{C}$. Its proof can be found in [94] (see also [151]). Recall that an Arakelian subset of $\mathbb{C}$ is a closed subset A such that $\mathbb{C}_\infty \setminus A$ is both connected and locally connected, and the *inner radius* of a set $A \subset \mathbb{C}$ is $\rho(A) := \sup\{r > 0 : \text{there exists a ball } B \text{ with radius } r \text{ such that } B \subset A\}$.

Theorem. *Assume that A is an unbounded subset of $\mathbb{C}$. Then the following properties are equivalent:*

(a) *There is an Arakelian subset $A_0 \subset \mathbb{C}$ such that $A \subset A_0$ and $\varrho(\mathbb{C} \setminus A_0) = \infty$.*

(b) *Given two functions $\varphi, \psi : [0, +\infty) \to (0, +\infty)$ such that φ is continuous and integrable on $[1, +\infty)$, and ψ is increasing, there exists a vector subspace $M = M(A, \varphi, \psi) \subset E$ satisfying that M is dense in E, each $f \in M \setminus \{0\}$ is Birkhoff-universal, $\lim_{\substack{z \to \infty \\ z \in A}} \exp(\varphi(|z|)|z|^{3/2})f(z) = 0$ and $\lim_{r \to +\infty} \frac{\max\{|f(z)| : |z| = r\}}{\psi(r)} = \infty$ for all $f \in M \setminus \{0\}$.*

In turn, the results of [94] complete some due to Calderón [149] about universality and the property of tending to zero on strips. Even the action of certain differential operators is proved to keep rapid decay in [94, 149]. To sum up, we can say that rapid decay (see Section 2.5) is compatible with Birkhoff-universality in order to generate lineability.

In the setting of harmonic functions, Armitage [15] in 1998 gave a direct proof of the following lineability result about functions with a prescribed growth condition.

Theorem. *Let $\mathcal{H}_N$ be the space of all harmonic functions $\mathbb{R}^N \to \mathbb{R}$, endowed with the compact-open topology. If $\alpha = (\alpha_1, \ldots, \alpha_N) \in \mathbb{N}_0^N$, denote $|\alpha| = \alpha_1 + \cdots + \alpha_N$ and let $\partial^\alpha = \frac{\partial^{|\alpha|}}{\partial x_1^{\alpha_1} \cdots \partial x_N^{\alpha_N}} : \mathcal{H}_N \to \mathcal{H}_N$ be the corresponding partial derivative operator. Let σ be the $(N-1)$-dimensional measure on the unit sphere S of $\mathbb{R}^N$, normalized so that $\sigma(S) = 1$. For every $h \in \mathcal{H}_N$ and every $\phi : (0, +\infty) \to (0, +\infty)$ with $\lim_{r \to +\infty} \phi(r) = +\infty$, denote $M_2(h, r) = \left(\int_S (h(rx))^2 \, d\sigma(x) \right)^{1/2}$ and $\mathcal{H}_N(\phi) = \{h \in \mathcal{H}_N : M_2(h, r) = O(\phi(r) r^{-(N-1)/2} e^r) \text{ as } r \to \infty\}$. Then the set $\mathcal{U}(\{\partial^\alpha : \alpha \in \mathbb{N}_0^N\})$ is denselineable in $\mathcal{H}_N$.*

An explicit construction of a common dense vector space of hypercyclic vectors for a countable family of weighted backward shifts acting on a Banach space supporting a Schauder basis is given by Seoane in [361].

Section 4.4. One can obtain dense-lineable sets of universal Taylor series satisfying, simultaneously, other universality properties. In particular, some kind of wild behavior near the boundary (see Section 2.6) is compatible with the property of being a universal Taylor series. For instance, Bernal, Bonilla, Calderón and Prado-Bassas [88] showed in 2009 that the family of universal Taylor series f having maximal cluster set $C(f, \gamma)$ along any curve $\gamma \subset \mathbb{D}$ tending to $\partial \mathbb{D}$ whose closure does not contain $\partial \mathbb{D}$ is dense-lineable in $H(\mathbb{D})$. Incidentally, in [98] it has been shown the maximal dense-lineability in $H(G)$ of the class of functions $f \in \mathcal{U}((C_{\varphi_n}))$ satisfying that boundary property, where G is a Jordan domain and (C_{φ_n}) is the sequence of composition operators generated by adequate holomorphic self-mappings $\varphi_n : G \to G$.

Charpentier and Mouze [160] have recently obtained, under certain conditions, the existence of dense subspaces and of closed infinite dimensional subspaces of universal Taylor series in $H(G)$ –where G is a simply connected domain– whose images under some regular matrix summability methods are automatically universal. Of course, for a domain G which is not necessarily $\mathbb{D}$, a point $\zeta \in G$ should be fixed in order to consider the Taylor series.

Section 4.5. It is worth mentioning that K.C. Chan [156] introduced in 1999 a new technique to study hypercyclic operators. His main idea is to endow $L(X)$ with the strongly operator topology (SOT), which is nothing but the topology of the pointwise convergence, under which $L(X)$ becomes a separable locally convex space if one assumes that X is a separable Banach space (note that $L(X)$ under the norm topology is not separable, unless X is finite dimensional). Given $T \in L(X)$, the left-multiplication operator $L_T : S \in L(X) \mapsto TS \in L(X)$ turns out to be SOT-continuous. In addition, it is proved in [156] (see also [309]) that L_T is SOT-hypercyclic on $L(X)$ if T satisfies the HCC (the reverse is proved by Martínez and Peris [299], who use tensor products). With the help of these properties, Chan and Taylor [157] prove Theorem 4.5.2 in a remarkable simpler way.

Regarding Theorem 4.5.8, in 2009 Bès and Martin [113] isolated geometric characteristics of operators under which the conclusion of the mentioned theorem holds. Specifically, for a member $T \in L(X, Y)$, where X and Y are Banach spaces, denote $C(T) = \inf\{\|TJ_W\| : W \in \mathcal{J}\}$, where $\mathcal{J}$ stands for the collection of all closed finite codimensional subspaces of X, and J_W is the canonical inclusion $W \hookrightarrow X$. Then it is proved in [113] that if $\{(T_{j,n})_{n\geq 1} : j \in \mathbb{N}\}$ is a countable collection of hereditarily densely universal sequences in $L(X, Y)$ satisfying $\sup\{C(T_{j,n}) : n \geq 1\} < \infty$ for every $j \in \mathbb{N}$, then $\bigcap_{j=1}^{\infty} \mathcal{U}((T_{j,n})_{n\geq 1})$ is spaceable.

As shown by Episkoposian and Müller in [198], Theorem 4.5.11 about partial sums of Fourier series has its counterpart in the setting of Walsh–Fourier series $\sum_{n=0}^{\infty} \hat{f}(n)W_n(x)$, where (W_n) denote the system of Walsh–Paley functions on the interval $[0, 1)$ defined by $W_0 := 1$ and $W_n(x) := \prod_{j=0}^{r}(\text{sign } \sin(2^{j+1}\pi x))^{\varepsilon_j}$ $(n \geq 1)$, where $\varepsilon_0, \ldots, \varepsilon_r \in 0, 1$ are the digits in the representation of n in the dyadic system, i.e. $n = \sum_{j=0}^{r} \varepsilon_j 2^j$.

In the special case of the space ω, H. Salas [356] obtained in 2011 common spaceability for countable many operators given by upper triangular matrices satisfying appropriate conditions. This extends results by Conejero and Bès [112].

A generalization of Theorem 4.5.13 to simply connected domains has been stated by Charpentier [158] in 2010. In fact, in [158] the Bayart *et al.* result [58] asserting the dense-lineability of $\mathcal{U}_A$ as soon as $\mathcal{U}_A \neq \varnothing$ (Theorem 4.4.2) is completed by showing that if X is a Banach space then $\mathcal{U}_A$ is spaceable as soon as $\mathcal{U}_A \neq \varnothing$. In 2011, Menet [301] has proved the assertion if X is a Fréchet space admitting a continuous norm, and the same author [302] has recently shown the same result for Fréchet spaces admitting a continuous seminorm p with $\text{codim}(\ker p) = \infty$. The spaceability of Padé universal series, where Taylor series are replaced by Padé approximants, is analyzed in [161]. In [88] and [98], respectively, the combined properties of maximality of cluster sets in each boundary point with either universality of Taylor series or compositional universality in $H(G)$ (where G is a Jordan domain in $\mathbb{C}$) considered therein are proved to give spaceability. As a related result, in [97] it is provided a large family of classical operators (for instance, differential or composition operators) $T : H(G) \to H(G)$ (with G a domain in $\mathbb{C}$) satisfying that, for any subset $A \subset G$ which is not relatively compact in G, the set $\{f \in H(G) : \overline{(Tf)(A)} = \mathbb{C}\}$ is spaceable. Nevertheless, the spaceability of the family of holomorphic monsters in G considered in Section 4.4 seems to be an open problem.

Existence of closed vector spaces of infinitely differentiable functions $\mathbb{R} \rightarrow \mathbb{R}$ (and other more restrictive functions) which are Fekete-universal –that is, whose Taylor sums approximate any given continuous function $h : \mathbb{R} \rightarrow \mathbb{R}$ with $h(0) = 0$– is analyzed in [159]. In this reference, a characterization of the spaceability of the set of universal series in the Fréchet space $\mathbb{K}^{\mathbb{N}}$ is also given.

An operator $T : H(\mathbb{C}) \rightarrow H(\mathbb{C})$ is said to be a convolution operator whenever T commutes with all translation operators $f \mapsto f(\cdot + a)$. It can be proved that T is a convolution operator if and only if $T = \Phi(D)$ for some entire function Φ of exponential type (see, e.g., [227]). But non-convolution operators have been also studied from the point of view of linear dynamics. For instance, Aron, Markose, Fernández and Hallack [32, 201] proved that the operator $T_{\lambda,b}$ on $H(\mathbb{C})$ given by $T_{\lambda,b}f(z) = f'(\lambda z + b)$ is hypercyclic whenever $|\lambda| \geq 1$, while León and Romero de la Rosa [285] have shown that this condition is also necessary. Moreover, it is shown in [285] that $HC(T_{\lambda,b})$ is spaceable if and only if $|\lambda| = 1$.

Another interesting aspect of the dynamics of operators is that of disjoint hypercyclicity. This notion was introduced, independently, in [80] and [117]. Given a topological vector space X and operators $T_1, \ldots, T_N \in L(X)$, these operators are said to be *disjoint hypercyclic* (or *d-hypercyclic*) provided that there is a vector $x_0 \in X$, called d-hypercyclic for $\{T_1, \ldots, T_N\}$, satisfying that the set $\{(T_1^n x_0, \ldots, T_N^n x_0) : n \geq 1\}$ is dense in X^N. Bès, Martin and Sanders have provided in [114] sufficient conditions for the spaceability of the family of d-hypercyclic vectors, assuming that X is a Banach space.

Section 4.6. Regarding Theorem 4.6.1, an interesting open problem is that of the existence of nontrivial algebras within $HC(P(D)) \cup \{0\}$, where P is a nonconstant polynomial.

Section 4.7. Concerning the comments following Theorem 4.7.1, it should be said that if T is a supercyclic operator on a locally convex space X then either $\sigma_p(T^*) = \varnothing$ or $\sigma_p(T^*) = \{\lambda\}$ for some $\lambda \neq 0$ [61, Chapter 1]. Moreover, from the statements of [248] (see also [311]) it can be derived that, if X is a Banach complex space and T is a supercyclic operator on X, then $SC(T)$ contains a T-invariant dense vector subspace if and only if the set of isolated points in $\sigma(T)$ that are not in $\sigma_e(T)$ is empty.

The extension of the concept of supercyclicity to a sequence (T_n) of operators is immediate: suffice it to consider the projective orbits $\{\lambda T_n x_0 : \lambda \in \mathbb{K}, n \geq 1\}$. As an analogue of Theorem 4.5.8, Zhang and Zhou [387] have obtained conditions in order that the family of common supercyclic vectors $\bigcap_{j=1}^{\infty} SC\left((T_{j,n})_{n\geq1}\right)$ of a countable family of sequences $(T_{j,n})_{n\geq1} \subset L(X)$ ($j \geq 1$) be spaceable, X being a Banach space.

Section 4.8. Menet [304] was able to establish the following criterion of nonexistence of closed subspaces in $FHC(T)$.

Theorem. *Let X be a separable infinite dimensional Banach space and $T \in L(X)$. If there exists $C > 0$ such that for any $K \geq 1$, any infinite dimensional closed subspace M of X, and any $x \in X$, there exists $y \in M$ such that $\|y\| \leq 1/K$ and $\sup_{k>K} \frac{\text{card}\{n \leq k: T^n(x+y) \geq C\}}{k+1} > 1 - \frac{1}{K}$ then $FHC(T)$ is not spaceable.*

The above theorem is used in [304] to present an example of a frequently hypercyclic weighted shift $B_w : \ell_p \to \ell_p$ $(1 \leq p < \infty)$ satisfying that $HC(B_w)$ is spaceable but $FHC(B_w)$ is not. Moreover, in [305] Menet succeeds in obtaining a characterization of those operators $T \in L(X)$ –with X a Fréchet space– such that $HC(T)$ is spaceable under the assumption that T satisfies the FHCC. Other related results involving sequences (T_n) of operators between two Fréchet spaces X and Y are provided in [115, 305]. In particular, criteria for spaceability of the set of upper frequently hypercyclic vectors of a operator T on a Fréchet space are provided. The notion of upper frequently hypercyclic vectors is the same as the one of frequently hypercyclic vectors just by changing lower densities to upper densities.

Section 4.9. An interesting notion that describes the dynamics of certain physical processes is that of semigroup of operators. A one-parameter family $(T_t)_{t \geq 0}$ of operators on a topological vector space X is called a *strongly continuous semigroup of operators*, or simply a C_0-*semigroup*, if the following three conditions are satisfied:

(i) $T_0 = I$,

(ii) $T_{t+s} = T_t T_s$ for all $t, s \geq 0$, and

(iii) $\lim_{s \to t} T_s x = T_t x$ for all $x \in X$ and $t \geq 0$.

Several corresponding results for C_0-semigroups –to which the notions of distributionally irregular vectors and distributional chaos can be extended in a natural way, as well as the notions of hypercyclicity and frequent hypercyclicity– are provided in [6] by Albanese, Barrachina, Mangino and Peris. The study of lineability for the set of hypercyclic vectors for a C_0-semigroup is virtually settled since Conejero, Müller and Peris [175] proved in 2007 that if X is an F-space, then $(T_t)_{t \geq 0}$ is hypercyclic if and only if each T_{t_0} $(t_0 > 0)$ is hypercyclic, and if and only if some T_{t_0} $(t_0 > 0)$ is hypercyclic. In this case, the set of hypercyclic vectors for $(T_t)_{t \geq 0}$ equals $HC(T_{t_0})$ for any $t_0 > 0$. The assertion holds if one replaces hypercyclicity by frequent hypercyclicity. Turning to distributional chaos, it seems that reasonable criteria for the spaceability of $DI(T)$ have not been furnished to date.

Chapter 5

Zeros of Polynomials in Banach Spaces

In this chapter, we concentrate on the existence and dimension of (possible) subspaces of $P^{-1}(0)$, where P is a $\mathbb{K}$-valued polynomial on a Banach space X. It is clear that we must make restrictions in order that there be a chance of something interesting emerging. Namely, the polynomials $P : \mathbb{K}^2 \to \mathbb{K}$, $P(x_1, x_2) = 1 + x_1^2 + x_2^2$, $Q : \mathbb{K}^2 \to \mathbb{K}$, $Q(x_1, x_2) = x_1^3 + x_2^3$ and $R : \mathbb{K}^2 \to \mathbb{K}$, $R(x_1, x_2) = x_1^2 + x_2^2$ illustrate the three most basic requirements:

- P shows that the polynomial in question must vanish at $0 \in X$;

- Comparing the polynomials P and R shows that the scalar field $\mathbb{K} = \mathbb{R}$ or $\mathbb{C}$ matters;

- Comparing the polynomials P and Q shows that the degree of the polynomial matters.

5.1   What one needs to know

Everyone knows what a polynomial P on $\mathbb{R}^n$ or $\mathbb{C}^n$ is, but what is a *polynomial on a Banach space?* In this preliminary section, we will review the definition of polynomial and introduce the necessary background. As we will see, this depends on the notion of symmetric multilinear mappings, and so we will review these and discuss the relation between polynomials and multilinear mappings. This will lead us to the *polarization formula*. Also, we briefly discuss tensor products and their relation to polynomials and multilinear mappings on Banach spaces.

We now give the formal definition of *polynomial on a Banach space X.* For it, we will first need the following.

Definition 5.1.1. Let X be a Banach space over scalar field K and $n \in \mathbb{N}$. By $\mathcal{L}(^n X)$ we denote the space of continuous n-linear forms

$$A : X \overbrace{\times \cdots \times}^{n} X \to \mathbb{K}.$$

Next we will define n-homogeneous polynomial in terms of $n-$linear forms.

Definition 5.1.2. *For a Banach space X over $\mathbb{K}$ and $n \in \mathbb{N}$, we say that $P : X \to \mathbb{K}$ is a continuous n-homogeneous polynomial if there is some associated $A \in \mathcal{L}(^n X)$ such that for all $x \in X$,*

$$P(x) = A(x, \cdots, x).$$

We denote by $\mathcal{P}(^n X)$ the space of such continuous n-homogeneous polynomials, and we let $\mathcal{P}(X)$ be the space of all finite sums of the form $\sum P_j$, where each $P_j \in \mathcal{P}(^j X)$. It is agreed that $\mathcal{L}(^0 X) = \mathcal{P}(^0 X)$ corresponds to the constant functions $X \to \mathbb{K}$. Also, observe that $\mathcal{L}(^1 X) = \mathcal{P}(^1 X) = X^*$.

There are a number of excellent sources on polynomials and multilinear forms in this context, such as the books of S. Dineen and J. Mujica [188,318]. In particular, the following important result, known as the *polarization formula*, can be found in either [188, Corollary 1.6] or [318, Proposition 3].

Proposition 5.1.3. *To any continuous n-homogeneous polynomial $P \in \mathcal{P}(^n X)$ there corresponds a* unique *symmetric continuous n-linear form $A \in \mathcal{L}(^n X)$ such that $P(x) = A(x, \cdots, x)$ for all $x \in X$. In fact, A is given by the following formula:*

$$A(x_1, ..., x_n) = \frac{1}{2^n n!} \sum_{\epsilon_i = \pm 1} \epsilon_1 \epsilon_2 \cdots \epsilon_n P(\sum_{i=1}^{m} \epsilon_i x_i).$$

A very simple example illustrating the link between this definition of polynomial and multilinear form is seen by taking a 2×2 real symmetric matrix $A = \begin{bmatrix} a & b \\ b & c \end{bmatrix}$. Then A gives rise to a 2-homogeneous polynomial $P : \mathbb{R}^2 \to \mathbb{R}$ by $P(x) = x A x^T$. Thus, our notion of polynomial is nothing more than an extension of the idea of *quadratic form*.

Since our interest here will only be on continuous multilinear forms and polynomials, the word *continuous* will subsequently be omitted in this chapter.

5.2 Zeros of polynomials: the results

The fundamental problem in this chapter will be to study the existence of vector subspaces contained in $P^{-1}(0)$, where $P \in \mathcal{P}(X)$ is a fixed polynomial.

We will begin with the case of complex polynomials on complex Banach spaces, discussing work of A. Plichko and A. Zagorodnyuk [334]. The main result here is that if X is an infinite dimensional complex Banach space and $P : X \to \mathbb{C}$ is a polynomial with $P(0) = 0$, then $P^{-1}(0)$ contains a separable, infinite dimensional subspace of X. This leads to several natural questions, such as the relation between two subspaces contained in some $P^{-1}(0)$ and to whether the dimension of maximal subspaces contained in the zero set of a polynomial remains the same for each such subspace.

Next, we will turn to the real case where the situation is completely different. Here, we'll describe results of the following type: If X is an infinite dimensional real Banach space and $P : X \to \mathbb{C}$ is an odd polynomial with $P(0) = 0$, then for every $n \in \mathbb{N}$, there is a subspace $Z_n \subset P^{-1}(0)$ of dimension n. On the other hand, we will see that in general, in this situation there is *no infinite dimensional subspace Z that is contained in $P^{-1}(0)$*.

Throughout, lineability is the only "issue," since spaceability is automatic: indeed, if a vector subspace $V \subset P^{-1}(0)$, then also $\overline{V} \subset P^{-1}(0)$.

Let us begin with a few simple examples.

Example 5.2.1. *It is trivial that the polynomial $P : \mathbb{R}^2 \to \mathbb{R}$ given by $P(x, y) = x^2 + y^2$ has only $\{(0,0)\}$ as its zero set. However, for the same polynomial $P : \mathbb{C}^2 \to \mathbb{C}$, $P(z, w) = z^2 + w^2$, a 1-dimensional complex line, $\mathrm{span}\{(z, iz)\}$, is contained in $P^{-1}(0)$.*

The fact that the polynomial is even is decisive, as the next triviality shows.

Example 5.2.2. *Next, consider $P : \mathbb{R}^2 \to \mathbb{R}$, $P(x, y) = x^3 + y^3$, for which $P^{-1}(0) = \mathrm{span}\{(x, -x)\}$.*

The above examples show that both the degree of the polynomial and the scalar field play important roles in this study.

Let us go on by "shooting down" a natural, but unfortunately incorrect, attempt at simplifying the problem. Namely, suppose that $P : X \to \mathbb{C}$ is a 2-homogeneous polynomial. Thus, P corresponds to a unique symmetric bilinear form $A : X \times X \to \mathbb{C}$, $A(x, x) = P(x)$ for all $x \in X$. Now, A in turn naturally corresponds to a unique linear form

$$L_A : X \bigotimes X \to \mathbb{C}$$

on the tensor product of X with itself. Here, L_A is determined by the condition $L_A(x \bigotimes y) = A(x, y)$. Since L_A is a linear functional, its null-space $L_A^{-1}(0)$ is a vector subspace of codimension one in $X \bigotimes X$, and it would be nice if from this subspace one could somehow extract a large subspace contained in $P^{-1}(0)$. As it happens, this simply does not happen: The problem is that elements in $L^{-1}(0)$ are finite combinations of the form $x_i \bigotimes y_i$, whereas what is required for $P^{-1}(0)$ are combinations of the form $x_i \bigotimes x_i$, each of which is in $P^{-1}(0)$.

Example 5.2.3. *For instance, let $P : \mathbb{R}^4 \to \mathbb{R}$ be defined by $P(x) = P(x_1, x_2, x_3, x_4) = x_1^2 + x_2^2 + x_3^2 + x_4^2$. The symmetric bilinear map associated to P is $A : \mathbb{R}^4 \times \mathbb{R}^4 \to \mathbb{R}$, $A(x, y) = x_1 y_1 + \cdots x_4 y_4$, and the linear form associated to A is $L_A : \mathbb{R}^4 \otimes \mathbb{R}^4 \to \mathbb{R}$, $L_A(x \otimes y) = \sum_{i=1}^{4} x_i y_i$. Now, the nullspace of L_A is non-trivial since, for example, $L_A(e_1 \otimes e_1 - e_2 \otimes e_2) = 0$, so that $e_1 \otimes e_1 - e_2 \otimes e_2 \in L_A^{-1}(0)$. On the other hand, $P^{-1}(0) = \{0\}$.*

Remark 5.2.4. *Several easy reductions are in order.*

1. *The $n = 0$ case: It is clear that we should restrict the constant term of P so that $P(0) = 0$.*

2. *The linear, $n = 1$, case: If $P : X \to \mathbb{K}$ is linear, then $P^{-1}(0)$ is a one co-dimensional hyperplane in X.*

3. *If we can prove a general result about "large" subspaces contained in $P_j^{-1}(0)$ for j-homogeneous polynomials P_j, then we will have the same type of result for arbitrary $P = P_1 + \cdots + P_k$. Indeed, for such a P, let $X_k \subset P_k^{-1}(0)$ be such a large subspace. Then we will be able to find a large subspace $X_{k-1} \subset P_{k-1}|_{X_1}^{-1}(0)$, and so on. Since we are dealing with a finite series $\sum_{j=1}^{k} P_j$, the process will terminate after a finite number of steps.*

4. *In rare instances, the argument in (3) above works for analytic functions with an infinite power series expansion. For instance, every holomorphic function $f : c_0(I) \to \mathbb{C}$ vanishes on an infinite (even non-separable) subspace of $c_0(I)$, provided that I is an uncountable index set.*

Let us illustrate the above comment with a simple example.

Example 5.2.5. *Let $P : \ell_2^{2n} \to \mathbb{C}$ be given by*

$$P(x) = <x, y_0> + \sum_{j=1}^{2n} x_j^2,$$

where $y_0 \in \ell_2$ is fixed and $<x, y_0>$ denotes the usual inner product on ℓ_2^n. Note that $P_1 : x \mapsto <x, y_0>$ is a linear mapping, and that $P_2 : x \mapsto \sum_{j=1}^{2n} x_j^2$ is a 2-homogeneous polynomial, associated to the symmetric bilinear mapping $(x, y) \in \ell_2^n \times \ell_2^n \mapsto \sum_{j=1}^{2n} x_j y_j$. (We remark that $(x, y) \mapsto <x, y>$ is not bilinear in the complex case, since this mapping is conjugate linear in the second variable.) Now, consider the n-dimensional vector space X_2 spanned by $\{e_1 + ie_2, e_3 + ie_4, ..., e_{2n-1} + ie_{2n}\}$. It is very easy to see that $P_2|_{X_2}$ is identically 0. Next, being linear, $P_1|_{X_2}$ has an $n - 1$-dimensional kernel, X_1. Hence, $P|_{X_1} \equiv 0$.

We will now see how this example can be extended to the case of general polynomials on arbitrary complex Banach spaces. The basic idea is due to A. Plichko and A. Zagorodnyuk [334, Theorem 5], although the approach that we follow is somewhat modified [34].

Theorem 5.2.6. *Let X be an infinite dimensional complex Banach space and let $P : X \to \mathbb{C}$ be a polynomial such that $P(0) = 0$. Then $P^{-1}(0)$ contains an infinite dimensional subspace of X.*

In other words, we have the lineability of the zero set of such polynomials. Our method of proof will be to first show the following corollary (see [34]).

Corollary 5.2.7. *Let n and d be fixed positive integers. Then there exists $m = m(n, d) \in \mathbb{N}$ such that for every n-homogeneous polynomial $P : \mathbb{C}^m \to \mathbb{C}$ there is a d-dimensional subspace contained in $P^{-1}(0)$.*

Restating the result, fix the degree of homogeneity n of a polynomial. Then for any desired dimension d of subspace, there is some number of variables m such that the zero set of *every* n-homogeneous polynomial in m variables contains a d-dimensional subspace.

It is important to note that Theorem 5.2.6 follows from Corollary 5.2.7 *and its proof*. We will show this right after the proof of Corollary 5.2.7. For this, we require an easy lemma.

Lemma 5.2.8. *Let $Q : \mathbb{C}^n \to \mathbb{C}$ be a non-constant polynomial, where $n \geq 2$. Then there is some point $z \in \mathbb{C}^n, z \neq 0$, such that $Q(z) = 0$.*

Proof. For some $j = 1, ..., n$, there is some $j \in \{1, ..., n\}$ and some $z_j \in \mathbb{C}, z_j \neq 0$, such that the polynomial

$$(z_1, ..., z_{j-1}, z_{j+1}, ..., z_n) \in \mathbb{C}^{n-1} \mapsto Q(z_1, ..., z_n) \in \mathbb{C}$$

is non-constant. By induction, the result follows. □

In fact, much more than the assertion of the above lemma is true. Namely, it is a classical result in several complex variables [238] that a non-constant entire function $f : \mathbb{C}^n \to \mathbb{C}$, $n \geq 2$, is either never 0 or else has an unbounded set of zeros. Given that any non-constant polynomial P on $\mathbb{C}^n$ has a zero, Lemma 5.2.8 is seen to be a very weak consequence of the classical result.

Our plan now is to prove Corollary 5.2.7. In Remark 5.2.10, we show how it implies Theorem 5.2.6.

Proof of Corollary 5.2.7. The proof is a double induction, first on the homogeneity of n and then on the dimension d of the desired subspace. In fact, our proof will really be a "pseudo-induction" on n, where by "pseudo" we mean that we will prove the cases of n-homogeneous polynomials for $n = 1, 2$ and 3. From these cases, it should be evident how to proceed in the general case.

As we already remarked, for $n = 1$ the result is trivial with $m = m(1, d) = d + 1$. The point here is that we're merely using the fact that the kernel of a linear form is a 1-codimensional.

Next, let's consider $n = 2$-homogeneous polynomials $Q : \mathbb{C}^m \to \mathbb{C}$ (where m is to be determined, depending on the degree of homogeneity $n = 2$ and desired dimension d). Let $A : \mathbb{C}^m \times \mathbb{C}^m \to \mathbb{C}$ be the symmetric bilinear form associated to Q, so that $Q(z) = A(z, z)$ for all $z \in \mathbb{C}^m$. The situation $d = 1$ is easy, since the above lemma ensures that provided $m \geq 2$, there is $z_1 \in \mathbb{C}^m$, $z_1 \neq 0$, such that $Q(z_1) = 0$. Since Q is homogeneous, the one-dimensional vector space generated by z_1, $[z_1]$, works.

Consider the set $\{z \in \mathbb{C}^m \mid A(z, z_1) = 0\}$, which is a 1-codimensional subspace of $\mathbb{C}^m$. Since it contains z_1, we can write $\{z \in \mathbb{C}^m \mid A(z, z_1) = 0\} = [z_1] \oplus \mathbb{C}^{m-2}$. If $m - 2 \geq 2$, i.e. if $m \geq 4$, then the lemma provides us with a non-zero point $z_2 \in \mathbb{C}^{m-2}$ with $Q(z_2) = 0$. (Note that we are regarding $\mathbb{C}^{m-2}$ as a subspace of $\{z \in \mathbb{C}^m \mid A(z, z_1) = 0\}$.) Since for any scalars a_1, a_2,

$$Q(a_1 z_1 + a_2 z_2) = a_1^2 Q(z_1) + 2a_1 a_2 A(z_2, z_1) + a_2^2 Q(z_2) = 0,$$

the 2-dimensional space $[z_1, z_2] \subset Q^{-1}(0)$ when $m \geq 4$.

Continuing, consider $\{z \in \mathbb{C}^{m-2} \mid A(z, z_2) = 0\}$, where again we consider

$$\{z \in \mathbb{C}^{m-2} \mid A(z, z_2) = 0\} \subset \mathbb{C}^{m-2} \subset \{z \in \mathbb{C}^m \mid A(z, z_1) = 0\} \subset \mathbb{C}^m.$$

Thus $\{z \in \mathbb{C}^{m-2} \mid A(z, z_2) = 0\} = [z_2] \oplus \mathbb{C}^{m-4}$, and so if $m - 4 \geq 2$, there is $z_3 \in \mathbb{C}^{m-4}$, $z \neq 0$, with $Q(z_3) = 0$. Repeating the argument of the preceding paragraph, $Q(a_1 z_1 + a_2 z_2 + a_3 z_3) = 0$ for all scalars a_0, a_1 and a_2, and so $[z_0, z_1, z_2] \subset Q^{-1}(0)$.

Arguing inductively, let us assume that we have found linearly independent $z_1, z_2, ..., z_k \in \mathbb{C}^{m-2(k-1)}$ such that $Q|_{[z_1,...,z_k]} \equiv 0$. If $m - 2(k - 1) \geq 2$, i.e. $m \geq 2k$, then Lemma 5.2.8 provides us with a non-zero $z_{k+1} \in \mathbb{C}^{m-2(k-1)}$ with $Q(z_{k+1}) = 0$. Following the above scheme, we show that Q is identically 0 on the $k + 2$-dimensional subspace $[z_1, ..., z_{k+1}]$, and this completes the argument for the case of $n = 2$ –homogeneous polynomials.

Let's briefly describe the $n = 3$ case. As before, Lemma 5.2.8 assures that there there is a non-zero point $z_1 \in \mathbb{C}^2$ such that $Q(z_1) = 0$. Hence, for any 3-homogeneous polynomial Q, $Q|_{[z_1]} \equiv 0$. Consider next the subspace $S(z_1, z_1) = \{z \in \mathbb{C}^m \mid A(z_1, z_1, z) = 0\}$. Since $S(z_1, z_1)$ is an $(m - 1)$-dimensional subspace of $\mathbb{C}^{m-1}$ that contains z_1, we can write $S(z_1, z_1) = [z_1] \oplus \mathbb{C}^{m-2}$ Applying induction, we use the $n = 2$ case to show that the 2-homogeneous polynomial $z \mapsto A(z_1, z, z)$ vanishes on an $[\frac{m-2}{2}]$-dimensional subspace $S(z_1)$ of $\mathbb{C}^{m-2}$. If it happens that $[\frac{m-2}{2}] \geq 2$, that is if it happens that $m \geq 6$, then there is $z_2 \in S(z_1)$, $z_2 \neq 0$, such that $Q(z_2) = 0$. We verify that $[z_1, z_2] \subset Q^{-1}(0)$. Indeed, for any scalars a_1, a_2, $Q(a_1 z_1 + a_2 z_2) = A(a_1 z_1 + a_2 z_2, a_1 z_1 + a_2 z_2, a_1 z_1 + a_2 z_2) = a_1^3 Q(z_1) + a_2^3 Q(z_2) + 3a_1^2 a_2 A(z_1, z_1, z_2) + 3a_1 a_2^2 A(z_1, z_2, z_2)$. The first two terms are 0 by our choice of z_1 and z_2. Next $A(z_1, z_1, z_2) = 0$ since A is symmetric and since $z_2 \in S(z_1) \subset S(z_1, z_1)$, and $A(z_1, z_2, z_2) = 0$ by our choice of $z_2 \in S(z_1)$.

Next, suppose that we have found d linearly independent vectors

$\{z_1, ..., z_d\} \subset \mathbb{C}^m$ whose span is contained in $P^{-1}(0)$. Furthermore, let us suppose that there is a subspace of $\mathbb{C}^m$ of the form

$$[z_1, ..., z_d] \bigoplus \mathbb{C}^{[\frac{m - d2^d}{2^d}]}$$

with the property that any three vectors u, v, w contained in it satisfy $A(u, v, w) = 0$. If $[\frac{m - d2^d}{2^d}] \geq 2$, that is if $m \geq 2^d(d + 2)$, then there is a non-zero vector $z_{d+1} \in \mathbb{C}^{[\frac{m - d2^d}{2^d}]}$ such that $P(z_{d+1}) = 0$. Let's simplify notation by setting $t = \frac{m - d2^d}{2^d} - 1$. Let $S(z_0, z_{d+1}) = \mathbb{C}^t$, where $z_0 = 0$. For $i = 1, ..., d + 1$, let $S(z_i, z_{d+1})$ be defined as

$$S(z_i, z_{d+1}) = \{z \in S(z_{i-1}, z_{d+1}) \mid A(z_i, z_{d+1}, z) = 0\}.$$

Therefore, $S(z_{d+1}, z_{d+1})$ is in general a $t - (d + 1)$ dimensional subspace of $\mathbb{C}^t$. Let $S(z_{d+1})$ be a further subspace having dimension at least $[\frac{t - (d+1)}{2}]$ on which the 2-homogeneous polynomial

$$z \mapsto A(z_{d+1}, z, z)$$

is identically 0. This is the induction step for the $n = 3$ case.

(Note that since

$$\frac{t - (n + 1)}{2} = \frac{\frac{m - d2^d}{2^d} - 1 - (d + 1)}{2} = \frac{m - (d + 1)2^{d+1}}{2^{d+1}},$$

we have shown that in fact $m(3, d) \leq 2^{d-1}(d + 1)$.)

As stated at the outset, the general argument follows along exactly the lines described above. $\qquad\square$

Several comments are in order concerning the above argument. First, as was noted it is evident that for $n = 1$, $m(1, d) = d + 1$. For the 2-homogeneous case, we have shown that $m(2, d) \leq 2d$. However, for $n \geq 3$, we have only crude estimates on an upper bound for $m(3, d)$. For instance, the above methods lead to the estimate that that for $d = 3$ and $n = 4$, $m = 2 \cdot 2^{27} \cdot 29 + 2$ works; that is any 4-homogeneous polynomial in m (the *horrible* number of the previous line) vanishes on a 3-dimensional subspace. It would be interesting, and it certainly should be possible, to get better results in this direction.

Second, concerning the specific case of 2-homogeneous polynomials, we leave it as an exercise to show that, in fact, $m(2, d) = 2d$. Note here that a 2-homogeneous polynomial $Q : \mathbb{C}^m \to \mathbb{C}$ can be associated to a quadratic form, i.e. a symmetric matrix $A = (a_{ij})_{i,j=1,...,m}$, which can be diagonalized. Thus, in some basis $Q(w) = \sum_w b_j w_j^2$, which can be handled as in Example 5.1.

Here is a restatement of the Corollary we have just proved.

Corollary 5.2.9. [34] *Given natural numbers n and m, there is a number $d = d(n, m) \in \mathbb{N}$ such that (i) $d(n, m) \to \infty$ as $m \to \infty$ and (ii) for every n-homogeneous polynomial $P : \mathbb{C}^m \to \mathbb{C}$, $P^{-1}(0)$ contains a d-dimensional subspace.*

Remark 5.2.10. *Several comments about the proof of Corollary 5.2.7 should be made. First, the argument is "incremental" in the sense that for a given n-homogeneous polynomial P, not only is the function $m(n, d)$ increasing with d but also the d-dimensional vector spaces contained in the zero set, $P^{-1}(0)$, are also increasing with d. The point about this is that if, instead of $\mathbb{C}^m$ for some m, we take an infinite dimensional space X, then the argument proves that there is an increasing sequence of d-dimensional subspaces contained in $P^{-1}(0)$. In other words, we have proved Theorem 5.2.6.*

Second, the above phenomenon is a purely complex feature of zero sets. By this we mean that we will soon see that there are real-valued odd-homogeneous polynomials P on infinite dimensional Banach spaces whose zero set contains vector spaces of all finite dimensions, although there is no infinite dimensional vector space contained in $P^{-1}(0)$. Thus, there is no hope to produce a proof along the lines of Corollary 5.2.7.

For non-separable complex Banach spaces, another natural question that arises is whether one can find very large subspaces of $P^{-1}(0)$ if $P : X \to \mathbb{C}$ is defined on a very large Banach space. Specifically, one can ask if the zero set of P contains a non-separable Banach space if P is defined on a non-separable Banach space. Although there are natural situations where the answer is affirmative (M. Fernández-Unzueta, [202]), it does not hold in general (A. Avilés and S. Todorcevic, [36]). We will briefly describe this situation below.

Another reasonable question comes from the following situation: Fix an n-homogeneous polynomial $P : X \to \mathbb{C}$ on a non-separable complex Banach space X. By Theorem 5.2.6, we know that there is an infinite dimensional subspace contained in the zero set of P. Hence, by a standard Zorn's lemma argument there is a maximal infinite dimensional subspace contained in the zero set of P. (Note: *maximal* rather than *largest*.) Now, there may be many, different infinite dimensional subspaces contained in $P^{-1}(0)$, and it seems reasonable to wonder whether maximal subspaces related to each of them are isomorphic, or at least of the same dimension. As the following result due to Avilés and Todorcevic [36] shows, the situation is more complicated than one might think.

Theorem 5.2.11. *Consider the complex Banach space $X = \ell_1(\omega_1)$. There is a 2-homogeneous polynomial $P : \ell_1(\omega_1) \to \mathbb{C}$ which has a maximal separable as well as a maximal non-separable subspace contained in $P^{-1}(0)$.*

We sketch the proof of this result, which makes use of the following somewhat counterintuitive result (see Exercise 5.2).

Proposition 5.2.12. *There is a collection $\mathcal{C} \subset \mathcal{P}(\mathbb{N})$ of infinite subsets of $\mathbb{N}$ having the following properties:*

(i) $\operatorname{card} \mathcal{C} = \mathfrak{c}$, *the cardinality of the real numbers, and*

(ii) *for any $A, B \in \mathcal{C}, A \cap B$ is a finite set.*

Proof of Theorem 5.2.11. Let $\mathcal{B}$ equal the disjoint union $\mathbb{N} \cup \mathcal{C}$, where $\mathcal{C}$ is the set that is described in the above proposition. Consider the non-separable Banach space $\ell_1(\mathcal{B})$ and the following polynomial $P : \ell_1(\mathcal{B}) \to \mathbb{C} :$ $P(x) \equiv \sum_{n \in \mathbb{N}, A \in \mathcal{C}} x_n x_A$, where $x = (x_i)_{i \in \mathbb{N} \cup \mathcal{C}}$. Since $|P(x)| \leq \|x\|_1^2$, P is a well-defined, continuous 2-homogeneous polynomial. Note that $P|_{\ell_1(\mathbb{N})} \equiv 0$ since in this case every $x_A = 0$. We claim that the separable subspace $\ell_1(\mathbb{N})$ is a *maximal* zero subspace of P. This will imply the conclusion of the result since, just like $\ell_1(\mathbb{N})$, P is identically 0 on the non-separable subspace $\ell_1(\mathcal{C})$ of $\ell_1(\mathcal{B})$.

So, looking for a contradiction, let us assume that $\ell_1(\mathbb{N})$ is not maximal. Thus, there must be a strictly larger subspace Y of $\ell_1(\mathcal{B})$, $\ell_1(\mathbb{N}) \subsetneqq Y \subset \ell_1(\mathcal{B})$ such that $P|_Y \equiv 0$. Let $y \in Y \backslash \ell_1(\mathbb{N})$, where without loss of generality y is of the form $y = \sum_{A \in \mathcal{C}} y_A \chi_A$ (where $y_A \in \mathbb{C}$ for each $A \in \mathcal{C}$). Since y is absolutely summable, there is some "coordinate" A so that $|y_A| = \max_{B \in \mathcal{C}} |y_B|$. There is a finite subset $\mathcal{F} \subset \mathcal{C}$ so that

$$\sum_{B \in \mathcal{C}, B \notin \mathcal{F}} |y_B| < |y_A|.$$

Now, by the properties of $\mathcal{C}$, there must be an integer $n \in \mathbb{N}$ such that $n \in A \backslash \cup_{B \in \mathcal{F}, B \neq A} B$.

Let $z \in \ell_1(\mathbb{N} \cup \mathcal{C}), z = e_n + y$. Observe that since $z \in Y$, we will have a contradiction once we show that $P(z) \neq 0$. Calculating,

$$P(z) = \sum_{B \in \mathcal{C}, n \in B} y_B = y_A + \sum_{B \in \mathcal{C}, B \notin \mathcal{F}} y_B,$$

since $B \neq A, B \in \mathcal{F}$ implies that $n \notin B$. Finally,

$$|\sum_{B \in \mathcal{C}, B \notin \mathcal{F}} y_B| \leq \sum_{B \in \mathcal{C}, B \notin \mathcal{F}} |y_B| < |y_A|,$$

and consequently $P(z) \neq 0$. Thus, $P|_Y \neq 0$, and the proof is complete. $\square$

Using a somewhat sophisticated index set Γ_2, Avilés and Todorcevic [36, Theorem 13] have even shown the following result, which we state without proof.

Theorem 5.2.13. *There is an uncountable index set Γ_2 and a 2-homogeneous polynomial $P : \ell_1(\Gamma_2) \to \mathbb{C}$ such that if $Y \subset P^{-1}(0)$ is a subspace then Y is separable.*

In other words, for $\ell_1(\Gamma_2)$, there is a 2-homogeneous polynomial P whose maximal zero-subspaces are all separable. By contrast, we now outline the argument of M. Fernández-Unzueta that shows that the situation is completely different when X is complex ℓ_∞. Specifically, we have the following result (see [202, Theorem 2.6]).

Theorem 5.2.14. *Let X be a complex Banach space that contains ℓ_∞. Let $P : X \to \mathbb{C}$ be a polynomial such that $P(0) = 0$. Then $P^{-1}(0)$ contains a non-separable subspace.*

We will indicate how the proof of this result proceeds. It clearly suffices to restrict to the case $X = \ell_\infty$. By Theorem 5.2.6, $P^{-1}(0)$ contains an infinite dimensional Banach space which, by a classical result (see, e.g., [290, Proposition 2.a.2]), contains a copy G of c_0. In fact, somewhat more is shown (see [202, Theorem 2.1]):

Theorem 5.2.15. *Let X be a complex Banach space that contains ℓ_∞. For each $j \in \mathbb{N}$, let $P_j : X \to \mathbb{C}$ be a polynomial with $P_j(0) = 0$. Suppose that $c_0 \subset P_j^{-1}(0)$ for all $j \in \mathbb{N}$. Then there is a non-separable subspace of X that is contained in every $P_j^{-1}(0)$, $j \in \mathbb{N}$.*

Curiously, the argument for Theorem 5.2.15 also makes use of the same set theoretical Proposition 5.2.12. Namely, the non-separable subspace can be constructed in a special, useful manner. Specifically, applying Proposition 5.2.12 we find an an uncountable index set $\mathcal{C}$ and, for each $\alpha \in \mathcal{C}$, an infinite $N_\alpha \subset \mathbb{N}$ such that $N_\alpha \cap N_\beta$ is finite ($\alpha, \beta \in \mathcal{C}$, $\alpha \neq \beta$). For each such α, let $x_\alpha \in \ell_\infty$ be given by

$$x_\alpha(j) = \begin{cases} 0 & \text{if } j \notin N_\alpha \\ 1 & \text{if } j \in N_\alpha. \end{cases}$$

Since N_α is infinite, $x_\alpha \in \ell_\infty \backslash c_0$. The non-separable subspace of Theorem 5.2.15 will be of the form span $\{x_\gamma \mid \gamma \in \Gamma\}$, where $\Gamma \subset \mathcal{C}$ has a countable complement.

Let us first assume that instead of a countable collection $(P_j)_j$ we have just one polynomial P that is n-homogeneous for some n. (See Exercise 5.3 to show how the case of a single polynomial P implies Theorem 5.2.15 for a countable family of P_j's. In fact, during the course of the discussion, we'll see that this Exercise is important since the induction argument for the nth step uses the full force of Theorem 5.2.15 for lower degree polynomials.)

The theorem will be proved once we have established the following result, which can be found in [202, Theorem 2.1].

Theorem 5.2.16. *Let $G \cong c_0$ be fixed. There is an uncountable set $\{x_\alpha : \alpha \in \mathcal{C}\} \subset \ell_\infty$ such that whenever $P : \ell_\infty \to \mathbb{C}$ is n-homogeneous with $G \subset P^{-1}(0)$, then there is a subset $\Gamma \subset \mathcal{C}$ with $\mathcal{C} \backslash \Gamma$ countable which satisfies the following conditions:*

$$F_\Gamma := \overline{\text{span}} \, \{x_\gamma \mid \gamma \in \Gamma\} \text{ is non-separable and } F_\Gamma \subset P^{-1}(0).$$

It is straightforward that this result implies Theorem 5.2.15.

Proof. (Sketch) The argument will show that each $x_\alpha \in \ell_\infty$ is just a sequence of 0's and 1's. Since Γ is uncountable and $\|x_\alpha - x'_\alpha\| = 1$ whenever $\alpha \neq \alpha'$, the non-separability of F_Γ will be assured. Also, from Proposition 5.1.3, to show that $F_\Gamma \subset P^{-1}(0)$, it will suffice to prove that for the associated symmetric n-linear form A, $A(x_1, x_2, ..., x_n) = 0$ for every $x_i \in F_\Gamma$, $i = 1, ..., n$.

To start the induction, suppose that $P = \varphi : \ell_\infty \to \mathbb{C}$ is a linear form (i.e. $n = 1$) such that $c_0 \subset \varphi^{-1}(0)$. Using Exercise 5.2, let C be an uncountable set, and for each $\alpha \in C$ let N_α be an infinite set such that $N_\alpha \cap N_\beta$ is finite whenever $\alpha \neq \beta$ and $\alpha, \beta \in C$. For each $\alpha \in C$, let $x_\alpha \in \ell_\infty$ be given by

$$x_\alpha(j) = \begin{cases} 0 & \text{if } j \notin N_\alpha \\ 1 & \text{if } j \in N_\alpha. \end{cases}$$

Since N_α is infinite, $x_\alpha \in \ell_\infty \backslash c_0$. By e.g. [154, Theorem 6.9], there is a subset $\Gamma \subset C$ such that $C \backslash \Gamma$ is countable and, furthermore, $\varphi(x_\gamma) = 0$ for every $\gamma \in \Gamma$. Consequently, $\varphi|_{\mathrm{span}\{x_\gamma \mid \gamma \in \Gamma\}} \equiv 0$. This proves the initial, $n = 1$, case.

Next, assume that the result holds for all polynomials P of degree $k < n$ for some $n > 1$. Suppose that $P : \ell_\infty \to \mathbb{C}$ is an n-homogeneous polynomial with $c_0 \subset P^{-1}(0)$. For each $k, 1 \leq k < n$ and each set of k positive integers, $i_1, ..., i_k$, consider the $(n - k)$-homogeneous polynomial $P_{i_1,...,i_k} : \ell_\infty \to \mathbb{C}$,

$$P_{i_1,...,i_k}(x) = A(e_{i_1}, \ldots, e_{i_k}, \overbrace{x, \ldots, x}^{n-k}),$$

where $A : X \times \cdots \times X \to \mathbb{C}$ is the symmetric n- linear form associated to P, $\{e_i \mid i \in \mathbb{N}\}$ is the standard basis of c_0 and $x \in \ell_\infty$. Applying the polarization formula, Proposition 5.1.3, each such $P_{i_1,...,i_k}$ vanishes on c_0. For each $i_1, ..., i_k$, we apply induction to get a set $\Gamma_{i_1,...,i_k} \subset C$ such that $C \backslash \Gamma_{i_1,...,i_k}$ is countable and such that $P_{i_1,...,i_k}$ is 0 on c_0. Since the set of such k- tuples of indices $i_{1.,,,.i_k}$ is countable, we conclude that there is an uncountable subset $\Gamma_1 \subset C$ such that $C \backslash \Gamma_1$ is countable and also

$$\overline{\mathrm{span}\{x_\gamma \mid \gamma \in \Gamma_1\}} \subset \bigcap_{i_1,...,i_k} P_{i_1,...,i_k}^{-1}(0).$$

Note that $\overline{\mathrm{span}\{x_\gamma \mid \gamma \in \Gamma_1\}}$ is non-separable.

We will now complete the proof of Theorem 5.2.16, making use of the fact that the set S is countable, where

$$S = \{\gamma \in \Gamma_1 \mid \text{for some } \alpha_2, ..., \alpha_k \in \Gamma_1, A(x_\gamma, x_{\alpha_2}, ..., x_{\alpha_k}) \neq 0\}.$$

However, we omit the rather technical proof of this fact. (For details, see [202].) Thus, by removing S from Γ_1, we are left with an uncountable set of indices, Γ_2 say, such that if $\gamma_k \in \Gamma_2$, $k = 1, ..., n$, then $A(x_{\gamma_1}, ..., x_{\gamma_n}) = 0$. It is evident that this implies that $P(x_\gamma) = 0$ for every $\gamma \in \Gamma_2$. Finally, let $x \in span\{x_\gamma \mid \gamma \in \Gamma_2\}$, $x = \sum_{j=1}^{k} \lambda_j x_{\gamma_j}$. Then

$$P(x) = \sum_{i_1,\ldots,i_n=1,\ldots,k} \lambda_{i_1} \cdots \lambda_{i_n} A(x_{\gamma_{i_1}}, \ldots, x_{\gamma_{i_n}}) = 0,$$

and this completes the argument. $\square$

As a form of "segue" between the complex and real parts of this chapter, we remark that the real analogue of Theorem 5.2.15 cannot hold in generality: Just take the 2-homogeneous polynomial

$$P : \ell_\infty \to \mathbb{R}, \ x = (x_n) \mapsto \sum_{n=1}^{\infty} (\frac{x_n}{n})^2,$$

to see that the result fails for *real* ℓ_∞. On the other hand, the argument in [202] for the complex case shows that, in addition, the following (see [202]) is true.

Theorem 5.2.17. *Let $P : \ell_\infty \to \mathbb{R}$ be a polynomial on the real Banach space ℓ_∞ with the property that $c_0 \subset P^{-1}(0)$. Then P vanishes on a non-separable subspace of ℓ_∞.*

We turn now to the "real story" about zeros of polynomials. We first consider the case of zeros of 2-homogeneous polynomials $P : X \to \mathbb{R}$. As we have already noted, it is to be expected that the discussion here should be quite short, given examples like $P : \mathbb{R}^n \to \mathbb{R}$, $P(x) = \sum_{j=1}^{n} x_j^2$. Such examples notwithstanding, we will see that interesting things can be said provided that the domain space X is quite "large." To give an example of what we're talking about, consider the space $X = c_0$ and the above polynomial $P :$ $c_0 \to \mathbb{R}$, $P(x) = \sum_{j=1}^{\infty} \left(\frac{x_j}{j}\right)^2$. It is easy to see that P really is a well-defined 2-homogeneous polynomial on c_0 whose only zero is $x = 0$. Replacing c_0 by its non-separable cousin $c_0(\Gamma)$ (where Γ is an index set of uncountable cardinality) does not yield an analogous example and, indeed, for every 2-homogeneous polynomial $P : c_0(\Gamma) \to \mathbb{R}$, $P^{-1}(0)$ contains an infinite dimensional subspace.

Much of the material in the next few pages comes from [21].

Definition 5.2.18. Let X be a real Banach space. An n-homogeneous polynomial $P : X \to \mathbb{R}$ is said to be *positive definite* if $P(x) \geq 0$ for all $x \in X$ and, moreover, $P(x) = 0$ if and only if $x = 0$.

Note that if P is positive definite, then $< x, y >:= A(x, y)$ defines an inner product on X. (Here, as usual, A is the unique symmetric bilinear form on $X \times X$ such that $A(x, x) = P(x)$ for all $x \in X$.)

The relation between positive definite 2-homogeneous polynomials P and the corresponding set of zeros is given in the following proposition.

Proposition 5.2.19. *Let X be a real Banach space. The following conditions are equivalent.*

(1) *There is a positive definite 2-homogeneous polynomial on X.*

(2) *There is a continuous, linear, one-to-one mapping $j : X \to H$, where H is a Hilbert space.*

(3) *There is a 2-homogeneous polynomial $P : X \to \mathbb{R}$ such that $P^{-1}(0) \subset V$ for some finite dimensional subspace $V \subset X$.*

Proof. (1) $\Rightarrow$ (2)   As noted above, if P is a positive definite 2-homogeneous polynomial on X, then the corresponding symmetric bilinear form A defines an inner product on X. With the renorming

$$x \in X \mapsto |||x||| = \sqrt{A(x,x)},$$

X is therefore a pre-Hilbert space with completion $(H, |||\cdot|||)$ say. The identity mapping $j : X \hookrightarrow H$ is clearly an injection. Moreover

$$|||j(x)||| = \sqrt{A(x,x)} = |P(x)|^{\frac{1}{2}} \leq \|P\|^{\frac{1}{2}} \|x\|,$$

which implies the continuity of j.

(2) $\Rightarrow$ (3)   This is very easy, since $P : x \in X \mapsto A(x,x)$ is the desired polynomial. Note that $P^{-1}(0) = \{0\}$.

(3) $\Rightarrow$ (1)   Suppose that $P : X \to \mathbb{R}$ is a 2-homogeneous polynomial whose zero set is contained in $V = \mathrm{span}\{v_1, \ldots, v_n\}$. For a contradiction, assume that there are points x and y in $X \backslash V$ such that $P(x) < 0 < P(y)$, and let $\gamma : [0,1] \to X \backslash V$ be a curve from $\gamma(0) = x$ to $y = \gamma(1)$. (Note that we are assuming that $\dim X = \infty$ in order to get the existence of such a curve γ which misses V; if $\dim X < \infty$, then the result is trivial.) Then, for some $z = \gamma(t_0) \notin V$, $P(z) = P \circ \gamma(t_0) = 0$, which contradicts our assumption (3).

Thus, $P(x)$ has the same sign, say $P(x) \geq 0$, for every $x \in X \backslash V$. To conclude, take a continuous linear projection $\Pi : X \to V$, $\Pi(x) = \sum_{j=1}^{n} \varphi_j(x) v_j$, where each $\varphi_j \in X^*$. Then the 2-homogeneous polynomial

$$Q : X \to \mathbb{R}, \ Q(x) = P(x) + \sum_{j=1}^{n} \varphi_j^2(x)$$

is the required positive definite 2-homogeneous polynomial on X. $\square$

One of the principal results in the case of zero-subspaces of real polynomials is the following.

Theorem 5.2.20. *Suppose that X is a real Banach space with the property that there are no 2-homogeneous polynomials on X that are positive definite. Then the zero set of every 2-homogeneous polynomial $P : X \to \mathbb{R}$ contains an infinite dimensional subspace. That is, there is an infinite dimensional subspace Z of X such that $Z \subset P^{-1}(0)$.*

One example of such a situation occurs when $X = \ell_p(I)$ where I is an uncountable index set and $p > 2$. In Proposition 5.2.27 we'll outline the proof of a characterization of those compact sets K for which $C(K)$ does have a 2-homogeneous positive definite polynomial.

As we will see, the proof of Theorem 5.2.20 is non-constructive, relying on Zorn's Lemma.

Proof. Let $P : X \to \mathbb{R}$ be an arbitrary 2-homogeneous polynomial on X. Let

$$S \equiv \{V \mid V \text{ is a subspace of } X \text{ and } P|_V \equiv 0\}.$$

Partially ordering S by inclusion, the standard Zorn's Lemma argument shows that S contains a maximal element V.

We will show that V cannot be finite dimensional. Suppose instead that $\{v_1, \ldots, v_n\}$ is a basis for V. Let $W = \cap_{j=1}^n \ker A(\cdot, v_j)$, where $A(\cdot, v_j) : X \to \mathbb{R}$, $x \rightsquigarrow A(x, v_j)$ and where, as usual, A is the unique symmetric bilinear form associated with P. It is easy that $W = \cap_{v \in V} \ker A(\cdot, v)$ and that W is of finite codimension in X. Obviously, W is a vector space which, we claim, contains the maximal subspace V. Indeed, let $y \in V$ be an arbitrary vector. Since V is a vector space, $x + y \in V$ for every $x \in V$. Now, we get

$$0 = P(x + y) = P(x) + 2A(x, y) + P(y) = 2A(x, y).$$

Restated, the above equality simply states that $y \in W$, as required.

Since we are assuming that $\dim V < \infty$, there is a subspace $Y \subset W$ such that $W = V \oplus Y$. By maximality, if $P(w) = 0$ for some $w \in W$, then $w \in V$. As a consequence, $P|_Y$ is either positive definite or negative definite. Without loss of generality, assume that $P(y) \geq 0$ for all $y \in Y$. Since $\dim V < \infty$, we can find $\varphi_1, \ldots, \varphi_n \in X^*$ such that the 2-homogeneous polynomial $P + \sum_{j=1}^n \varphi_j^2$ is positive definite on W. Since W is of finite codimension in X, it follows that there is a positive definite 2-homogeneous polynomial on X. But this contradicts the equivalence (1) $\iff$ (3) of Proposition 5.2.19, since it implies that X does admit a 2-homogeneous positive definite polynomial. Therefore $\dim V = \infty$, which concludes the proof. $\square$

The argument in Theorem 5.2.20 will now be "recycled" to provide a dichotomy result. For it, we will need to assume that X is of type 2, and we will use the following important *extension theorem* of B. Maurey which we state in the special, simpler context that is needed here (see [186, Theorem 12.22]).

Theorem 5.2.21. *Let $W \subset X$ be a subspace of the Banach space X, where we further assume that X is type 2. Then any continuous linear operator from W to a Hilbert space H extends to a continuous linear operator from X to H.*

Using the above result of Maurey, we're now ready to state and prove our dichotomy.

Theorem 5.2.22. *Suppose that X is a Banach space having type 2. Then exactly one of the following holds:*

(1) *There is a positive definite 2-homogeneous polynomial on X.*

(2) *Every 2-homogeneous polynomial on X vanishes on a non-separable subspace of X.*

Proof. Let us suppose that the first condition fails, and let us then show that to every 2-homogeneous polynomial $P : X \to \mathbb{R}$ one can find a non-separable subspace that is contained in $P^{-1}(0)$. Following the scheme of the argument in Theorem 5.2.20, let V be a maximal subspace of X having the property that $P|_V \equiv 0$.

We claim that V is non-separable. Otherwise, assume that V is separable and that $\{v_j \mid j \in \mathbb{N}\}$ is a countable dense set in V of norm one vectors. By a small modification of the argument in Theorem 5.2.20, we see that there is a subspace $W \subset X$ such that W can be written as the *algebraic* direct sum of V and an algebraic complement, $Y : W = V \bigoplus_a Y$. Without loss of generality, $P|_W$ is positive definite on Y. Since we are assuming that V is separable, there is a countable collection $\{\varphi_j \mid j \in \mathbb{N}\}$ such that $\sum_{j=1}^{\infty} \varphi_j^2$ is a positive definite 2-homogeneous polynomial on V. Thus, $P + \sum_{j=1}^{\infty} \varphi_j^2$ is a positive definite 2-homogeneous polynomial on $V \bigoplus_a Y = W$.

By the equivalence (1) $\iff$ (2) of Proposition 5.2.19, we see that there is a continuous linear injection i from W to some Hilbert space H. Using our hypothesis together with Maurey's Theorem 5.2.21, we get that there is a continuous linear extension $\tilde{i} : X \to H$. To obtain the desired contradiction, all that is necessary is to notice that we have a continuous linear mapping $j : X \to H \bigoplus_2 \ell_2$, given by

$$ j(x) = \left(\tilde{i}(x), \left(\frac{A_{v_j}(x)}{j^2 \|A_{v_j}\|} \right)_{j=1}^{\infty} \right). $$

It is not difficult to verify that j is also injective, and another application of Proposition 5.2.19 implies that there is a positive definite 2-homogeneous polynomial on X, which contradicts our assumption. $\square$

Remark 5.2.23. *1. One should compare the above results, valid in the real case for 2-homogeneous polynomials, with work of Avilés and Todorcevic [36] and Fernández-Unzueta [202] in the case of complex polynomials defined on "very large" Banach spaces.*

2. It is an open problem whether the conclusion of Theorem 5.2.22 holds if one omits the assumption that X is type 2. It seems natural to conjecture that the result still holds in this generality, and that the appeal to Maurey's theorem is more a weakness in the inventiveness of the writer than a "defect" in the theory itself.

We briefly turn to n-homogeneous polynomials for higher homogeneities, $n = 3$ and $n = 4$. Our first result in this direction concerns 4-homogeneous polynomials which, as we will see, gives us information about the 3-homogeneous case as well. Note that if a Banach space X does not admit a positive definite $2m$-homogeneous polynomial then there exists no positive definite 2-homogeneous polynomial on X either.

Theorem 5.2.24. *Suppose that the real Banach space X does not admit a positive definite 4-homogeneous polynomial. Let $(\psi_k)_{k=1}^{\infty}$ be an arbitrary sequence in X^*. Then, for any countable family $(P_j)_{j=1}^{\infty}$ of 2-homogeneous polynomials on X, there exists a non-separable subspace $Z \subset \bigcap_{k=1}^{\infty} \psi_k^{-1}(0)$ such that $P_j|_Z \equiv 0$ for all $j \in \mathbb{N}$.*

An example of such a space is $X = \ell_p(I)$, where I is uncountable and $p > 4$. The point here is that every 4-homogeneous polynomial on $\ell_p(I)$ can be approximated, uniformly on the ball of $\ell_p(I)$, by elements in the algebra generated by functionals in $\ell_p(I)^*$. Since each such functional vanishes off a countable subset of I, the same must be true for elements in the algebra, and in fact for uniform limits of such elements. In particular, any 4-homogeneous polynomial $P : \ell_p(I) \to \mathbb{R}$ is such that $P|_{\ell_p(J)} \equiv 0$ for an uncountable subset $J \subset I$. (See, for example, [179] for further clarification of this type of situation.)

Proof. As we will see, the proof of Theorem 5.2.24 is similar to earlier arguments. As before, let $V \subset X$ be a maximal subspace of $\bigcap_{k=1}^{\infty} \psi_k^{-1}(0)$ such that $P_j|_V \equiv 0$ for every $j \in \mathbb{N}$. We claim that V is non-separable and so, for a contradiction, assume that V is separable with countable dense set $\{v_i \mid i \in \mathbb{N}\}$. Let

$$W = \bigcap_{k=1}^{\infty} \psi_k^{-1}(0) \cap \bigcap_{i=1}^{\infty} \bigcap_{j=1}^{\infty} ker\left(A_j(\cdot, v_i)\right),$$

where, as usual, A_j is the unique symmetric bilinear form associated with the 2-homogeneous polynomial P_j.

One shows that $V \subset W$ so that, once again, we can write W as an algebraic direct sum $W = V \bigoplus_a Y$ for some subspace $Y \subset W$. Now, the 4-homogeneous polynomial on W,

$$Q = \sum_{j=1}^{\infty} \frac{P_j{}^2}{j^2 \|P_j\|^2},$$

is positive definite on Y, the reason being that if $Q(w) = 0$ for some $w \in W$, then our choice of V dictates that w must be in Y. Using the assumption that V is separable, there is a countable sequence $(\varphi_i)_{i=1}^{\infty} \subset X^*$ so that the 4-homogeneous polynomial

$$\sum_{j=1}^{\infty} \frac{P_j{}^2}{j^2 \|P_j\|^2} + \sum_{i=1}^{\infty} \varphi_i^4$$

is positive definite on W.

Finally, the 4-homogeneous polynomial

$$\sum_{j=1}^{\infty} \frac{P_j{}^2}{j^2\|P_j\|^2} + \sum_{i=1}^{\infty} \varphi_i^4 + \sum_{i=1}^{\infty}\sum_{j=1}^{\infty} \frac{(A_j(\cdot,v_i))^4}{i^2 j^2\|A_j(\cdot,v_i)\|^4} + \sum_{k=1}^{\infty} \frac{\psi_k^4}{k^2\|\psi_k\|^4}$$

is a 4-homogeneous positive definite polynomial on X, which contradicts our assumption. Thus, V must be non-separable, and the proof is complete. $\qquad\square$

Here is an important, albeit self-evident, special case of the above result.

Corollary 5.2.25. *If the real Banach space X does not admit a positive definite 4-homogeneous polynomial, then any 2-homogeneous polynomial on X vanishes on a non-separable subspace of X.*

The next corollary is somewhat less trivial.

Corollary 5.2.26. *If X does not admit a positive definite 4-homogeneous polynomial, then every 3-homogeneous polynomial on X is identically 0 on a non-separable subspace of X.*

Proof. Fix a 3-homogeneous polynomial $P : X \to \mathbb{R}$ with associated symmetric trilinear form A. As by now is usual, let V be a maximal subspace of X having the property that $P|_V \equiv 0$. Suppose that V is separable, with countable dense set $\{v_i \mid i \in \mathbb{N}\}$. For each i,j, let $A_{v_i,v_j} \in X^*$ be given by

$$A_{v_i,v_j}(x) = A(v_i,v_j,x).$$

Also, for each i, denote by $Q_i : X \to \mathbb{R}$ the 2-homogeneous polynomial $Q_i(x) = A(v_i,x,x)$.

By Theorem 5.2.24, there is a non-separable space W contained in the intersection

$$\bigcap_{i,j} A_{v_i,v_j}^{-1}(0) \cap \bigcap_i Q_i^{-1}(0)$$

(noting that the set of zeros of each $A_{v_i,v_j}(\cdot)$ are of course the same as the set of zeros of the 2-homogeneous polynomial $A_{v_i,v_j}^2(\cdot)$). Suppose that $y \in W$ is such that $P(y) = 0$. Let $\lambda \in \mathbb{R}$ be arbitrary and $x = \sum \alpha_i v_i$ any finite linear combination of $\{v_i\}$. Then

$$\begin{aligned} P(x+\lambda y) &= A(x+\lambda y, x+\lambda y, x+\lambda y) \\ &= P(x) + 3A(x,x,\lambda y) + 3A(x,\lambda y,\lambda y) + \lambda^3 P(y) \\ &= P(x) + 3\lambda A\left(\sum \alpha_i v_i, \sum \alpha_i v_i, y\right) \\ &\quad + 3\lambda^2 A\left(\sum \alpha_i v_i, y, y\right) + \lambda^3 P(y). \end{aligned}$$

The first and fourth terms in the above expression are each 0 since $x \in V$ and, by assumption, $P(y) = 0$. Therefore,

$$P(x + \lambda y) = 3\lambda \sum_{i,j} \alpha_i \alpha_j A(v_i, v_i, y) + 3\lambda^2 \sum_i A(v_i, y, y)$$

$$= 3\lambda \sum_{i,j} \alpha_i \alpha_j A_{v_i, v_j}(y) + 3\lambda^2 \sum_i Q_{v_i}(y).$$

Now, $y \in W$ and so each term in the above two sums is 0. Summarizing, for any $y \in W$ with $P(y) = 0$, we have shown that $P(x + \lambda y) = 0$ for every scalar λ and every finite linear combination $x \in \text{span}\{v_i\}$. By continuity of P, we conclude that $P(x + \lambda y) = 0$ for every $y \in W$ and $x \in V$.

From this and the maximality of V, it follows that the restriction of P to W, $P|_W$, has the property that all its zeros are contained in V. Since W is non-separable and since we are assuming that V is separable, we can decompose W as $W = (V \cap W) \bigoplus_a Y$, where Y is a non-separable subspace of W. But the above work shows that every zero of $P|_W$ is contained in V. The result is that the 3-homogeneous polynomial $P|_Y$ has $0 \in Y$ as its only zero which, of course, is impossible. Thus, our assumption that V is separable leads to a contradiction, and the proof is complete. $\qquad\square$

As promised, we conclude this section on positive definite polynomials by sketching the argument for the following result.

Theorem 5.2.27. *For a compact Hausdorff space K, there is a positive definite 2-homogeneous polynomial on $\mathcal{C}(K)$ if and only if there is a strictly positive measure on K.*

Proof. The proof is based on an appeal to Proposition 5.2.19, together with fundamental results on 2-summing operators. Namely, the continuous injection $j : \mathcal{C}(K) \to H$, where H is a Hilbert space, must be 2-summing [186, Theorem 3.5]. Therefore, by an application of the Pietsch factorization theorem (see, e.g., [186, Corollary 2.15]), we can decompose j as $j = i \circ j_2$, where $j_2 : \mathcal{C}(K) \hookrightarrow \mathcal{L}_2(K, \mu)$ is the canonical inclusion, for a finite regular Borel measure μ, and $i : \mathcal{L}_2(K, \mu) \to H$ is an isomorphism into (see diagram (5.2)). Since j is $1-1$, so is j_2, and this forces μ to be strictly positive (see [186, p. 42]).

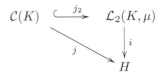

Finally, the reverse implication is straightforward: If μ is a strictly positive measure on K, then $P : \mathcal{C}(K) \to \mathbb{R}$, $P(f) = \int_K f^2 d\mu$, is the required positive definite 2-homogeneous polynomial. $\qquad\square$

We now turn to the work of P. Hájek et al., on zero subspaces of polynomials $P : X \to \mathbb{R}$, where X is a real Banach space and P is a homogeneous polynomial of *odd* degree. Roughly speaking, we will discuss two general, perhaps surprising, results about the existence and size of subspaces of $P^{-1}(0)$ in this case. First, we'll show that a result similar to Corollary 5.2.7 holds in the context of *real* Banach space X and *odd*-homogeneous polynomial $P : X \to \mathbb{R}$. Namely, we'll show that given an odd integer n and a desired dimension d, there is $m = m(n, d)$ such that every n-homogeneous polynomial $P : \mathbb{R}^m \to \mathbb{R}$ vanishes on a subspace of dimension d. One easy consequence of this result is that if X is a real, infinite dimensional Banach space and $P : X \to \mathbb{R}$ is an odd-homogeneous polynomial, then for every $d \in \mathbb{N}$, $P^{-1}(0)$ will contain a d-dimensional subspace. However, we will also show whenever X is a real separable infinite dimensional Banach space, then for every odd n, there is an n-homogeneous polynomial $P : X \to \mathbb{R}$ such that $P^{-1}(0)$ does not contain any infinite dimensional subspace. We therefore see that the real analogue of the Plichko–Zagordonyuk result, Theorem 5.2.6, is false. As a matter of fact, in [29, Theorem 4] the following result can be found.

Theorem 5.2.28. *Given an odd integer n and a dimension d, if m satisfies*

$$m > d! \, (\log_2 m)^d \binom{d + n - 1}{d - 1} \qquad (*),$$

then the following holds:

For any $P : \mathbb{R}^m \to \mathbb{R}$ which is n-homogeneous, $P^{-1}(0)$ contains a d-dimensional vector space.

This result is merely the latest (perhaps) in a series of papers on this topic. For instance, B. J. Birch [120] showed that if $\mathbb{K}$ is an algebraic number field, $h > 1, m > 1$ are natural numbers, and $r_1, ..., r_h$ are h odd positive integers, then there is a number $N = \Psi(r_1, ..., r_h, m, \mathbb{K})$ such that any h homogeneous polynomials of degrees $r_1, ..., r_h$ in N variables must vanish on an m-dimensional subspace. It should be noted that in [27], one can find a result similar to Theorem 5.2.28 in the case of 3-homogeneous odd polynomials P on $\mathbb{R}^N$, with a different bound on m.

The proof of Theorem 5.2.28 requires several observations. To begin, let $\{e_1, ..., e_k\}$ denote the standard basis of $\mathbb{R}^k$. The following lemma can be found in, for instance, [342].

Lemma 5.2.29. *Given natural numbers k and n, there is a subset $S(k, n) \subset \mathbb{R}^k$ with the following two properties:*

(a) $e_j \in S(k, n)$ *for each $j = 1, ..., k$, and*

(b) *Given an n-homogeneous polynomial $P : \mathbb{R}^k \to \mathbb{R}$, if P vanishes on $S(k, n)$, then $P \equiv 0$.*

For instance, it is a trivial exercise that $S(2,2) = \{e_1, e_2, e_1 + e_2\}$ works. In fact, $\operatorname{card}\left(S(k,n)\right) = \begin{pmatrix} k+n-1 \\ k-1 \end{pmatrix}$.

The next lemma is somewhat less trivial. For it, consider any $k, \ell \in \mathbb{N}, k \le \ell$, and let $\mathcal{A} = (A_1, ..., A_k)$ be pairwise disjoint subsets of $\{1, ..., \ell\}$ whose union, $\cup_{i=1}^{k} A_i$, equals $\{1, ..., \ell\}$. Let us agree to say that any such partition of $\{1, ..., \ell\}$ has *rank k*.

Lemma 5.2.30. *Given* $k \le \ell$, *there exist an integer* $p(k,\ell) \in \mathbb{N}$, $p(k,\ell) \le k!(\log_2(\ell))^k$ *and a collection* $\{\mathcal{A}_I \mid I = 1, ..., p(k,\ell)\}$ *of* $p(k,\ell)$ *rank k partitions of* $\{1, ..., \ell\}$ *with the following property:*

For any set $B \subset \{1, ..., \ell\}$ *having* k *elements, there is some* $\mathcal{A}_I = (A_1, ..., A_k)$ *such that* $B \cap A_i \ne \emptyset$ *for each* $i = 1, ..., k$.

Proof. The proof will be by induction on the rank, k. To simplify the argument, we will assume that $\ell = 2^n$ for some $n \in \mathbb{N}$. It is straightforward that any system of partitions that "works" for $\{1, ..., 2^n\}$ will also work for $\{1, ..., \ell\}$ whenever $2^{n-1} < \ell \le 2^n$.

When $k = 1$, the situation is trivial: There is only one partition $\mathcal{A}$, namely $\mathcal{A} = \{1, ..., \ell\}$, and of course the conclusion is satisfied. Let's move on to the $k = 2$ case. For any $j = 1, ..., n$, let $\mathcal{A}_j = (A_1^j, A_2^j)$, where

$$A_1^j = \{i \in \{1, ..., \ell\} \mid [\frac{i-1}{2^{n-j}}] \text{ is even}\} \quad \text{and} \quad A_2^j = \{i \in \{1, ..., \ell\} \mid [\frac{i-1}{2^{n-j}}] \text{ is odd}\},$$

where $[r]$ denotes the integer part of r. Thus, for example, when $n = 1$ then $\mathcal{A}_1 = (\{1\}, \{2\})$. When $n = 2$, $\mathcal{A}_1 = (\{1,2\}, \{3,4\})$ and $\mathcal{A}_2 = (\{1,3\}, \{2,4\})$, etc. It is straightforward that for any $B \subset \{1, ..., 2^n\}$, $|B| = k$, there is some $\mathcal{A}_j$ such that $B \cap A_j^1 \ne \emptyset \ne B \cap A_j^2$. Observe that there are n such rank 2 partitions, so that $p(2, 2^n) = n$.

Instead of continuing for general k, let's consider the $k = 3$ case more closely. For $j \in \{1, ..., n\}$, let A_1^j be as above, noting that $|A_1^j| = 2^{n-1}$. By the $k = 2$ case, there exist $p(2, 2^{n-1})$ partitions of rank 2, say $(\mathcal{A}_I^{j,1})_{I=1}^{p(2, 2^{n-1})} = (A_1^{j,1,I}, A_2^{j,1,I})_{I=1}^{p(2, 2^{n-1})}$, with the following property:

If $B \subset A_1^j$, $|B| = 2$, *then for some* $I \in \{1, ..., p(2, 2^{n-1})\}$, B *intersects both* $A_1^{j,1,I}$ *and* $A_2^{j,1,I}$ *in one point.*

It follows that if $B \subset A_1^j$, $|B| = 2$, for some j, then the triple $(A_1^{j,1,I}, A_2^{j,1,I}, A_2^j)$ satisfies the condition of Lemma 5.2.8. In a similar manner, if $B \subset A_2^j$, $|B| = 2$, then we can "break" A_2^j up into two pieces, $A_1^{j,2,I}, A_2^{j,2,I}$, so that the triple $(A_1^j, A_1^{j,2,I}, A_2^{j,2,I})$ works in the lemma. Thus, the requirement of the lemma is satisfied with $p(3, 2^n)$ rank 3 partitions, where $p(3, 2^n) = 2p(2, 2^n)p(2, 2^{n-1})$. Now by the $k = 2$ case, $p(2, 2^m) = m$ for any m, and so $p(3, 2^n) = 2n(n-1)$. For general $k \ge 2$, we continue, obtaining that $p(k+1, 2^n) \le kp(2, 2^n)p(2, 2^{n-1}) \le k!(n-1)^k n$, which completes the argument. □

We are almost ready to prove Theorem 5.2.28. However, we need one more not surprising, and not difficult, lemma.

Lemma 5.2.31. *Let $Q \ne 0$ be an n-homogeneous polynomial on a k-dimensional space X. Assume that $\{e_1, ..., e_k\}$ is a basis for X. Then there is a basis $\{\tilde{e}_1, ..., \tilde{e}_k\}$ on X with respect to which Q may be written as*

$$Q(x) = Q(\sum_{j=1}^{k} x_j \tilde{e}_j) = \sum_{|\alpha|=n} a_\alpha x^\alpha,$$

where each "pure" coefficient $a_{(n,0,...,0)}, ..., a_{(0,...,0,n)}$ is non-zero.

Observe that since $Q \ne 0$, there is some direction x_1 in X such that $\frac{\partial^n Q(0,...,0)}{\partial x_1^n} \ne 0$. Next, if for example $Q(x_1, x_2) = x_1^2 + x_1 x_2$, then the change of variables $x_1 \to C(u_1 + u_2), x_2 \to u_2$ will, for large C, provide us with a set of coordinates $\{u_1, u_2\}$ such that $\frac{\partial Q^2(0,0)}{\partial u_j^2} \ne 0$ for both $j=1$ and $j=2$. We leave the details of the general argument to the reader (Exercise 5.10).

Proof of Theorem 5.2.28. The argument will make use of the following version of the Borsuk Antipodal theorem (see, e.g., Lloyd [291, Theorem 3.2.7]): Let $Q : \mathbb{R}^N \to \mathbb{R}^m$ be a continuous function, where $N > m$. Then there is a non-zero point $x \in \mathbb{R}^N$ such that $Q(x) = Q(-x)$. In our case, Q will be an odd-homogeneous polynomial, and so $Q(x) = -Q(x)$; that is $Q(x) = 0$.

To start, fix any non-zero n-homogeneous polynomial $P : \mathbb{R}^m \to \mathbb{R}$, where n is odd and m satisfies condition (*) of the statement. Applying Lemma 5.2.31, we may assume that the basis $\{e_j \mid j = 1, ..., m\}$ for $\mathbb{R}^m$ is such that all the monomial, "pure" coefficient terms of P, namely all $\frac{\partial^n P(0)}{\partial x_j^n}$, are non-zero. Let's now apply Lemma 5.2.30, which yields a collection

$$\{A_I \mid I = 1, ..., p(d, m)\}$$

consisting of rank d partitions of the set $\{1, ..., m\}$. Also, using Lemma 5.2.29, choose a set

$$S(d, n) = \left\{ v_j \mid j = 1, ..., card\ S(d, n) = \binom{d+n-1}{d-1} \right\} \subset \mathbb{R}^m$$

containing each $e_i, 1 \le i \le m$.

Let $\{e_{I,J}\}$ be a basis for the space $\mathbb{R}^{p(d,m)\binom{d+n-1}{d-1}}$.

In order to apply the Borsuk Antipodal theorem, we will need an n-homogeneous polynomial $\tilde{Q} : \mathbb{R}^m \to \mathbb{R}^{p(d,m)\binom{d+n-1}{d-1}}$. (Note that it is here that our rather ugly, but necessary, hypothesis (*) in Theorem 5.2.28 will be needed.) In order to define $\tilde{Q}$, we'll need some notation. First, for any subset $A \subset \{1, ..., \ell\}$, let $P_A : \mathbb{R}^\ell \to \mathbb{R}^\ell$ be given by $P_A(x) = P_A(\sum_{i=1}^\ell x_i e_i) = \sum_{i \in A} x_i e_i$. That is, P_A is the projection of $\mathbb{R}^\ell$ onto the subspace spanned by the coordinates in A. For $v = (v_1, ..., v_d) \in \mathbb{R}^d$ and a partition $\mathcal{A} = \{A_1, ..., A_k\}$ of $\{1, ..., \ell\}$ of rank k, set

$$v \cdot \mathcal{A} : \mathbb{R}^\ell \to \mathbb{R}^\ell, \ v \cdot \mathcal{A}(x) = \sum_{j=1}^k v_j P_{A_j}(x).$$

Let $\tilde{Q} : \mathbb{R}^m \to \mathbb{R}^{p(d,m)\binom{d+n-1}{d-1}}$ be the n-homogeneous polynomial defined by

$$\tilde{Q}(x) = \sum_I \sum_J Q(v_J \cdot \mathcal{A}_I(x)) e_{I,J}.$$

By the Borsuk Antipodal theorem, there is some non-zero vector $v^0 = (x_1^0, ..., x_m^0) \in \mathbb{R}^m$ such that $\tilde{Q}(x^0) = 0$. Let's call supp $x^0 = B$. All that we have to do at this point is to show that B has more than d coordinates. Indeed, assuming this to be true, let's finish the proof. So, for some $I \in \{1, ..., p(d,m)\}$, $\mathcal{A}_I = (A_1, ..., A_d)$ has the property that for each $i, 1 \leq i \leq d$, the set B meets A_i in at least one point. As a consequence, $P_{A_i}(x^0) \neq 0$. Let's call $x^i = P_{A_i}(x^0)$. Set $R : \mathbb{R}^d \to \mathbb{R}$,

$$R((t_1, .., t_d)) = Q(\sum_{i=1}^k t_i x^i).$$

From the definition and the fact that $\tilde{Q}(x_0) = 0$, it follows that $R(v^j) = Q(v^J \cdot \mathcal{A}_i(x^0)) = 0$ for all $J, 1 \leq J \leq N(d,n)$. But Lemma 5.2.29 now shows that $R \equiv 0$ on $\mathbb{R}^d$. In other words, $Q \equiv 0$ on the d-dimensional space spanned by $\{x^1, ..., x^d\}$.

In order to complete the proof, we only need show that card (B) must be at least $d+1$. So, suppose that card $(B) \leq d$. By Lemma 5.2.30, there is some $\mathcal{A}_I = (A_1, ..., A_d)$ such that the intersection $B \cap A_i$ has no more than one element. Since $B \neq \emptyset$, there must be some i so that $B \cap A_i$ has exactly one element, say $\{t\}$. Using Lemma 5.2.29 we may choose J, $1 \leq J \leq N(d,n)$, so that $v^J = (0, ..., 0, 1, 0, ..., 0) = e_t$. Now, by definition, $v^J \cdot \mathcal{A}_I(x^0) = x_t^0$, and so $Q(v^J \cdot \mathcal{A}_I(x^0)) = Q(x_t^0)$ must be 0 since $\tilde{Q}(x^0) = 0$. Hence, the odd polynomial Q evaluated at $v^J \cdot \mathcal{A}_I(x^0)$ must be 0. On the other hand, Lemma 5.2.31 enabled us to choose e_t so that $Q(e_t) \neq 0$. This contradiction shows that card B cannot be $\leq k$, and the proof is finished.   □

We conclude this chapter with the promised proof that the real version of the result of A. Plichko and A. Zagorodnyuk, Theorem 5.2.6, is false. One might compare the following result with the results of Avilés and Todorcevic [36] about maximal "size" of zero-subspaces for 2-homogeneous polynomials on the complex Banach spaces of type $\ell_1(I)$.

Theorem 5.2.32. *Let X be a separable real Banach space and let $n \in \mathbb{N}$ be an odd integer, $n \geq 3$. Then there exists an n-homogeneous polynomial $P : X \to \mathbb{R}$ such that $P^{-1}(0)$ does not contain an infinite dimensional subspace.*

In fact, the argument will show considerably more. Namely, we will indicate how to find an n-homogeneous polynomial $P : X \to \mathbb{R}$ such that for any non-zero vector x such that $P(x) = 0$, there is $N = N(x) \in \mathbb{N}$ such that if Y is a subspace of X such that $x \in Y \subset P^{-1}(0)$, then dim $Y \leq N$.

Proof. We sketch the argument. First, it suffices to prove the result for $X = c_0$ and $n = 3$. To see this, suppose that we have found a 3-homogeneous polynomial $P : c_0 \to \mathbb{R}$ such that if $P(x) = 0$ for some $0 \neq x \in c_0$, then the dimension of every zero subspace of $P^{-1}(0)$ that contains x is bounded by some fixed $N(x)$. Let $\{\varphi_j \mid j \in \mathbb{N}\} \subset B_{X^*}$ be such that if for all $j, \phi_j(x) = 0$, then $x = 0$. Let $T : X \to c_0$ be the injection given as follows:

$$T(x) = \left(\frac{\phi_j(x)}{j} \right)_{j=1}^{\infty}.$$

Consider $Q : X \to \mathbb{R}$, $Q(x) = P \circ T(x)$, which is a 3-homogeneous polynomial on X. Suppose that $Q(z_0) = 0$ for some non-zero $z_0 \in X$ and that z_0 is contained in a subspace $Z \subset X$ which in turn is contained in $Q^{-1}(0)$. Thus, $T(z_0)$ is contained in the subspace $T(Z) \subset P^{-1}(0) \subset c_0$, whose dimension is bounded by $N(T(z_0))$. Hence, dim $Z \leq N(T(z_0))$. Finally, if $n = 3 + 2k \in \mathbb{N}$ is arbitrary, the same argument works with the polynomial $x \in X \rightsquigarrow Q(x)[\sum_{j=1}^{\infty} \frac{1}{2^j} \varphi_j(x)^{2k}]$.

So, our work consists in proving the assertion for some 3-homogeneous polynomial $P : c_0 \to \mathbb{R}$. We will indicate why a polynomial P that is given by

$$P(x) = \sum_{k=1}^{\infty} \sum_{i=k+1}^{\infty} \alpha_k^i x_i^2, \text{ where } x = (x_i) \in c_0,$$

satisfies our requirements. To do this, the first step is to observe that we can easily choose the coefficients $\alpha_k^i > 0$ in such a way that the polynomial is well-defined; indeed, any choice such that $\sum_k \sum_{i=k+1}^{\infty} \alpha_k^i < \infty$ will do here. (In fact, more technical requirements are needed on these coefficients if we want P to satisfy the assertion.)

Next, fix $0 \neq x \in c_0$ such that $P(x) = 0$. Clearly, there is no loss in assuming that $\|x\| \leq 1$. Consider the polynomial $R : c_0 \to \mathbb{R}$,

$$R(y) = P(x + y) = \sum_{k=1}^{\infty} \sum_{i=k+1}^{\infty} \alpha_k^i (x_i + y_i)^2.$$

So, R is a (non-homogeneous) polynomial of degree three, which we write as

$$R = R_0 + R_1 + R_2 + R_3,$$

with $R_0 = R(0) = P(y)$ being the constant term, ..., R_3 being the 3-homogeneous term in R. In particular,

$$R_2(y) = R_2((y_i)) = \sum_{k=1}^{\infty} x_k \sum_{i=k+1}^{\infty} \alpha_k^i y_i^2 + \sum_{k=1}^{\infty} y_k \sum_{i=k+1}^{\infty} 2\alpha_k^i x_i y_i.$$

One can find $N \in \mathbb{N}$ such that, if $Z = \overline{\text{span}}\{e_i \mid i > N\}$, then $R(z) > 0$ for every $0 \neq z \in Z$. (The not difficult, albeit technical, details showing the existence of such an N can be found in [30].) Assuming this to be the case, observe first that $\text{codim}(Z) = N$. For fixed $z \in Z, z \neq 0$, consider the non-zero third degree polynomial $\lambda \in \mathbb{R} \rightsquigarrow R(\lambda z) \in \mathbb{R}$, which we write as $R(\lambda z) = \sum_{\ell=0}^{3} \lambda^\ell R_\ell(z)$. There is some λ such that $P(x + \lambda z) = R(\lambda z) \neq 0$. Therefore, for any non-zero vector $z \in Z$, there is some $\lambda \in \mathbb{R}$ such that $P(x + \lambda z) \neq 0$. In other words, if $Y \subset c_0$ is any vector space contained in $P^{-1}(0)$, then $Z \cap Y = \{0\}$. Hence, $\dim(Y) \leq N$, as required. $\qquad\square$

5.3   Exercises

Exercise 5.1. This simple exercise deals with quadratic forms and symmetric matrices A. The basic result (Sylvester's Law of Inertia) is that the triple (n_+, n_-, n_0) of positive, negative and 0 eigenvalues of A remains the same under any congruence $A \to C^T A C$ of A (where C is a non-singular matrix).

Let A be a symmetric $n \times n$ matrix with associated quadratic form (i.e. 2-homogeneous polynomial P). Derive a formula for the largest possible dimension of a subspace contained in $P^{-1}(0)$ in terms of (n_+, n_-, n_0).

Exercise 5.2. There is a collection $\mathcal{C} \subset \mathcal{P}(\mathbb{N})$ of infinite subsets of $\mathbb{N}$ having the following properties:

  (i) $\text{card } \mathcal{C} = c$, the cardinality of the real numbers;

  (ii) For any $A, B \in \mathcal{C}, A \cap B$ is a finite set.

Hint: For each $r \in \mathbb{R}$, choose an infinite sequence $S_r \subset \mathbb{Q}$ that converges to r. Considered as sets, each S_r is countably infinite and any two such sets have finite intersection. Show that this implies the existence of such a family $\mathcal{C}$.

Exercise 5.3. Let A be an uncountable index set and let G be a subspace of ℓ_∞. Suppose that the following result holds: *There is an uncountable collection of vectors, $\{x_\alpha\}_{\alpha \in A}$, such that for every n-homogeneous polynomial $P : \ell_\infty \to \mathbb{C}$, we have $G \subset P^{-1}(0)$. Then there is a subset $\Gamma \subset A$ with* countable complement such that the non-separable space $\overline{\text{span}}\,\{x_\gamma \mid \gamma \in \Gamma\}$ is contained in $P^{-1}(0)$.

Then the same result holds if the single polynomial P is replaced by a countable family of polynomials. Specifically, given a countable set of homogeneous polynomials $\{P_i \mid i \in \mathbb{N}\}$ (of varying homogeneity) such that $G \subset \cap_{i=1}^\infty P_i^{-1}(0)$, one can find a subset $\Gamma \subset A$ with $A\backslash\Gamma$ countable so that the non-separable space

$$\overline{\text{span}}\,\{x_\gamma \mid \gamma \in \Gamma\}$$

is contained in each $P_i^{-1}(0)$.

Exercise 5.4. In each case, decide whether every 2-homogeneous polynomial $P : X \to \mathbb{R}$ on the real Banach space X vanishes on an infinite dimensional subspace:

(a) X is separable.

(b) $X = \mathcal{C}(K)$ for some compact set K.

(c) $X = \ell_\infty$.

(d) $X = \ell_p(I)$ for some index set I.

Exercise 5.5. Prove that for a 2-homogeneous polynomial $P : X \to \mathbb{R}$ to be positive definite, either of the following conditions is necessary and sufficient:

(a) P satisfies the parallelogram law:

$$P(x+y) + P(x-y) = 2(P(x) + P(y)), \quad \text{for all } x, y \in X.$$

(b) For every $x, y \in X$ such that $x \neq \pm y$,

$$|A(x,y)| < \frac{1}{2}(P(x) + P(y)).$$

As a consequence, show that if P is positive definite, then $\|P\| = \|A\|$.

Exercise 5.6. Show how the Zorn's Lemma argument, given in the first paragraph of the proof of Theorem 5.2.20, can be applied to show that every $\mathbb{C}$-valued polynomial on an infinite dimensional complex Banach space X is constant on an infinite dimensional subspace of X.

Exercise 5.7. Let X be a real Banach space that does not admit a positive definite n-homogeneous polynomial for any n and let $P : X \to \mathbb{R}$ be a polynomial satisfying $P(0) = 0$. Then there is a non-separable subspace of X on which $P \equiv 0$. There are several known examples of spaces X that satisfy the hypothesis. The simplest such examples are $X = c_0(I)$ and $X = \ell_p(I)$, $p > 2$, for some uncountable index set I.

Exercise 5.8. Show that Corollaries 5.2.7 and 5.2.9 are equivalent.

Exercise 5.9. Show that Theorem 5.2.28 implies that given any positive integer k and any odd integer n, there is $m \in \mathbb{N}$ such that every odd polynomial $Q : \mathbb{R}^m \to \mathbb{R}$ of degree at most n vanishes on a k-dimensional subspace.

Exercise 5.10. Prove Lemma 5.2.31, which is needed for Theorem 5.2.28.

Exercise 5.11. To what extent is the assumption of separability of X "vital" in Theorem 5.2.32?

Chapter 6

Miscellaneous

As we have been showing up to now, the topic of lineability has reached many different fields of Mathematics, from Linear Chaos to Real and Complex Analysis, passing through Set Theory and Linear and Multilinear Algebra, or even Operator Theory, Topology, Measure Theory and, very recently, even in Probability Theory ([200]). This chapter focuses on a number of disconnected topics that we also have considered interesting and that are (probably) too specialized to be included in any of the previous chapters.

6.1   Series in classical Banach spaces

Firstly, we would like to recall the following notion, that shall be employed on several occasions in this chapter and, particularly, in this section. A family $\{A_\alpha : \alpha \in I\}$ of infinite subsets of $\mathbb{N}$ is called *almost disjoint* if $A_\alpha \cap A_\beta$ is finite whenever $\alpha, \beta \in I$ and $\alpha \neq \beta$. The usual procedure to generate such a family is the following. Let us denote I the irrationals in $[0,1]$ and $\{q_n : n \in \mathbb{N}\}$ denotes $[0,1] \cap \mathbb{Q}$. For every $\alpha \in I$ we choose a subsequence $(q_{n_k})_k$ of $\{q_n : n \in \mathbb{N}\}$ so that $\lim_{k\to\infty} q_{n_k} = \alpha$ and we define $A_\alpha = \{n_k : k \in \mathbb{N}\}$. By construction we obtain that $\{A_\alpha : \alpha \in I\}$ is an almost disjoint family of subsets of $\mathbb{N}$.

If V denotes the set of conditionally convergent series then, clearly, $V \cup \{0\}$ is not a vector space in $CS(\mathbb{K})$, the set of convergent series. The following theorem shows that the set of conditionally convergent series is $\mathfrak{c}$-lineable.

Theorem 6.1.1. *The set* $CS(\mathbb{K})$ *contains a vector space* E *satisfying the following properties:*

(i) *Every* $x \in E \setminus \{0\}$ *is a conditionally convergent series.*

(ii) $dim(E) = \mathfrak{c}$.

(iii) $span\{E \cup c_{00}\}$ *is an algebra and its elements are either elements of* c_{00} *or conditionally convergent series.*

Proof. Let us fix any conditionally convergent series $\sum_i a_i$ such that $a_i \neq 0$ for every $i \in \mathbb{N}$. Consider the family $(A_\alpha)_{\alpha \in I}$ of almost disjoint subsets of $\mathbb{N}$. We have that $card(I) = \mathfrak{c}$. For every $\alpha \in I$ we define x^α given by $x_i^\alpha = a_n$ if $i = n$th element of A_α and $x_i^\alpha = 0$ otherwise. For every $\alpha \in I$ we have that the series $\sum_i x_i^\alpha$ has a conditionally convergent subseries and, therefore, it is conditionally convergent. Let $E = span\{x^\alpha : \alpha \in I\}$. We have that $\{x^\alpha : \alpha \in I\}$ is a linearly independent family, and so $dim(E) = \mathfrak{c}$.

Suppose that $\{\lambda_1, \dots, \lambda_n\} \subset \mathbb{K} \setminus \{0\}$ and $\{\alpha_1, \dots, \alpha_n\} \subset I$. We now see that $z = \lambda_1 x^{\alpha_1} + \dots + \lambda_n x^{\alpha_n}$ is a conditionally convergent series. Indeed, it is easy to check that there exists $A \subset A_{\alpha_1}$, infinite, so that $A_{\alpha_1} \setminus A$ is finite, and satisfying that $A \cap (A_{\alpha_2} \cup \dots \cup A_{\alpha_n}) = \emptyset$. Then $\sum_{i \in A} z_i = \sum_{i \in A} \lambda_i x_i^\alpha$ is conditionally convergent and, therefore, so is z.

To check that $span(E \cup c_{00})$ is an algebra it suffices to notice that

$$(\lambda_1 x^{\alpha_1} + \dots + \lambda_n x^{\alpha_n})(\mu_1 x^{\beta_1} + \dots + \mu_m x^{\beta_m}) \in c_{00}$$

if $\{\lambda_1, \dots, \lambda_n, \mu_1, \dots, \mu_m\} \subset \mathbb{K} \setminus \{0\}$ and $\{\alpha_1, \dots, \alpha_n, \beta_1, \dots, \beta_m\} \subset I$. $\square$

6.2   Dirichlet series

Convergence and divergence of Dirichlet series is also a fruitful source of challenging questions (see, e.g., [197, 338]). Recall that a Dirichlet series is any series of the form $\sum_{k=1}^{\infty} a_k k^{-s}$, where s is complex and a_n is a complex sequence (the case $a_n = 1$ is precisely the case of the Riemann zeta function).

Let $\mathcal{H}^\infty$ be the set of analytic functions defined on $\mathbb{C}_+ :=$ $\{s \in \mathbb{C} : Re(s) > 0)\}$, which are bounded in $\mathbb{C}_+$ and can be represented by a Dirichlet series $f(s) = \sum_{k=1}^{\infty} a_k k^{-s}$, convergent in some half-plane. It is well known that when endowed with the sup norm $\mathcal{H}^\infty$ is a Banach space and, moreover, an algebra.

If $f(s) = \sum_{k=1}^{\infty} a_k k^{-s}$ is a function in $\mathcal{H}^{\infty}$ that converges in some half-plane, a theorem of Bohr asserts that f in fact converges in $\mathbb{C}_+$. But, what do we know about the convergence in the boundary $i\mathbb{R}$? This question, due to Hedenmalm [243], was answered by Bayart, Konyagin and Quéffelec [59] by asserting that there is a Dirichlet series $f(s) = \sum_{k=1}^{\infty} a_k k^{-s}$ such that $\sum_{k=1}^{\infty} a_k k^{-it}$ diverges for all real number t. The lineability and spaceability of the set of such series (Dirichlet series that are bounded in the right half-plane and diverge everywhere in the imaginary axis) were studied in [54] and [53]. In [63] the algebrability of this set was proved.

6.3 Non-convergent Fourier series

Now we shall focus on continuous functions whose Fourier series expansion diverges on a set of Lebesgue measure zero.

The convergence of Fourier series has been studied thoroughly in the past. It came as a considerable surprise when Du Bois-Reymond produced an example of a continuous function $f : \mathbb{T} \to \mathbb{C}$ whose Fourier series is divergent at one point (see [277, pp. 67–73] for a modern reference.) This result can be improved by means of an example of a continuous function whose Fourier series expansion diverges on a set of measure zero ([269, p. 58].) This last result is the best possible, since the Fourier expansion of every continuous function converges almost everywhere (by the famous theorem of Carleson; see, e.g., [277, p. 75]). Moreover, by means of the Baire category theorem, one can show that there exists a G_δ dense subset $E \subset \mathbb{T}$ such that the set of continuous functions whose Fourier expansion diverges on this set is a G_δ dense subset of $\mathcal{C}(\mathbb{T})$ ([351, p. 102]). Also, a remarkable result by Bayart [54] shows that if $\mathcal{F}_E \subset \mathcal{C}(\mathbb{T})$ is the set of continuous functions whose Fourier series expansion diverges on a set of measure zero, E, then $\mathcal{F}_E$ is dense-lineable. This result is, actually, a consequence of the following one due to Aron, Pérez and Seoane [33].

Theorem 6.3.1. *Let $E \subset \mathbb{T}$ be a set of measure zero. Let $\mathcal{F}_E \subset \mathcal{C}(\mathbb{T})$ be the set of continuous functions whose Fourier series expansion diverges at every point $t \in E$. Then $\mathcal{F}_E$ is dense-algebrable.*

The proof of this result is highly technical and we shall provide its proof omitting a few technicalities that the interested reader can check in [33, Theorem 2.1] for a more detailed proof.

In order to begin, we need several different tools, amongst them a couple of technical lemmas and the construction of a special ordering and a double sequence.

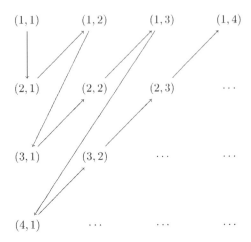

FIGURE 6.1: The order defined in (6.1).

By [269, pp. 57–58], given a set of measure zero $E \subset \mathbb{T}$ we can find a sequence $(\tilde{H}_k)_k$ of trigonometric polynomials, a sequence $(n_k)_k$ of positive integers and a sequence $(E_k)_k$ of measurable subsets of $\mathbb{T}$ such that every $t \in E$ belongs to infinitely many E_k's and such that:

1. $|\tilde{H}_k(s)| \leq 1$ for all $s \in \mathbb{T}$.

2. $|S_{n_k}(\tilde{H}_k, t)| \geq 2^{k^2}$ for every $t \in E_k$.

Without loss of generality we can always assume that $\tilde{H}_k = \sum_{r=0}^{\tilde{a}_k} \widehat{\tilde{H}_k}(r)e^{irt}$, that is, $\mathrm{supp}(\tilde{H}_k) \subset [0, \tilde{a}_k]$, with $(\tilde{a}_k)$ increasing, $\tilde{a}_k > n_k$ for every $k \in \mathbb{N}$. Next, define $H_k(t) = 1 + e^{i(\tilde{a}_k+1)t}\tilde{H}_k(t)$ and let us call $a_k = 2\tilde{a}_k + 1$ (to have $\mathrm{supp}(H_k) \subset [0, a_k]$).

Let $(Q_j)_j$ be a sequence of trigonometric polynomials that is dense in $\mathcal{C}(\mathbb{T})$ and such that $\mathrm{supp}(Q_j) \subset [-q_j, q_j]$. We may clearly suppose that the sequence $(q_j)_j$ is increasing, and we let $b_j = \max\{a_j, q_j\}$.

Next, consider the following well order $\prec$ on $\mathbb{N} \times \mathbb{N}$:

$$(j, k) \prec (j', k') \iff \begin{cases} j + k < j' + k' \quad \text{or} \\ j + k = j' + k' \text{ and } k < k' \end{cases} \tag{6.1}$$

This order can be represented by Figure 6.1, in which each arrow connects a pair with its immediate successor.

With this order we can define by recursion the following double sequence

$(p_k^j)_{j,k}$:

$$p_1^1 = b_1 + 1$$
$$p_k^j = 3kb_k + 2\max\{k, k'\}p_{k'}^{j'} + 1,$$

where (j', k') is the immediate predecessor of (j, k). Of course, $p_k^j < p_{k'}^{j'}$ if and only if $(j, k) \prec (j', k')$. Moreover $(p_k^j)_{j,k}$ is increasing in both indices and verifies, trivially, the following properties:

(P1) If $(j, k) \preceq (j', k')$ then $b_k < p_{k'}^{j'}$.

(P2) $p_d^i > 3db_d$ for every i, d.

(P3) If $(j, k) \prec (i, d)$, then $p_d^i > 2dp_k^j + 3db_d$.

(P4) If $(i, d) \prec (j, k)$, then $p_k^j > 2dp_d^i$.

We now use the above elements to state a technical lemma that will be necessary for the proof of the main result. We omit its proof (we posed it as an Exercise at the end of this chapter, Exercise 6.5).

Lemma 6.3.2. *Let us suppose that*

$$p_d^{i_1} + \ldots + p_d^{i_n} + u = p_{k_1}^{j_1} + \ldots + p_{k_s}^{j_s} + v \tag{6.2}$$

and:

(i) $s \leq n \leq d$.

(ii) $p_d^{i_1} \geq p_d^{i_2} \geq \cdots \geq p_d^{i_n}$ *(that is, $i_1 \geq \cdots \geq i_n$).*

(iii) $p_{k_1}^{j_1} \geq p_{k_2}^{j_2} \geq \cdots \geq p_{k_s}^{j_s}$.

(iv) $|u| \leq db_d$.

(v) $-db_d \leq v \leq b_{k_1} + \ldots + b_{k_s} + db_d$.

Then $n = s$, $i_r = j_r$, $k_r = d$ for every r and $u = v$.

Next, for each $j \geq 1$ and $m \geq 1$, let us define the function

$$f_j^m(t) = \sum_{k=1}^{m} 2^{-k} \cdot e^{ip_k^j t} H_k(t).$$

Thanks to the Weierstrass M-test we have that, for each $j \in \mathbb{N}$, the sequence $(f_j^m)_m$ converges uniformly to a continuous function $f_j \in \mathcal{C}(\mathbb{T})$ with $\|f_j\| \leq 2$. Define $g_j = \frac{1}{j}f_j + Q_j$ and let $\mathcal{A}$ be the algebra generated by $\{g_j\}_j$.

Since $(Q_j)_j$ is a dense sequence in $\mathcal{C}(\mathbb{T})$, it is clear that $(g_j)_j$ (and hence $\mathcal{A}$) is also dense in $\mathcal{C}(\mathbb{T})$. This immediately implies that $\mathcal{A}$ is infinite dimensional.

We will see that every $g \in \mathcal{A} \setminus \{0\}$ has a Fourier series divergent at any $t \in E$. So let us take a generic element of $\mathcal{A}$:

$$g = \sum_{j=1}^{N} \alpha_j g_{i_1^j} \cdots g_{i_{s_j}^j}$$

where $s_1 \geq \cdots \geq s_N$ (s_j is the number of functions that we multiply in each summand of g) and $i_1^j \geq \cdots \geq i_{s_j}^j$ for every j.

The heart of the matter is the next technical lemma:

Lemma 6.3.3. *If $d \geq \max\{s_1, i_1^1, \ldots, i_1^N\}$ then*

(a) $\hat{g}(p_d^{i_1^1} + \cdots + p_d^{i_{s_1}^1} + u) = \alpha_1 \frac{1}{i_1^1 \cdots i_{s_1}^1} 2^{-s_1 d} \widehat{H_d^{s_1}}(u)$ *for $0 \leq u \leq s_1 a_d$.*

(b) $\hat{g}(u) = 0$ *for $u \leq -(p_d^{i_1^1} + \cdots + p_d^{i_{s_1}^1})$.*

(c) *If $l > i_1^1$, then $\hat{g}_l(p_d^{i_1^1} + \cdots + p_d^{i_{s_1}^1} + u) = 0$ for every $0 \leq u \leq s_1 a_d$ and hence $g_l \neq g$.*

Now, after the previous lemmas and considerations, we can give the proof of Theorem 6.3.1:

Let us call $\beta_1 = \alpha_1 \frac{1}{i_1^1 \cdots i_{s_1}^1}$ and take any $d \geq \max\{s_1, i_1^1, \ldots, i_1^N\}$. It follows that

$$\left| S_{p_d^{i_1^1} + \cdots + p_d^{i_{s_1}^1} + \tilde{a}_d + 1 + n_d}(g, t) - S_{p_d^{i_1^1} + \cdots + p_d^{i_{s_1}^1} + \tilde{a}_d}(g, t) \right| =$$

$$= \left| \sum_{-(p_d^{i_1^1} + \cdots + p_d^{i_{s_1}^1} + \tilde{a}_d + 1 + n_d)}^{-(p_d^{i_1^1} + \cdots + p_d^{i_{s_1}^1} + \tilde{a}_d + 1)} \hat{g}(u)e^{iut} + \sum_{p_d^{i_1^1} + \cdots + p_d^{i_{s_1}^1} + \tilde{a}_d + 1}^{p_d^{i_1^1} + \cdots + p_d^{i_{s_1}^1} + \tilde{a}_d + 1 + n_d} \hat{g}(u)e^{iut} \right|. \quad (6.3)$$

By Lemma 6.3.3 (b), the first term of (6.3) is zero. Thus, by Lemma 6.3.3 (a), (6.3) is equal to

$$\beta_1 2^{-s_1 d} \left| \sum_{u=\tilde{a}_d + 1}^{\tilde{a}_d + 1 + n_d} \widehat{H_d^{s_1}}(u)e^{iut} \right|.$$

But, by the definition of H_d, we have that

$$H_d^{s_1}(t) = \sum_{k=0}^{s_1} \binom{s_1}{k} e^{ik(\tilde{a}_d + 1)t} \tilde{H}_d^k(t)$$

and, since $\text{supp}(\tilde{H}_d^k) \subset [0, \infty)$ for any k, we have that for $u \in [\tilde{a}_d + 1, \tilde{a}_d + 1 + n_d]$,

$$\widehat{H_d^{s_1}}(u) = s_1 \widehat{\tilde{H}_d}(u - (\tilde{a}_d + 1)) \quad \text{(recall that } n_d \leq \tilde{a}_d\text{)}.$$

Therefore (6.3) equals

$$\beta_1 s_1 2^{-s_1 d} \left| \sum_{u=0}^{n_d} \widehat{\tilde{H}_d}(u) e^{iut} \right| = \beta_1 s_1 2^{-s_1 d} \left| S_{n_d}(\tilde{H}_d, t) \right| \geq \beta_1 s_1 2^{d^2 - s_1 d},$$

for every $t \in E_d$. Since any $t \in E$ belongs to infinitely many E_d's, we have that

$$\limsup_n |S_n(g, t)| = \infty \quad \text{for every } t \in E.$$

To finish the proof of Theorem 6.3.1 it only remains to prove that $\mathcal{A}$ is infinitely generated. Now, if $\mathcal{A}$ were finitely generated, there would exist an $l \in \mathbb{N}$ such that every g_h, $h \in \mathbb{N}$, can be expressed as

$$g_h = \sum_{j=1}^{N} \alpha_j g_{i_1^j} \cdots g_{i_{s_j}^j} \tag{6.4}$$

with $s_1 \geq \cdots \geq s_N$ and $l > i_1^j \geq \cdots \geq i_{s_j}^j$ for every j. Taking $h = l$ in (6.4) this contradicts Lemma 6.3.3 (c), which finishes the proof of Theorem 6.3.1.

Also, let us recall a very interesting result by F. Bayart [54] stating that (under certain conditions, [54, Theorem 3]) if $X \subset L_1(\mathbb{T})$ is a Banach space and $E \subset \mathbb{T}$ then there exists a dense subspace X_0 of X such that, for every $f \in X_0 \setminus \{0\}$ and every $t \in E$, the Fourier series $(S_n(f, t))_{n \geq 0}$ is unboundedly divergent.

In the special case in which E is *countable*, even universality properties of Fourier series of continuous functions are found to be compatible with lineability; see Section 4.5 and [84].

6.4   Norm-attaining functionals

Although, as we said in the Preface, this monograph would surely be missing plenty of information, we would not want to leave without mentioning the classical problem of the lineability of the set of *norm-attaining functionals*. Recall the well known theorem by Bishop and Phelps (1961, [122]), stating that for any Banach space X, the set $\mathcal{NA}(X) := \{\varphi \in X^* : \varphi \text{ attains its norm}\}$ is a dense subset of X^*.

It is natural to ask about the *lineability* of the set $\mathcal{NA}(X)$. For many specific spaces X, the answer is simple. For example,

$$\mathcal{NA}(c_0) = \{x = (x_n) \in \ell_1 : x_n = 0 \text{ for all sufficiently large } n\}.$$

In this case, $\mathcal{NA}(c_0)$ is itself a dense vector subspace of $c_0^* = \ell_1$. However, note that $\mathcal{NA}(c_0)$ contains no infinite dimensional closed subspace, so here we have lineability without spaceability. On the other hand,

$$\mathcal{NA}(\ell_1) = \{y = (y_n) \in \ell_\infty : \|y\|_\infty = \max_n |y_n|\}$$

is not a subspace. Indeed, both $y = (-1, 0, 0, \dots)$ and $y' = (1, \frac{1}{2}, \frac{2}{3}, \dots, \frac{n}{n+1}, \dots)$ are in $\mathcal{NA}(\ell_1)$ although $y + y' \notin \mathcal{NA}(\ell_1)$. Furthermore, $\mathcal{NA}(\ell_1)$ contains both dense vector subspaces and infinite dimensional closed subspaces. This raises the question:

> *Does every dual space contain a 2-dimensional subspace consisting of norm-attaining functionals?*

This question (first posed by Godefroy in [226]) has been recently studied in [43] by Bandyopadhyay and Godefroy, who provided, among other results, conditions that ensure that $\mathcal{NA}(X)$ is not spaceable; see also the recent works [1, 214] for more results on the linear structure of $\mathcal{NA}(X)$ and its complement. Very recently, García-Pacheco and Puglisi [218] showed that every Banach space admitting an infinite dimensional separable quotient can be equivalently renormed in such a way that the set of its norm attaining functionals contains an infinite dimensional linear subspace. However, it was not until 2015 when the general problem of the 2-lineability of the set of norm-attaining functionals was solved *in the negative* by M. Rmoutil. Rmoutil's result has a highly technical proof and we refer the interested reader to [343] for the complete details of it.

6.5   Annulling functions and sequences with finitely many zeros

As we already mentioned in Section 1.4, a function $f \in \mathcal{C}[0, 1]$ is said to be an annulling function on $[a, b] \subset [0, 1]$ if f has infinitely many zeros in $[a, b]$ (see [196, Definition 2.1]). Recently, it was proved that this set is spaceable, [196, Corollary 3.8]. However, the techniques employed in the proof are highly technical and require some tools that are far from the purpose of this book. We state the result by Enflo, Gurariy and Seoane [196, Corollary 3.8] for the interested reader.

Theorem 6.5.1. *For every infinite dimensional subspace X of $\mathcal{C}[0, 1]$ there is an infinite dimensional subspace Y of X and a sequence $\{t_k\}_{k\in\mathbb{N}} \subset [0, 1]$, of pairwise different elements, such that $y(t_k) = 0$ for every $k \in \mathbb{N}$ and every $y \in Y$. In particular, for every infinite dimensional subspace X of $\mathcal{C}[0, 1]$ the set of functions in X having infinitely many zeros in $[0, 1]$ is spaceable in X.*

On a totally different framework, but related to the study of the amount of zeros of functions on a given interval, let us recall a question originally posed by Aron and Gurariy in 2003, where they asked whether there exists an infinite dimensional subspace of ℓ_∞ every nonzero element of which has only a finite number of zero coordinates. If we denote by P the set of odd prime numbers and we call $x_p = \left(\frac{1}{p}, \frac{1}{p^2}, \frac{1}{p^3}, \frac{1}{p^4}, \cdots\right) \in \ell_\infty$ $(p \in P)$ then it is easy to see that any nontrivial finite linear combination of $\{x_p : p \in P\}$ satisfies the desired property. Some partial answers to the original problem were recently given (for other sequence spaces) by García-Pacheco, Pérez-Eslava and Seoane in ([217], 2010). In this paper the authors proved, among other results in this direction, the following:

Theorem 6.5.2. *Let X be an infinite dimensional Banach space with a normalized Schauder basis $(e_n)_{n \in \mathbb{N}}$. There exists a linear space $V \subset X$ such that:*

(a) *If $a = \sum_{n=1}^{\infty} a\,[n]\,e_n \in V \setminus \{0\}$ then $\mathrm{card}\,\{n \in \mathbb{N} : a\,[n] = 0\} < \infty$.*

(b) *If $a, b \in V$ then $\sum_{n=1}^{\infty} a_n b_n e_n \in V$.*

(c) *V is dense and not barrelled.*

However, recently ([153]) Cariello and Seoane gave a negative answer to the original question posed by Aron and Gurariy. Actually, if M denotes the subset of ℓ_∞ formed by sequences having only a finite number of zero coordinates, then they showed more than that:

1. They answered the question not just for ℓ_∞ but for all the Banach ℓ_p spaces (i.e. for all the range $p \in [1, \infty]$).

2. They also showed that, inside every infinite dimensional closed subspace of ℓ_p $(p \in [1, \infty])$, there are basic sequences spanning an infinite dimensional closed subspace X, every element of which has infinitely many zeros (i.e. we study the complementary of the set M above as well and, also, for all ranges of $p \in [1, \infty]$).

3. Also, the previous closed subspace X is complemented in ℓ_p for $p \in [1, \infty)$.

4. They also showed the counterpart of this question. If we remove the assumption of being *closed* then the answer is *positive*.

5. At the same time, if we start with a closed algebra (and not just a closed subspace!) in ℓ_p for $p \in [1, \infty]$, they also showed that there exists a closed subalgebra every element of which has infinitely many zeros.

Let us reproduce here below the proof of item 4 above. In order to do that, let us denote by $Z(X)$ the subset of X formed by sequences having only a finite number of zero coordinates.

Proposition 6.5.3. *The set $Z(X)$ is maximal algebrable for $X = c_0$ or ℓ_p, $p \in [1, \infty]$.*

Proof. For every real number $p \in {]}0, 1{[}$ denote

$$x_p = \left(p^1, p^2, p^3, \dots\right),$$

and let $V = \operatorname{span}\{x_p : p \in {]}0, 1{[}\}$. Notice that $V \subset X$, for $X = c_0$ or ℓ_p, $p \in [1, \infty]$.

Next, take any $x \in V \setminus \{0\}$. We shall show that $x \in Z(X)$. We can write x as

$$x = \sum_{j=1}^{N} \lambda_j x_{p_j},$$

with $N \in \mathbb{N}$, $p_j \in {]}0, 1{[}$ for every $j \in \{1, 2, \dots, N\}$, $p_N > p_{N-1} > \dots > p_1$, and $(\lambda_j)_{j=1}^{N} \subset \mathbb{C}$. Let us suppose that there exists an increasing sequence of positive integers $(m_k)_{k \in \mathbb{N}}$ such that $x(m_k) = 0$ for every $k \in \mathbb{N}$. Then, we have

$$0 = \sum_{j=1}^{N} \lambda_j p_j^{m_k}$$

for every $k \in \mathbb{N}$. Dividing the last identity by $p_N^{m_k}$, we obtain (for every $k \in \mathbb{N}$)

$$0 = \sum_{j=1}^{N-1} \lambda_j \left(\frac{p_j}{p_N}\right)^{m_k} + \lambda_N. \tag{6.5}$$

Now, since $0 < \dfrac{p_j}{p_N} < 1$ for every $j \in \{1, 2, \dots, N-1\}$ and $\lim_{k \to \infty} m_k = \infty$, we have $\lim_{k \to \infty} \left(\dfrac{p_j}{p_N}\right)^{m_k} = 0$. Thus, $\lambda_N = 0$ in equation (6.5). By induction, we can easily obtain $\lambda_j = 0$ for every $j \in \{1, 2, \dots, N\}$. This is a contradiction, since $x \neq 0$.

This argument also shows that the set $\{x_p : p \in {]}0, 1{]}\}$ is linearly independent, so V is $\mathfrak{c}$-dimensional (where $\mathfrak{c}$ stands for the continuum) and, thus, $Z(X)$ is maximal lineable for $X = c_0$ or ℓ_p, $p \in [1, \infty]$.

Now let $x_p, x_q \in \{x_r, \ r \in {]}0, 1{[} \}$. Notice that the coordinatewise product of x_p and x_q is $x_{pq} \in \{x_r, \ r \in {]}0, 1{[} \}$. Therefore the algebra generated by $\{x_r, \ r \in {]}0, 1{[} \}$ is the subspace generated by $\{x_r, \ r \in {]}0, 1{[} \}$ which is V.

Consider any countable subset $W \subset V$. The subalgebra generated by W is a vector space generated by finite products of elements of W, but the set of finite products of elements belonging to a countable set is still countable. Therefore the subalgebra generated by W has countable dimension and, thus, W cannot be a set of generators for the algebra V, since $\dim(V)$ is uncountable. Therefore any set of generators of V is uncountable. $\qquad \square$

The following result is a straightforward consequence of Proposition 6.5.3.

Corollary 6.5.4. *The set $Z(X)$ is maximal lineable for $X = c_0$ or ℓ_p, $p \in [1, \infty]$.*

The proofs of items 1, 2, 3 and 5 are much more complex, and we refer the interested reader to [153] for the details.

6.6 Sierpiński-Zygmund functions

This section shall focus on a very special class of real valued functions, the so-called *Sierpiński-Zygmund functions*. First of all, let us recall that, as a consequence of the classic Luzin's theorem, we have that for every measurable function $f \colon \mathbb{R} \to \mathbb{R}$, there is a measurable set $S \subset \mathbb{R}$ (of infinite measure) such that $f|_S$ is continuous. Certainly, a natural question would be whether similar results could be obtained for arbitrary functions (not necessarily measurable). That is, given any arbitrary function $f \colon \mathbb{R} \to \mathbb{R}$, can one find a "large" subset $S \subset \mathbb{R}$ for which $f|_S$ is continuous? In 1922, Blumberg [123] provided an affirmative answer to this question:

Theorem 6.6.1. *Let $f \colon \mathbb{R} \to \mathbb{R}$ be an arbitrary function. There exists a dense subset $S \subset \mathbb{R}$ such that the function $f|_S$ is continuous.*

Blumberg's proof of his theorem (see, e.g., [272, p. 154]) shows that the set S above is, actually, countable. We could wonder whether we can choose the subset S in Blumberg's theorem to be uncountable. A (partial) negative answer was given in 1923 [368] by Sierpiński and Zygmund.

Theorem 6.6.2. *There exists a function $f \colon \mathbb{R} \to \mathbb{R}$ such that, for any set $Z \subset \mathbb{R}$ of cardinality $\mathfrak{c}$, the restriction $f|_Z$ is not a Borel map (and, in particular, not continuous).*

Now, using a standard terminology we shall say that a function $f \colon \mathbb{R} \to \mathbb{R}$ is a *Sierpiński-Zygmund function* if it satisfies the condition in Sierpiński-Zygmund's theorem, and we denote

$$\mathcal{SZ} = \{ f \colon \mathbb{R} \to \mathbb{R} : f \text{ is a Sierpiński-Zygmund function} \}.$$

Let us recall some known results about the class $\mathcal{SZ}$ (see [212]).

It is known that if the Continuum Hypothesis (CH) holds then the restriction of a function in $\mathcal{SZ}$ to any uncountable set can not be continuous (see, e.g., [272, pp. 165, 166]). Also, CH is necessary in this frame. Shinoda proved in 1973 [365] that if Martin's Axiom and the negation of CH hold then, for every $f \colon \mathbb{R} \to \mathbb{R}$, there exists an uncountable set $Z \subset \mathbb{R}$ such that

$f|_Z$ is continuous. The functions in $\mathcal{SZ}$ are never measurable and, although it is possible to construct them being injective, they are nowhere monotone in a very strong way: their restriction to any set of cardinality $\mathfrak{c}$ is not monotone. In 1997, Balcerzak, Ciesielski and Natkaniec showed in [40] that, assuming the set-theoretical condition $\operatorname{cov}(\mathcal{M}) = \mathfrak{c}$ (which is true under Martin's Axiom or CH), there exists a Darboux function that is in $\mathcal{SZ}$ as well. They prove also that there exists a model of ZFC (Zermelo-Fraenkel-Axiom of Choice) in which there are no such functions (see also [170, 171, 335]). Later, Gámez, Muñoz, Sánchez and Seoane (2010) proved in [209, Theorems 5.6 and 5.10] that the set $\mathcal{SZ}$ is $\mathfrak{c}^+$-lineable and also $\mathfrak{c}$-algebrable. As a consequence, assuming that $\mathfrak{c}^+ = 2^{\mathfrak{c}}$ (which follows, for instance, from the Generalized Continuum Hypothesis or GCH), $\mathcal{SZ}$ would be $2^{\mathfrak{c}}$-lineable. Also, in 2010 [210, Corollary 2.11], Gámez, Muñoz and Seoane proved that $\mathcal{SZ}$ is actually $d_{\mathfrak{c}}$-lineable, where $d_{\mathfrak{c}}$ is a cardinal invariant defined as

$$d_{\mathfrak{c}} = \min\{\operatorname{card} F \;:\; F \subset \mathbb{R}^{\mathbb{R}}, \, (\forall \varphi \in \mathbb{R}^{\mathbb{R}}) \, (\exists f \in F) \, (\operatorname{card}(f \cap \varphi) = \mathfrak{c})\}.$$

This cardinal can take as value any regular cardinal between $\mathfrak{c}^+$ and $2^{\mathfrak{c}}$, depending on the set-theoretical axioms assumed. Later, in [50, Theorem 2.6], Bartoszewicz, Głąb, Pellegrino and Seoane showed that $\mathcal{SZ}$ is actually strongly κ-algebrable for some $\mathfrak{c}^+ \leq \kappa \leq 2^{\mathfrak{c}}$ if there is in $\mathfrak{c}$ an almost disjoint family of cardinality κ (see Definition 6.6.3 below and, for a particular case, the beginning of Section 6.1). Assuming either Martin's Axiom, or CH, or $2^{<\mathfrak{c}} = \mathfrak{c}$, this κ can be chosen to be $2^{\mathfrak{c}}$, so we would have that $\mathcal{SZ}$ is strongly $2^{\mathfrak{c}}$-algebrable.

Definition 6.6.3. Let S be a set of cardinality κ. We say that a family $\mathcal{F} \subset \mathcal{P}(S)$ is an *almost disjoint family* in S if the following conditions hold:

(1) If $A \in \mathcal{F}$ then $\operatorname{card} A = \kappa$.

(2) If $A, B \in \mathcal{F}$, $A \neq B$, then $\operatorname{card}(A \cap B) < \kappa$.

Until very recently, it was not known whether any additional set-theoretical assumptions were needed or not in order to show the $2^{\mathfrak{c}}$-strongly algebrability (and the $2^{\mathfrak{c}}$-lineability) of $\mathcal{SZ}$. Nevertheless, in [212] Gámez and Seoane showed the following:

Theorem 6.6.4. *Let κ be a cardinal number such that $\mathfrak{c}^+ \leq \kappa \leq 2^{\mathfrak{c}}$. The following are equivalent:*

(a) *$\mathcal{SZ}$ is strongly κ-algebrable.*

(b) *$\mathcal{SZ}$ is κ-algebrable.*

(c) *$\mathcal{SZ}$ is κ-lineable.*

(d) *There exists an additive group $\mathcal{G} \subset \mathcal{SZ} \cup \{0\}$ of size (i.e., cardinality) κ.*

(e) *There exists in $\mathfrak{c}$ an almost disjoint family of cardinality κ.*

Now, we review a series of results on almost disjoint families, all of which can be found in [280]. On the one hand, recall that under ZFC there is an almost disjoint family of cardinality $\mathfrak{c} = 2^{\aleph_0}$ in $\aleph_0$. On the other, the existence of an almost disjoint family of cardinality $2^{\aleph_1}$ in $\aleph_1$ is undecidable. Also, and under the set-theoretical assumption $2^{<\mathfrak{c}} = \mathfrak{c}$, there exists an almost disjoint family of cardinality $2^{\mathfrak{c}}$ in $\mathfrak{c}$.

Let us point out that, from these previous results, we infer that it is consistent with ZFC that $\mathcal{SZ} \cup \{0\}$ contains a vector space of dimension $2^{\mathfrak{c}}$. By means of the *forcing technique*, in [212] the authors proved that the contrary is also consistent, that is:

Theorem 6.6.5. *The $2^{\mathfrak{c}}$-lineability (maximal lineability) of the set of Sierpiński-Zygmund functions is undecidable.*

We shall now present a proof of the previous result. We warn the reader about the technicalities of it, since the use of forcing is, indeed, compulsory. Although this would be the only time throughout this monograph in which we shall employ this technique, we believe that it is illustrative of how complex a lineability problem can actually get. Let us show that, in some model of ZFC, there is no almost disjoint family in $\mathfrak{c}$ whose cardinality is $2^{\mathfrak{c}}$.

Proof of Theorem 6.6.5. Let us take a model M of $ZFC + GCH$ as ground model. Let $\mathbb{P} \in M$ be an Easton forcing obtained from an index function $E(\aleph_0) = \aleph_2$, $E(\aleph_1) = \aleph_4$; see [280, Ch. VIII, §4]. (This forcing is equivalent to the iteration of $\mathrm{Fn}(\aleph_4 \times \aleph_1, 2, \aleph_1)$ and $\mathrm{Fn}(\aleph_2 \times \aleph_0, 2, \aleph_0)$ ([280, Lemma VIII 4.3]).)

Let G be a generic filter for $\mathbb{P}$. In the generic extension $M[G]$ we have ([280, Theorem VII 4.7]) that $\mathfrak{c} = 2^{\aleph_0} = \aleph_2$, $2^{\aleph_1} = \aleph_4$ and also $2^{\mathfrak{c}} = 2^{\aleph_2} = \aleph_4$. We shall see that in this generic extension, there is no almost disjoint family of cardinality $\aleph_4$ in ω_2. Indeed, suppose that some $p \in \mathbb{P}$ forces the existence of a family of $\aleph_4$ almost disjoint subsets of ω_2. Then there would be $\mathbb{P}$-names $\dot{E}_\alpha$ for $\alpha < \omega_4$ such that p forces that each $\dot{E}_\alpha \subset \omega_2$ and that $\mathrm{card}(\dot{E}_\alpha \cap \dot{E}_\beta) < \aleph_2$, whenever $\alpha < \beta$. By [280, Lemma VIII 4.4], $\mathbb{P}$ has the $\aleph_2$-cc. Therefore, using [280, Lemma VIII 5.6], whenever $\alpha < \beta$, there is a $\gamma_{\alpha,\beta} < \omega_2$ such that p forces that $\dot{E}_\alpha \cap \dot{E}_\beta \subset \gamma_{\alpha,\beta}$.

Next, using the $(2^{\aleph_2})^+ \to (\aleph_3)^2_{\aleph_2}$ instance of the Erdös-Rado Partition theorem (see [280, p. 290 (B1)] or [288, Theorem 3.13]), which is equivalent to $\aleph_4 \to (\aleph_3)^2_{\aleph_2}$ because GCH holds in M, we have that there exists a subset $H \subset \omega_2$ such that $\mathrm{card}\, H = \aleph_3$ and $\gamma < \omega_2$ such that p forces that $\dot{E}_\alpha \cap \dot{E}_\beta \subset \gamma$ whenever $\alpha, \beta \in H$, $\alpha < \beta$. If for every $\alpha \in H$ we define $F_\alpha = \dot{E}_\alpha \setminus \gamma$ and $\dot{F}_\alpha$ is a $\mathbb{P}$-name for it, we have:

1. For every $\alpha \in H$, p forces that $\dot{F}_\alpha \subset \omega_2$.

2. For every $\alpha \in H$, p forces that $\dot{F}_\alpha \neq \varnothing$, because $\mathrm{card}\, \dot{E}_\alpha = \aleph_2$ and $\mathrm{card}\, \gamma < \aleph_2$.

3. If $\alpha, \beta \in H$, $\alpha < \beta$, then p forces that $\dot{F}_\alpha \cap \dot{F}_\beta = \varnothing$.

Thus, we get a contradiction, because in $M[G]$ the family $\{F_\alpha\}_{\alpha \in H}$ is a *pairwise disjoint* family of $\aleph_3$ many elements in ω_2.   □

This would be the first time in which one encounters a (highly nontrivial!) undecidable proposition in this theory of lineability and spaceability.

6.7   Non-Lipschitz functions with bounded gradient

A standard result from Real Analysis states that, for any interval I, a differentiable function $f : I \longrightarrow \mathbb{R}$ is Lipschitz if and only if it has bounded derivative. One could think if the result still holds under weaker conditions. In [383] it is provided an example of a continuous non-Lipschitz function, which is differentiable almost everywhere and has bounded derivative almost everywhere. Recently, Jímenez-Rodríguez, Muñoz and Seoane [257, Theorems 2.1 and 3.1] proved the following results.

Theorem 6.7.1.   (a) *The set of continuous functions on $[0, 1]$ which are a.e. differentiable, with a.e. bounded derivative and not Lipschitz, is $\mathfrak{c}$-lineable.*

(b) *The set of differentiable functions $f : \mathbb{R} \longrightarrow \mathbb{R}^2$ that do not enjoy the classical Mean Value theorem is $\mathfrak{c}$-lineable.*

(c) *The set of differentiable functions $f : D \to \mathbb{R}$ with bounded gradient, non-Lipschitz and therefore not satisfying the classical Mean Value theorem is $\mathfrak{c}$-lineable* [257, Theorem 3.1], *where*

$$D = \{(x, y) \in \mathbb{R}^2 : x^2 + y^2 < 1\} \setminus \{(x, y) \in \mathbb{R}^2 : x = 0 \text{ and } y > 0\},$$

which is a path connected, nonconvex set.

We remark that the above results can be improved to dense-lineability by means of [26, Theorem 2.2 and Remark 2.5]. Also, just recently [256], item (a) above has been improved by showing that c_0 is isometrically isomorphic to a subspace of Cantor-Lebesgue functions, that is, continuous non-Lipschitz functions $f : [0, 1] \to \mathbb{R}$ with $f' = 0$ a.e. This, in particular, gives spaceability in $\mathcal{C}[0, 1]$ of the set defined in (a).

Let us illustrate the above Theorem 6.7.1 by seeing the proof of its part (c).

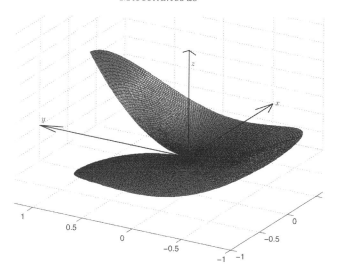

FIGURE 6.2: The function $f_2 : D \to \mathbb{R}$ from Theorem 6.7.1(c).

Proof of Theorem 6.7.1 (c). Define, for every $\lambda > 1$, $f_\lambda : D \longrightarrow \mathbb{R}$ as

$$
f_\lambda(x,y) = \begin{cases}
y^\lambda \arctan\left(\dfrac{\lambda y}{x}\right) & \text{if } x < 0, \\[2mm]
y^\lambda \left[\arctan\left(\dfrac{\lambda y}{x}\right) + \pi\right] & \text{if } x > 0, \\[2mm]
\dfrac{\pi}{2} y^\lambda & \text{if } x = 0.
\end{cases}
$$

We shall prove that the set $H = \{f_\lambda : \lambda > 1\}$ is linearly independent and that span(H) is in the set we are studying (see Figure 6.2 for a sketch of these functions).

Assume first $g(x,y) = \sum_{i=1}^{k} \alpha_i f_{\lambda_i}(x,y) = 0$ for $\lambda_1, \ldots, \lambda_k > 1$ and $\alpha_1, \ldots, \alpha_k \in \mathbb{R}$. Then, for every $-1 < y < 0$ we obtain

$$
0 = g(0, y) = \frac{\pi}{2} \sum_{i=1}^{k} \alpha_i y^{\lambda_i},
$$

which allows us to conclude $\alpha_i = 0$ for every $1 \le i \le k$, since $\{y^\lambda : \lambda > 1\}$ is a linearly independent set over $(-1, 0)$. To check now that g (assumed $\alpha_i \ne 0$ for all $i = 1, \ldots, k$) is in our set we shall work with the functions f_λ alone, since the arguments we are going to use can be easily extended to finite linear combinations. Let us obtain first $\frac{\partial f_\lambda}{\partial x}(x_0, y_0)$. If $x_0 \ne 0$ we can simply differentiate f_λ to get $\frac{\partial f_\lambda}{\partial x}(x_0, y_0) = -\lambda \frac{y_0^{\lambda+1}}{x_0^2 + (\lambda y_0)^2}$.

For $x_0 = 0$, we get

$$
\lim_{t \to 0^+} \frac{f_\lambda(t, y_0) - f_\lambda(0, y_0)}{t} = \lim_{t \to 0^+} \frac{y_0^\lambda \left[\frac{\pi}{2} + \arctan \left(\frac{\lambda y_0}{t} \right) \right]}{t}
$$

$$
= \lim_{t \to 0^+} \frac{y_0^\lambda (-y_0 \lambda)}{t^2 + (\lambda y_0)^2} = -\frac{y_0^{\lambda - 1}}{\lambda},
$$

and identical calculations lead to the same value for $\lim_{t \to 0^-} \frac{f_\lambda(t, y_0) - f_\lambda(0, y_0)}{t}$.

Hence, we obtain

$$
\frac{\partial f_\lambda}{\partial x}(x_0, y_0) = \begin{cases} -\lambda \dfrac{y_0^{\lambda + 1}}{x_0^2 + (\lambda y_0)^2} & \text{if } x \neq 0, \\[2mm] -\dfrac{y_0^{\lambda - 1}}{\lambda} & \text{if } x_0 = 0. \end{cases}
$$

Analogously one can obtain

$$
\frac{\partial f_\lambda}{\partial y}(x_0, y_0) = \begin{cases} \lambda y_0^{\lambda - 1} \left(\arctan \left(\dfrac{\lambda y_0}{x_0} \right) + \dfrac{x_0 y_0}{x_0^2 + (\lambda y_0)^2} \right) & \text{if } x < 0, \\[3mm] \lambda y_0^{\lambda - 1} \left(\arctan \left(\dfrac{\lambda y_0}{x_0} \right) + \dfrac{x_0 y_0}{x_0^2 + (\lambda y_0)^2} + \pi \right) & \text{if } x > 0, \\[3mm] \dfrac{\pi}{2} \lambda y_0^{\lambda - 1} & \text{if } x = 0. \end{cases}
$$

Thus, the partial derivatives exist. Let us now see that they are continuous (which would prove $f_\lambda \in \mathcal{C}^1(\mathbb{R}^2, \mathbb{R})$), for which we shall only have to focus on the points of the form $(0, y_0)$. Indeed,

$$
\lim_{(x,y) \to (0, y_0)} \left| \frac{\partial f_\lambda}{\partial x}(x, y) - \frac{\partial f_\lambda}{\partial x}(0, y_0) \right| = \lim_{(x,y) \to (0, y_0)} \left| \frac{y_0^{\lambda - 1}}{\lambda} - \frac{\lambda y^{\lambda + 1}}{x_0^2 + (\lambda y_0)^2} \right| = 0.
$$

In a similar fashion,

$$
\lim_{(x,y) \to (0, y_0)} \left| \frac{\partial f_\lambda}{\partial y}(x, y) - \frac{\partial f_\lambda}{\partial y}(0, y_0) \right| = 0,
$$

from which it follows that the partial derivatives are continuous.

Assume now $x_0 \neq 0$. Then $\left| \frac{\partial f_\lambda}{\partial x}(x_0, y_0) \right| \leq \lambda |y_0|^{\lambda - 1} \leq \lambda$. Similarly $\left| \frac{\partial f_\lambda}{\partial x}(0, y_0) \right| \leq \frac{1}{\lambda} < 1 < \lambda$, which gives us $\left| \frac{\partial f_\lambda}{\partial x}(x_0, y_0) \right| \leq \lambda$ for every $(x_0, y_0) \in D$. In an analogous way, $\left| \frac{\partial f_\lambda}{\partial y}(x_0, y_0) \right| \leq \lambda \frac{1 + 3\pi}{2}$ for all $(x_0, y_0) \in D$, and hence we deduce that Df_λ is bounded on D.

Finally, suppose that f_λ is Lipschitz with constant $K > 0$. Thus, given $(x, y), (\hat{x}, \hat{y}) \in D$ we have

$$
|f_\lambda(\hat{x}, \hat{y}) - f_\lambda(x, y)| \leq K \| (\hat{x} - x, \hat{y} - y) \|_2.
$$

Now, if we fix $\hat{y} = y > 0$ and force $x > 0$ and $\hat{x} < 0$ we obtain

$$|f_\lambda(\hat{x}, y) - f_\lambda(x, y)| = \left| y^\lambda \left[\arctan\left(\frac{\lambda y}{\hat{x}}\right) - \arctan\left(\frac{\lambda y}{x}\right) - \pi \right] \right|$$
$$\leq K\|(\hat{x} - x, 0)\|_2.$$

But $|f_\lambda(\hat{x}, y) - f_\lambda(x, y)| \xrightarrow[\hat{x} \to 0, x \to 0]{} 2\pi|y|^\lambda \neq 0$ for $y \neq 0$ and $K\|(\hat{x} - x, 0)\|_2 \xrightarrow[\hat{x} \to 0, x \to 0]{} 0$, which makes it impossible for f_λ to be Lipschitz. $\qquad \square$

To finish this section, let us recall that Balcerzak, Bartoszewicz and Filipczak [39] have shown the dense strongly $\mathfrak{c}$-algebrability in $\mathcal{C}[0, 1]$ of the smaller set of strongly singular functions. Recall that a function $f : [0, 1] \to \mathbb{R}$ is called strongly singular if $f \in CBV[0, 1]$ (the space of bounded variation continuous functions on $[0, 1]$), $f' = 0$ a.e. and f is not constant on any subinterval of $[0, 1]$. The spaceability of the last set (with nonseparable subspace) in the space $CBV[0, 1]$ (endowed with the norm $\|f\| = \sup_{[0,1]} +\text{Variation}_{[0,1]} f$) is also shown in [39]. See also the section Notes and Remarks of Chapter 7 and Exercise 7.4.

6.8 The Denjoy-Clarkson property

Let us recall that the derivative of any real valued function enjoys the Denjoy-Clarkson property: if $u : \mathbb{R} \to \mathbb{R}$ is everywhere differentiable, then the counterimage through u' of any open subset of $\mathbb{R}$ is either empty or has positive Lebesgue measure. The extension of this result to several (real) variables is known as the Weil Gradient Problem [378] and, after being an open problem for almost 40 years, was eventually solved in the negative for $\mathbb{R}^2$ by Buczolich [147]. His example was simplified in some subsequent articles, in particular very recently by Deville and Matheron [182]. They constructed an everywhere differentiable function on $Q = [0, 1]^n$ and then extended it through $\mathbb{Z}^n$-periodicity to the whole of $\mathbb{R}^n$ obtaining a bounded, everywhere differentiable function $f : \mathbb{R}^n \to \mathbb{R}$ such that

1. f and ∇f vanish on the boundary of Q,

2. $\|\nabla f\| = 1$ almost everywhere in $\mathbb{R}^n$ and $\|\nabla f(x)\| \leq 1$ for all $x \in \mathbb{R}^n$.

Thus it is clear that f fails the Denjoy-Clarkson property, since $(\nabla f)^{-1}(B(0, 1))$ is a nonempty set of zero Lebesgue measure.

In [213] the authors employed the above function (which they called the Deville-Matheron function) in order to show the following result:

Theorem 6.8.1. *For every $n \geq 2$ there exists an infinite dimensional Banach* space of differentiable functions on $\mathbb{R}^n$ which (except for 0) fail the Denjoy-Clarkson property.

Proof. For the sake of simplicity we will work with $\mathbb{R}^2$, but everything is also valid for $\mathbb{R}^n$ with $n \geq 2$.

For every $(k,l) \in \mathbb{Z}^2$ let $Q_{k,l} = [k, k+1] \times [l, l+1]$ and $f_{k,l} : \mathbb{R}^2 \to \mathbb{R}$ defined as the restriction of f on $Q_{k,l}$ and 0 everywhere else. Of course, now $(\nabla f_{k,l})^{-1}(B(0,1))$ has infinite Lebesgue measure, but the function still fails the Denjoy-Clarkson property. For this we just need to show that there exist an $0 < \alpha < 1$ and a point $a_\alpha \in Q_{k,l}$ with $\|\nabla f_{k,l}(a_\alpha)\| = \alpha$.

For the sake of completeness we give a very simple proof of the fact that for every $\alpha \in [0, \|f\|_\infty]$, there exists a point $a_\alpha \in Q_{k,l}$ with $\|\nabla f_{k,l}(a_\alpha)\| = \alpha$. This is all we need to know about the intermediate values of the gradient in order to reach the conclusion of our theorem. By Weierstrass' theorem and connectedness we have $|f|(Q) = [0, \|f\|_\infty]$. Hence, by changing f by $-f$ if necessary, we have $[0, \|f\|_\infty] \subset f(Q)$. Let $\alpha \in (0, \|f\|_\infty]$. Consider $b \in Q_{k,l}$ with $f_{k,l}(b) = \alpha$. Of course, b does not belong to the boundary of $Q_{k,l}$. Define the differentiable function $h : Q_{k,l} \to \mathbb{R}$ by

$$h(x,y) = f_{k,l}(x,y) + \alpha(x-k).$$

Since $h(b) > f_{k,l}(b) = \alpha = \max_{\partial Q_{k,l}} h$, it follows that h attains its maximum in the interior of $Q_{k,l}$. Thus, there exists a point a_α with $0 = \nabla h(a_\alpha) = \nabla f_{k,l}(a_\alpha) + (\alpha, 0)$ and so $\|\nabla f(a_\alpha)\| = \alpha$.

Now we choose $0 < \delta < \min\{\alpha, 1-\alpha\}$ and we have that

$$(\nabla f_{k,l})^{-1}(B(\nabla f(a_\alpha), \delta)) \subset (\nabla f)^{-1}(B(0,1)) \cap Q$$

is a nonempty set of zero Lebesgue measure.

We define $\Phi : c_0(\mathbb{N}^2) \to C_b(\mathbb{R}^2)$ which associates to every double sequence $d = (d_{k,l})_{k,l}$ the function $\Phi_d(x,y) = \sum d_{k,l} f_{k,l}(x,y)$. Since $f_{k,l} = \nabla f_{k,l} = 0$ on $\mathbb{R}^2 \setminus \mathrm{int}(Q_{k,l})$, the pieces glue together well and so Φ_d is an everywhere differentiable function. Clearly we have

$$\nabla \Phi_d(x,y) = \sum d_{k,l} \nabla f_{k,l}(x,y)$$

and so $\|\nabla \Phi_d(x,y)\| \in \{|d_{k,l}|, (k,l) \in \mathbb{N}^2\}$ for almost all $(x,y) \in \mathbb{R}^2$. Given $d \neq 0$, there exists $d_{k_0,l_0} \neq 0$. Choose $\alpha \in (0, \|f\|_\infty]$ such that $\alpha |d_{k_0,l_0}| \notin K = \{|d_{k,l}| : (k,l) \in \mathbb{N}^2\} \cup \{0\}$. Since $|d_{k_0,l_0}|\alpha$ does not belong to the compact set K the distance δ from $|d_{k_0,l_0}|\alpha$ to K is positive. Consider $a_\alpha \in Q_{k_0,l_0}$ with $\|\nabla f_{k_0,l_0}(a_\alpha)\| = \alpha$. Then $\|\nabla \Phi_d(a_\alpha)\| = |d_{k_0,l_0}|\alpha$. If $x \in (\nabla \Phi_d)^{-1}(B(\nabla \Phi_d(a_\alpha), \delta))$ then $\|\nabla \Phi_d(x)\| \neq |d_{k,l}|$ for all $(k,l) \in \mathbb{N}^2$ and so $(\nabla \Phi_d)^{-1}(B(\nabla \Phi_d(a_\alpha), \delta))$ is a nonempty set of zero Lebesgue measure.

All this shows that $\Phi(c_0(\mathbb{N}^2))$ is a vector space of differentiable functions which (except for 0) do not have the Denjoy-Clarkson property. But $\Phi(c_0(\mathbb{N}^2))$

is also a closed subspace of $(C_b(\mathbb{R}^2), \|\cdot\|_\infty)$ since $\|\Phi_d\|_\infty = \|(d_{k,l})_{k,l}\|_\infty \|f\|_\infty$ for all $d \in c_0(\mathbb{N}^2)$, and so $\big(\Phi(c_0(\mathbb{N}^2)), \|.\|_\infty\big)$ is an infinite dimensional Banach space isometric to $c_0(\mathbb{N}^2)$. $\qquad\square$

The following remark is also considered by the authors in [213].

Remark 6.8.2. *1. The proof can be simplified if, instead of using the simple intermediate value result for gradients, one uses the much stronger Darboux property for gradients; see [297].*

2. *The above result can be reformulated in the following way. If we denote by $NDC(\mathbb{R}^n)$ the set of everywhere differentiable functions on $\mathbb{R}^n$ which do not enjoy the Denjoy-Clarkson property then $NDC(\mathbb{R}^n)$ is $\mathfrak{c}$-spaceable.*

3. *Also, it is possible to construct another infinite dimensional Banach space of differentiable functions on $\mathbb{R}^n$ which (except for 0) fail the Denjoy-Clarkson property (and which contains the Deville-Matheron function). In order to do so, it suffices with replacing $c_0(\mathbb{N}^n)$ in the above proof with the Banach space $c(\mathbb{N}^n)$ of all convergent sequences in $\mathbb{R}$ endowed with the supremum norm. In that case f is the image of the sequence $(1, 1, \ldots)$.*

4. *Finally, one can also construct a non separable infinite dimensional normed space E of differentiable functions on $\mathbb{R}^n$ which (except for 0) fail the Denjoy-Clarkson property and such that E contains the Deville-Matheron function as well. Indeed, consider the non separable norm space $X = \mathrm{span}\{\chi_P : P \subset \mathbb{N}^n\}$, endowed with the supremum norm, instead of $c_0(\mathbb{N}^n)$ in the proof above. Our method cannot be extended to $\ell_\infty(\mathbb{N}^n)$, the completion of X.*

6.9 Exercises

Exercise 6.1. Use the same technique as in Theorem 6.1.1 to show that there exists a vector space $E \subset BS(\mathbb{K})$ (the set of all series with bounded partial sums) such that:

(i) Every $x \in E \setminus \{0\}$ is a divergent series.

(ii) $\dim(E) = c$ and E is non-separable.

(iii) $\mathrm{span}\{E \cup c_{00}\}$ is an algebra and every element of it is either a divergent series or is an element of c_{00}.

Exercise 6.2. Use the same technique as in Theorem 6.1.1 to show that there exists a vector space $E \subset \ell_\infty$ such that:

(i) $\dim(E) = \mathfrak{c}$.

(ii) Every $x \in E \setminus \{0\}$ is a divergent sequence.

(iii) $E \oplus c_0$ is an algebra.

(iv) Every element in $\overline{E} + c_0$ is either a divergent sequence or a sequence in c_0.

Exercise 6.3. If X is a Banach space and $\sum_i x_i$ is a series in X, we say that $\sum_i x_i$ is *unconditionally convergent* (*UC*) if, for every permutation π of $\mathbb{N}$, we have that $\sum_{i=1}^{\infty} x_{\pi(i)}$ converges. We say that $\sum_i x_i$ is *weakly unconditionally Cauchy* (*WUC*) if $\sum_{i=1}^{\infty} |f(x_i)| < \infty$ for every $f \in X^*$, the dual space of X. It is also known that [119, 184] if X is a Banach space, then there exists a WUC series in X which is convergent but which is not unconditionally convergent if and only if X has a copy of c_0. It is a well known fact that every infinite dimensional Banach space has a series $\sum_i x_i$ which is unconditionally convergent and so that $\sum_i \|x_i\| = \infty$ [195]. Use the same technique as in Theorem 6.1.1 in order to show the following:

(a) There exists a vector space $E \subset \ell_1^{\omega}(c_0)$ (the space of all weakly unconditionally Cauchy series in c_0) satisfying:

   (i) $\dim(E) = \mathfrak{c}$.
   (ii) If $x \in E \setminus \{0\}$ then $\sum_i x_i$ is not weakly convergent.

(b) Let X be an infinite dimensional Banach space. Then there exists a vector subspace E of $UC(X)$ such that $\dim(E) = \mathfrak{c}$, and if $x \in E \setminus \{0\}$ then $\sum_i \|x_i\| = \infty$.

Notice that from this exercise it follows that X has a copy of c_0 if and only if there exists a vector subspace E of $\ell_1^{\omega}(X)$ with $\dim(E) = \mathfrak{c}$, so that every non-zero element of E is a non-weakly convergent series.

Exercise 6.4. Is any of the sets from Theorem 6.7.1, (b) or (c), algebrable?

Exercise 6.5. Prove Lemma 6.3.2.

Exercise 6.6. Show Lemma 6.3.3.

Exercise 6.7. Show that the family of Dirichlet series $\sum_{k=1}^{\infty} a_k k^{-s}$ that are absolutely convergent in the open half-plane $\mathrm{Re}\, s > 0$ but divergent at $s = 0$ is lineable.
Hint: Split the series $\sum_{k=1}^{\infty} k^{-s-1}$ into infinitely many appropriate series.

Exercise 6.8. Denote by $\mathcal{F}$ the family consisting of all polynomials $P : \mathbb{R}^2 \to \mathbb{R}$ for which the origin $(0,0)$ is a saddle point, that is, $(0,0)$ is a critical point of P but it is not a relative extremum.

(a) Verify that $\mathcal{F} \cup \{0\}$ is not a vector space.

(b) Prove that $\mathcal{F}$ is lineable.

Hint: Concerning (b), consider the functions $P_n(x,y) = x^{2n} - y^{2n}$ ($n = 1, 2, \dots$).

Exercise 6.9. Let $f \in \mathrm{ES}(\mathbb{R})$. In Section 1.5 it was shown that $\mathrm{ES}(\mathbb{R})$ is $2^{\mathfrak{c}}$-lineable. Now, show (directly) that the set $\mathrm{ES}(\mathbb{R})$ is $\aleph_0$-lineable by taking odd powers of f.
Hint: Consider the set $S = \{f^{2k+1} : k \in \mathbb{N}\}$ and check that $\mathrm{span}(S) \subset \mathrm{ES}(\mathbb{R}) \cup \{0\}$. The solution can be found in [360].

Exercise 6.10. Let V be a topological vector space (over $\mathbb{K} = \mathbb{R}$ or $\mathbb{C}$) and α be a cardinal number with $\alpha < \dim(V)$. Provide a dense α-lineable subset of V that is not β-lineable for any cardinal number $\beta > \alpha$.

Exercise 6.11. Show that the linear space generated by the two linearly independent functions $\{\sin x, \cos x\}$ can not be used to study the 2-lineability of the set of continuous functions on $[0, 2\pi]$ which attain their maximum at exactly one point (of course, this set is not 2-lineable as we saw in Section 1.7, Theorem 1.7.8).

Exercise 6.12. Let $\mathcal{E}$ be the vector space of all entire functions $f : \mathbb{R} \to \mathbb{R}$, and let $\mathcal{A}$ be the subfamily of $\mathcal{E}$ consisting of those functions f such that

$$\lim_{x \to +\infty} \frac{|f(ax)|}{1 + |f(x)|} = \infty \quad \text{for all } a > 1.$$

Prove that $\mathcal{A}$ is maximal lineable in $\mathcal{E}$.
Hint: Consider the functions $e^{\alpha x}$ ($\alpha > 0$).

Chapter 7

General Techniques

The reader who has been able to reach this place after having read the previous chapters might have realized that the proofs of a remarkable number of results follow a relatively small number of patterns. With this in mind, the purpose of this chapter is to collect several general techniques that generate or improve lineability. As a matter of fact, not many of such general criteria are known up to the moment. This is why this chapter is rather short. It should be said that many of the statements given in this book could have been demonstrated by using the mentioned criteria but, since it is mainly addressed to beginners, we have preferred to give more natural proofs for each statement and then, in a final chapter, to extract the deep essence of the proofs in a few results. There is, in addition, a historical reason for this. Namely, most lineability results appeared in an independent and isolated way; it was several years later when the (few) general techniques arose.

7.1   What one needs to know

Recall, again, that an *F-space* X is a completely metrizable topological vector space. In particular, the conclusion of the Baire category theorem holds. As a consequence, if X has infinite dimension, then it cannot be filled up with countably many subspaces of finite dimension, from which it follows that $\dim(X)$ is uncountable in this case. Moreover, and again as a consequence of the Baire category theorem, one has that if Y is a vector subspace of X that is of second category in X then $Y = X$.

Another elementary but important fact, which in turn is a consequence of the Hahn–Banach theorem, is that if X is a locally convex space then any finite dimensional subspace Z is complemented, that is, there exists a closed vector subspace $Y \subset X$ such that $X = Y \oplus Z$, the direct sum of Y and Z. This means that $X = Y + Z$ and $Y \cap Z = \varnothing$. For this, see for instance Rudin's book [352, Chapter 4]. For a general vector space X, if $X = Y \oplus Z$ then $\dim(Y)$ is said to be the codimension of Z. This codimension equals the dimension of the quotient X/Y.

Recall that, in a vector space X, a subset A is called *absolutely convex* if $cx \in A$ for all $x \in A$ and all scalars c with $|c| \leq 1$. And A is said to be *absorbing* provided that, given $x_0 \in X$, there is a scalar $c > 0$ such that $x \in cA$. A locally convex space X is said to be *barrelled* whenever every absorbing closed absolutely convex subset of X is a neighborhood of 0. It can be proved that every Baire locally convex space (in particular, every Fréchet space) is barrelled; see for instance the Schaefer–Wolff book [359]. Moreover, every closed subspace Y with countable codimension of a barrelled space X is also barrelled; see [358].

Nikolskii's theorem of characterization of *basic sequences* has been already used in the context of Banach spaces; see for instance Sections 3.5 and 4.7. But it also works in the context of *F-spaces,* as shown in [267, Theorem 5.1.8, p. 67]. Recall that an F-space is a complete metrizable topological vector space, and that every F-space X supports an F-norm $\| \cdot \|$ such that the mapping $d(x, y) := \|x - y\|$ is a distance generating its topology; see, e.g., [266, pp. 2–5]. Recall also that an F-norm on a vector space X is a functional $\| \cdot \| : X \to [0, \infty)$ satisfying, for all $x, y \in X$ and $\lambda \in \mathbb{K}$, the following properties: $\|x + y\| \leq \|x\| + \|y\|$; $\|\lambda x\| \leq \|x\|$ if $|\lambda| \leq 1$; $\|\lambda x\| \to 0$ if $\lambda \to 0$; $\|x\| = 0$ only if $x = 0$. A sequence $(x_n) \subset X$ is called a basic sequence whenever for every vector $x \in \overline{\mathrm{span}}\{x_n : n \geq 1\}$ there is a unique sequence of scalars (a_n) such that

$$x = \sum_{n=1}^{\infty} a_n x_n.$$

Then Nikolskii's theorem asserts that a sequence $(x_n) \subset X \setminus \{0\}$ is basic if and only if there is a constant $\alpha \in [1, \infty)$ such that, for every pair $r, s \in \mathbb{N}$ with $s \geq r$ and every finite sequence of scalars $a_1, \ldots, a_s$, one has

$$\Big\| \sum_{n=1}^{r} a_n x_n \Big\| \leq \alpha \Big\| \sum_{n=1}^{s} a_n x_n \Big\|.$$

The Weierstrass approximation theorem asserting that any continuous function $[a, b] \to \mathbb{R}$ can be uniformly approximated by polynomials has a very useful extension, namely, the *Weierstrass-Stone theorem* (see, e.g., [352]), which tells us the following. Assume that S is a compact topological space and that the space $\mathcal{C}(S)$ of all continuous functions $S \to \mathbb{R}$ is endowed with the uniform distance. Suppose that $\mathcal{A} \subset C(S)$ is an algebra such that given

$x_0 \in S$ there is $F \in \mathcal{A}$ with $F(x_0) \neq 0$ and such that, in addition, given a pair of distinct points $x_0, x_1 \in S$, there is $F \in \mathcal{A}$ with $F(x_0) \neq F(x_1)$. Then $\mathcal{A}$ is dense in $C(S)$.

7.2 The negative side

In this book we have found many examples of sets that, being topologically large, are also algebraically large. Nevertheless, although being a residual set is "symptomatic" of containing large subspaces, there is clearly no immediate implication. Indeed, we have got several examples:

- The set $\mathcal{M} = \{f \in \mathcal{C}([0,1]) : f$ attains its maximum at exactly one point of $[0,1]\}$ is also residual but, again, if V is a vector space contained in $\mathcal{M} \cup \{0\}$, then $\dim(V) \leq 1$; see Section 1.7.

- The set $\mathcal{U}((\Delta_{\alpha_n}))$ of universal sequence of diagonal operators given in Section 4.3 is residual on $H(\mathbb{C})$ as soon as $\{(a_{k,n})_{k\geq 0} : n \in \mathbb{N}\}$ is dense in $\mathbb{C}^{\mathbb{N}_0}$, but it is just 1-lineable.

- Banakh *et al.* proved in [42] that, if $E(x)$ denotes the set

$$\left\{a \in \mathbb{R} : \text{ exists } A \subset \mathbb{N} \text{ such that } \sum_{n \in A} x(n) = a\right\}$$

for every $x = (x(n)) \in \ell_1$, then the set of all $x \in \ell_1$ for which $E(x)$ is homeomorphic to the Cantor set is residual and strongly $\mathfrak{c}$-algebrable, but not spaceable.

In spite of this, one might believe that the above ones are pathological examples. This is far from being true, as the following result of [26] shows. Note that it allows us to conclude that, in addition, there is no relation between spaceability and dense-lineability.

Theorem 7.2.1. (a) *Let X be an infinite dimensional locally convex space. There exists a subset $M \subset X$ such that M is spaceable and dense, although it is not dense-lineable.*

(b) *Let X be an infinite dimensional F-space. There exists a subset $M \subset X$ which is lineable and dense, but which is not spaceable. If X is separable, then M can also be chosen to be dense-lineable.*

Proof. (a) Take any vector $u \in X$ and let $Z := \text{span}(\{u\}) = \{c\,u : c \in \mathbb{K}\}$. Since Z has dimension $1 < \infty$, there is an algebraic complement Y in X for Z, satisfying that Y is a closed subspace. Take

$$M = Y + \{\lambda\,u : \lambda \in \mathbb{Q}_0\},$$

where $\mathbb{Q}_0$ denotes the set $\mathbb{Q}$ of rational numbers if $\mathbb{K} = \mathbb{R}$, and the set $\mathbb{Q} + i\mathbb{Q}$ if $\mathbb{K} = \mathbb{C}$.

We will now show that M enjoys the required properties. Since $Y \subset M$ and Y is infinite dimensional (because X is infinite dimensional and $\dim(Z) = 1$), the set M is spaceable. The set M is also dense. Indeed, $\{\lambda u : \lambda \mathbb{Q}_0\}$ is dense in Z, so $M = Y + \mathbb{Q}_0$ is dense in $Y + Z = X$.

The set M is not dense-lineable. Indeed, suppose that W is a dense vector subspace of X contained in $M \cup \{0\} = M$. Take $w \in W$ and write

$$w = y + q\,u$$

with $y \in Y$ and $q \in \mathbb{Q}_0$. Our aim is to show that $q = 0$. Suppose not, assuming that $q \neq 0$. Then $\frac{\pi}{q} w \in W \subset M$; that is,

$$\frac{\pi}{q} w = y' + q'\,u$$

with $y' \in Y$ and $q' \in \mathbb{Q}_0$. Since the sum $X = Y + Z$ is direct, we infer that

$$y' = \frac{\pi}{q}\,y \quad \text{and} \quad \frac{\pi}{q} q = q',$$

obtaining the contradiction that $\pi \in \mathbb{Q}_0$. Thus, $q = 0$. Finally, $W \subset Y$, which (again) is a contradiction because Y is closed and proper.

(b) Let us consider a Hamel basis $\mathcal{B}$ for X and a countably infinite subset $\{b_n : n \in \mathbb{N}\}$ of $\mathcal{B}$. We write $X = Y \oplus Z$ where $Y = \text{span}\{b_n : n \in \mathbb{N}\}$ and $Z = \text{span}(\mathcal{B} \setminus \{b_n : n \in \mathbb{N}\})$. Our candidate will be

$$M = Y + \text{span}_{\mathbb{Q}_0}(\mathcal{B} \setminus \{b_n : n \in \mathbb{N}\}),$$

where $\text{span}_{\mathbb{Q}_0}(A)$ denotes the $\mathbb{Q}_0$-linear span of the set A. We claim that M satisfies the required properties. First, M is lineable, since $Y \subset M$. And M is also dense. To see this, note that $\text{span}_{\mathbb{Q}_0}(\mathcal{B} \setminus \{b_n : n \in \mathbb{N}\})$ is dense in Z, so $M = Y + \text{span}_{\mathbb{Q}_0}(\mathcal{B} \setminus \{b_n : n \in \mathbb{N}\})$ is dense in $Y + Z = X$.

Finally, M is not spaceable. Indeed, suppose that W is an infinite-dimensional closed vector subspace of X contained in $M \cup \{0\} = M$. Now, we proceed similarly as we did in part (a). Take $w \in W$ and write

$$w = y + q_1 b_1' + \cdots + q_r b_r'$$

with $y \in Y$, $q_1, \ldots, q_r \in \mathbb{Q}_0$ and $b_1', \ldots, b_r' \in \mathcal{B} \setminus \{b_n : n \in \mathbb{N}\}$. Our aim is to show that $q_j = 0$ for every $j \in \{1, 2, \ldots, r\}$. Suppose not, assuming that $q_1 \neq 0$ without loss of generality. Then $\frac{\pi}{q_1} w \in W \subset M$; that is,

$$\frac{\pi}{q_1} w = y' + q_1' b_1' + \cdots + q_s' b_s',$$

with $s \geq r$, $y' \in Y$, $q'_1, \ldots, q'_s \in \mathbb{Q}_0$, and $b'_1, \ldots, b'_s \in \mathcal{B} \setminus \{b_n : n \in \mathbb{N}\}$. We therefore have that

$$y' = \frac{\pi}{q_1} y,$$

and

$$\pi b'_1 + \frac{q_2 \pi}{q_1} b'_2 + \cdots + \frac{q_n \pi}{q_1} b'_n = q'_1 b'_1 + \cdots + q'_s b'_s,$$

obtaining again the contradiction that $\pi \in \mathbb{Q}_0$. Thus, all q_j's are 0. Consequently, $W \subset Y$, but this is impossible because the cardinality of any Hamel basis of W is uncountable.

To finish the proof, notice that if X is separable, then we can choose Y to be dense in X, and therefore M is dense-lineable. $\qquad\square$

Notice that, in part (a), the set M satisfies $\lambda(M) = \dim(X)$ or, in other words, M is maximal spaceable. Indeed, $M \supset Y$ and Y is the complementary subspace of a straight line.

7.3 When lineability implies dense-lineability

In view of Theorem 7.2.1, it is natural to ask for criteria telling us how to get dense-lineability from the mere lineability. A few of them have appeared recently, and are presented in the next theorem. The idea which is in the core of these results is to obtain the desired dense subspace by adding small vectors coming from a known lineable set to the vectors of a dense subset. According to [26], if A and B are subsets of a vector space X, then A is said to be *stronger than* B provided that $A + B \subset A$.

Theorem 7.3.1. *Assume that X is a topological vector space. Let $A \subset X$. Suppose that there exists a subset $B \subset X$ such that A is stronger than B and B is dense-lineable. We have:*

(a) *If α is a cardinal number, A is α-lineable and X has an open basis $\mathcal{B}$ for its topology such that $\mathrm{card}(\mathcal{B}) \leq \alpha$, then A is dense-lineable. If, in addition, $A \cap B = \varnothing$, then $A \cup \{0\}$ contains a dense vector space D with $\dim(D) = \alpha$.*

(b) *If X is metrizable and separable and α is an infinite cardinal number such that A is α-lineable, and $A \cap B = \varnothing$, then $A \cup \{0\}$ contains a dense vector space D with $\dim(D) = \alpha$.*

(c) *If the origin possesses a fundamental system $\mathcal{U}$ of neighborhoods with $\mathrm{card}(\mathcal{U}) \leq \dim(X)$, A is maximal lineable and $A \cap B = \varnothing$, then A is maximal dense-lineable. In particular, the same conclusion follows if X is metrizable, A is maximal lineable and $A \cap B = \varnothing$.*

Proof. Observe that (b) is derived from (a) because if X is metrizable and separable then it is second countable; hence, it has a countable open basis $\mathcal{B}$ for its topology. Therefore $\mathrm{card}(\mathcal{B}) = \mathrm{card}(\mathbb{N}) = \aleph_0 \leq \alpha$ because α is infinite, and (a) applies.

Let us show that (c) is also a consequence of (a). For this, assume that A, B and $\mathcal{U}$ are as in the hypothesis of (c). Let C denote a dense countable subset of $\mathbb{K}$, and let $\{u_i\}_{i \in I}$ be an algebraic basis of X, so that $\mathrm{card}(I) = \dim(X)$. Denote by $\mathcal{P}_f(I)$ the family of nonempty finite subsets of I. Since $\mathrm{card}(\mathcal{U}) \leq \dim(X)$, we must have that $\dim(X)$ is not finite, hence $\mathrm{card}(\mathcal{P}_f(I)) = \mathrm{card}(I) = \dim(X) \geq \mathrm{card}(C)$. Moreover, $\mathrm{card}(I^F) = \mathrm{card}(I)$ for any nonempty finite set F, and $\mathrm{card}(C \times I) = \mathrm{card}(I)$. Now, it is easy to see that the family

$$\mathcal{B} := \left\{ U + \sum_{i \in F} \alpha_i u_i : U \in \mathcal{U},\ \alpha_i \in C \text{ for all } i \in F,\ F \in \mathcal{P}_f(I) \right\}$$

is an open basis for the topology of X. We have that

$$\mathrm{card}(\mathcal{B}) \leq \mathrm{card}\left(\mathcal{U} \times \bigcup_{F \in \mathcal{P}_f(I)} (C \times I)^F \right) = \mathrm{card}\left(\mathcal{U} \times \bigcup_{F \in \mathcal{P}_f(I)} I^F \right)$$
$$\leq \mathrm{card}(\mathcal{U} \times \mathcal{P}_f(I) \times I) = \mathrm{card}(\mathcal{U} \times I \times I)$$
$$= \max\{\mathrm{card}(\mathcal{U}), \mathrm{card}(I)\} = \dim(X).$$

Since A is $\dim(X)$-lineable, by applying (a) again we obtain the first part of (c). As for the second part, simply observe that if X is metrizable then $\mathcal{U}$ can be chosen countable, so $\mathrm{card}(\mathcal{U}) \leq \dim(X)$ if $\dim(X)$ is infinite. If $\dim(X)$ is finite then the conclusion is evident because $A \cup \{0\} = X$; indeed, every vector subspace M of a finite dimensional vector space X such that $\dim(M) = \dim(X)$ must equal X.

Thus, our only task is to prove (a). Suppose that A is α-lineable and that $\mathrm{card}(\mathcal{B}) \leq \alpha$ for some open basis $\mathcal{B}$ of X. We are also assuming that $A + B \subset A$ and B is dense-lineable. It follows that there exist vector spaces A_1, B_1 such that $A_1 \subset A \cup \{0\}$, $B_1 \subset B \cup \{0\}$, B_1 is dense in X and $\dim(A_1) = \alpha \geq \mathrm{card}(\mathcal{B})$. Hence there are sets I, J, vectors a_i ($i \in I$) and open sets U_j ($j \in J$), such that $\mathrm{card}(I) = \alpha$, $\{a_i\}_{i \in I}$ is a linearly independent system contained in A_1, $\mathcal{B} = \{U_j\}_{j \in J}$ and there exists a surjective mapping $\varphi : I \to J$. By density, we can assign to each $j \in J$ a vector $b_j \in U_j \cap B_1$. Fix $j \in J$. As $U_j - b_j$ is a neighborhood of 0 and multiplication by scalars is continuous on X, for each $i \in \varphi^{-1}(\{j\})$ there is $\varepsilon_i > 0$ satisfying $\varepsilon_i a_i \in U_j - b_j$, or $\varepsilon_i a_i + b_j \in U_j$. Define

$$D := \mathrm{span}\ \{\varepsilon_i a_i + b_{\varphi(i)} : i \in I\}.$$

Then D is a vector subspace of X. Since φ is surjective, we can pick for each $j \in J$ and index $i(j) \in I$ with $\varphi(i(j)) = j$. As $\{U_j\}_{j \in J}$ is an open basis and

$v_{i(j)} a_{i(j)} + b_j \in U_j$ $(j \in J)$, these vectors form a dense subset of X. But D contains these vectors, so D is also dense. Furthermore, if $x \in D$ then there are $p \in \mathbb{N}$, $(\lambda_1, \ldots, \lambda_p) \in \mathbb{K}^p \setminus \{(0, \ldots, 0)\}$ and $i_1, \ldots, i_p \in I$ with

$$x = \lambda_1 \varepsilon_{i_1} a_{i_1} + \cdots + \lambda_p \varepsilon_{i_p} a_{i_p} + \lambda_1 b_{\varphi(i_1)} + \cdots + \lambda_p b_{\varphi(i_p)}.$$

Define $u := \lambda_1 \varepsilon_{i_1} a_{i_1} + \cdots + \lambda_p \varepsilon_{i_p} a_{i_p}$ and $y := \lambda_1 b_{\varphi(i_1)} + \cdots + \lambda_p b_{\varphi(i_p)}$. Then $y \in B_1 \subset B \cup \{0\}$, and $u \in A_1 \setminus \{0\}$ because of the linear independence of the a_i's. Hence $u \in A$ and

$$x = u + y \in A + (B \cup \{0\}) \subset A \cup A = A.$$

Consequently, $D \setminus \{0\} \subset A$ and A is dense-lineable.

Finally, we suppose further that $A \cap B = \varnothing$. We want to prove that $\dim(D) = \alpha$ or, that is the same, the vectors

$$x_i := \varepsilon_i a_i + b_{\varphi(i)} \quad (i \in I)$$

are linearly independent. With this aim, consider a $p \in \mathbb{N}$ and two p-tuples $(\lambda_1, \ldots, \lambda_p) \in \mathbb{K}^p$ and $(i_1, \ldots, i_p) \in I^p$ such that $\sum_{j=1}^{p} \lambda_j x_{i_j} = 0$. Assume, by way of contradiction, that $(\lambda_1, \ldots, \lambda_p) \neq (0, \ldots, 0)$. Then

$$u + y = 0,$$

where u and y are as in the preceding paragraph. Hence $y \in A$ (because $y = -u \in A_1 \setminus \{0\} \subset A$) and $y \in B$ (because $y = -u \neq 0$, so $y \in B_1 \setminus \{0\} \subset B$), which implies $A \cap B \neq \varnothing$. This contradicts the assumption $A \cap B = \varnothing$, and we are done. $\qquad\qquad\square$

Corollary 7.3.2. *Let X be a topological vector space. Suppose that Γ is a family of vector subspaces of X such that $\bigcap_{S \in \Gamma} S$ is dense in X. We have:*

(a) *If $\bigcap_{S \in \Gamma}(X \setminus S)$ is α-lineable and X has an open basis $\mathcal{B}$ for its topology such that $\mathrm{card}(\mathcal{B}) \leq \alpha$ then $\bigcap_{S \in \Gamma}(X \setminus S)$ is dense-lineable and, moreover, it contains a dense vector space D with $\dim(D) = \alpha$.*

(b) *If X is metrizable and separable and α is an infinite cardinal number such that $\bigcap_{S \in \Gamma}(X \setminus S)$ is α-lineable, then $\bigcap_{S \in \Gamma}(X \setminus S)$ contains, except for zero, a dense vector space D with $\dim(D) = \alpha$.*

(c) *If the origin possesses a fundamental system $\mathcal{U}$ of neighborhoods with $\mathrm{card}(\mathcal{U}) \leq \dim(X)$ then $\bigcap_{S \in \Gamma}(X \setminus S)$ is maximal dense-lineable. The same conclusion holds if X is metrizable and $\bigcap_{S \in \Gamma}(X \setminus S)$ is maximal lineable.*

Proof. In order to apply Theorem 7.3.1, it is enough to check that $A := \bigcap_{S \in \Gamma}(X \setminus S)$ is stronger than $B := \bigcap_{S \in \Gamma} S$, that B is dense-lineable and that $A \cap B = \varnothing$. The last property is obvious, whereas the dense-lineability of B is trivial in view of its denseness and the fact that B is itself a vector space. As for the property $A + B \subset A$, consider $x \in A$, $y \in B$ and $z := x + y$. If $z \notin A$ then there exists $S \in \Gamma$ with $z \in S$. Then

$$x = z + (-y) \subset S - B \subset S - S = S$$

as S is a vector subspace. Thus $x \notin A$, a contradiction, which concludes the proof.   □

Theorem 7.3.1 (or its corollary) can be applied in many situations, so as to yield quick proofs. As an example, assume that X is a separable F-space and that T is an operator on X satisfying the hypotheses of Theorem 4.5.2. Suppose, in addition, that there is a dense subset $D \subset X$ such that $(T^n x)$ converges for every $x \in D$. Then $HC(T)$ is maximal dense-lineable in X (this was the content of Exercise 4.10). Indeed, take $\Gamma = \{S\}$ in Corollary 7.3.2, where S is the set of points of X having convergent (T_n)-orbit. The reader can find further applications in the Exercises section of this chapter.

In the case of the complement of a subspace, there is nothing to add to lineability in order to get dense-lineability in the separable case, as the following simple assertion shows.

Theorem 7.3.3. *Let X be a metrizable separable topological vector space and Y be a vector subspace of X. If $X \setminus Y$ is lineable then $X \setminus Y$ is dense-lineable. Consequently, both properties of lineability and dense-lineability for $X \setminus Y$ are equivalent provided that X has infinite dimension.*

Proof. It is evident that $X \setminus Y$ is lineable if and only if Y has infinite algebraic codimension. The assumptions imply that X has a countable open basis $\{G_n : n \geq 1\}$. Assume that $X \setminus Y$ is lineable. In particular, $Y \subsetneq X$. Then $Y^0 = \varnothing$, hence $X \setminus Y$ is dense. Therefore there is $x_1 \in G_1 \setminus Y$. Since $\operatorname{codim}(Y) = \infty$, we have $\operatorname{span}(Y \cup \{x_1\}) \subsetneq X$. Then $(\operatorname{span}(Y \cup \{x_1\}))^0 = \varnothing$. It follows that there exists $x_2 \in G_2 \setminus \operatorname{span}(Y \cup \{x_1\})$. With this procedure, we get recursively a sequence of vectors $\{x_n\}_{n \geq 1}$ satisfying

$$x_n \in G_n \setminus \operatorname{span}(Y \cup \{x_1, \ldots, x_{n-1}\}) \quad (n \geq 1).$$

In particular, the set $\{x_n : n \geq 1\}$ is dense. Now, if we define $M := \operatorname{span}\{x_n : n \geq 1\}$ then M is a dense vector space and $M \setminus \{0\} \subset X \setminus Y$.   □

Theorem 7.3.3 is easy to apply. For instance, the set $A := \{f \in \mathcal{C}(\mathbb{R}) : f$ is unbounded$\}$ is dense-lineable because $A = X \setminus Y$ with $X = \mathcal{C}(\mathbb{R})$, $Y = \{$bounded continuous functions $\mathbb{R} \to \mathbb{R}\}$, Y is a vector subspace and A is lineable; indeed, A contains the vector space of all non-zero polynomials P with $P(0) = 0$.

A partial complement of Theorem 7.3.3 is possible in the non-separable case. Namely, by assuming the Continuum Hypothesis, the next assertion is easily obtained.

Proposition 7.3.4. *Let* X *be a non-separable F-space and* Y *be a closed separable vector subspace of* X. *Then* $X \setminus Y$ *is maximal lineable.*

Proof. Indeed, let Z be a vector space that is an algebraic complement of Y, so that $Z \setminus \{0\} \subset X \setminus Y$. Note that

$$\dim(Y) \leq \mathfrak{c} \leq \dim(X) = \dim(Y) + \dim(Z).$$

If $\dim(Z) \leq \aleph_0$ then Z, and so $X \ (= Y + Z)$, would be separable (a contradiction). Hence $\dim(Z) \geq \mathfrak{c}$, which implies $\dim(Z) = \dim(X)$, and we are done. □

7.4 General results about spaceability

As in dense-lineability, most spaceability proofs on specific settings have been done directly and constructively. One has to go back to Wilansky ([381], 1975) to find what might have been the first general criterium for Banach spaces. The following improved version of his result, where X is allowed to be a Fréchet space, is ascribed by Kitson and Timoney [275, Theorem 2.2] to Kalton.

Theorem 7.4.1. *If* Y *is a closed vector subspace of a Fréchet space* X, *then* $X \setminus Y$ *is spaceable if and only if* Y *has infinite codimension.*

Proof. We first need the following result: If E is an infinite dimensional Fréchet space over the field $\mathbb{K}$ such that the weak topology coincides with the Fréchet topology, then X is isomorphic to $\mathbb{K}^{\mathbb{N}}$ (with the product topology). Let us prove it. As the topology of E is Fréchet, hence there is a countable basis of zero neighborhoods in E, it must be that there are countably many linear functionals determining the weak topology. That means that the dual topological space E^* of E must have countable algebraic dimension. Let $\{\phi_1, \phi_2, \dots\}$ be an algebraic basis for E^* and consider the map $\pi : E \to \mathbb{K}^{\mathbb{N}}$ given by

$$\pi(y) = (\phi_j(y))_{j \in \mathbb{N}},$$

which is linear, continuous, injective, a homeomorphism onto its range since X has the weak topology, hence $\pi(X)$ is complete and closed in $\mathbb{K}^{\mathbb{N}}$. By the Hahn–Banach theorem and the linear independence of the φ_j, $\pi(E) = \mathbb{K}^{\mathbb{N}}$.

Now, assume that X is a Fréchet space and that Y is a closed vector subspace. It is clear that spaceability of $X \setminus Y$ implies infinite codimension of Y. Also, as finite dimensional vector subspaces are always complemented, we need only consider the situation where both Y and the quotient X/Y are infinite dimensional Fréchet spaces. We first consider the situation where the (subspace) topology on Y coincides with the weak topology of Y. In this case, by the assertion proved in the first paragraph, one gets that Y is isomorphic to $\mathbb{K}^{\mathbb{N}}$. The Hahn–Banach theorem implies that Y is complemented in X, and then the kernel of the projection is a closed subspace contained in $(X \setminus Y) \cup \{0\}$. From this one obtains that $X \setminus Y$ is spaceable in this case.

Assume then that the topology of Y is not the same as the weak topology. Since the topology of Y has a zero neighborhood basis of closed convex sets (the closed unit balls in continuous seminorms), it has a zero neighborhood basis of weakly closed sets because if A is a convex subset of a locally convex spaces, then A is closed if and only if it is weakly closed. By the new assumption, there is a net $(y_\alpha)_\alpha \subset Y$ such that $y_\alpha \to 0$ in the weak topology but not in the F-space topology. Hence there is a strong neighborhood U of 0 such that $y_\alpha \notin U$ for all α. By a result of Kalton [263, Theorem 3.2], there is a basic sequence (y_n) in Y that is contained in the complement of U. This basic sequence is then bounded away from 0. Applying the considerations above to the quotient X/Y, we can conclude that X/Y contains a basic sequence $(x_n + Y)$. From Kalton [263, Lemma 4.3], it follows that there is a sequence $(t_n) \subset (0, \infty)$ such that $(y_n + t_n x_n)$ is a basic sequence in X. Now, define

$$Z := \overline{\operatorname{span}} \{ y_n + t_n\, x_n : n = 1, 2, \dots \}.$$

If $x \in Z \cap Y$, then there are scalars λ_n with $x = \sum_{n=1}^\infty \lambda_n (y_n + t_n x_n)$. Considering the quotient, we have $0 = x + Y = \sum_{n=1}^\infty \lambda_n t_n (x_n + Y)$ and so $\lambda_n = 0$ for all $n \in \mathbb{N}$. Thus $x = 0$ and $Z \cap Y = \{0\}$. Hence $Z \subset (X \setminus Y) \cup \{0\}$, which shows that $X \setminus Y$ is spaceable. $\square$

The last result should be compared to Theorem 7.3.3. The authors of [275] exploited it to obtain the following assertion. As in Theorem 7.4.1, a number of high-level techniques are used to obtain it. Yet we have decided to present a proof for the double sake of completeness and intrinsic interest of such results.

Theorem 7.4.2. *Let Z_n $(n \in \mathbb{N})$ be Banach spaces and X a Fréchet space. Let $T_n : Z_n \to X$ be continuous linear mappings and Y the linear span of $\bigcup_{n=1}^\infty T_n(Z_n)$. If Y is not closed in X then the complement $X \setminus Y$ is spaceable.*

Proof. First, Y must have infinite codimension. To see this, consider the family of product spaces $W_n = \oplus_{j=1}^n Z_j$ $(n \geq 1)$ endowed with the product topology. Then each W_n is a Fréchet space. Define $S_n : W_n \to X$ by $S_n(z_1, \dots, z_n) = \sum_{j=1}^n T_j(z_j)$. Then

$$Y = \bigcup_{n=1}^\infty S_n(W_n).$$

Assume, by way of contradiction, that Y has a finite dimensional complement F in X. Let $q : X \to X/F$ be the quotient map. Then the union of the ranges $q \circ S_n(W_n)$ $(n = 1, 2, \dots)$ is all of X/F. Since X/F is a Fréchet space, from the Baire category theorem it follows that there is $m \geq 1$ such that

$(q \circ S_m)(W_m)$ is of second category in Y. Since $S_m(W_m)$ is a vector subspace, we get $(q \circ S_m)(W_m) = X/F$, and this entails $S_m(W_m) = Y$. Consequently, Y is the range of a Fréchet space operator and Y has closed complementary subspace F. It follows that Y is closed, a contradiction. Therefore Y has infinite codimension.

Next, it is sufficient to deal with the case where X is separable. Since Y is not closed, there is a sequence $\{y_k\}_{k=1}^{\infty}$ with $\lim_{k\to\infty} y_k = x \notin Y$. Let $\widetilde{X}$ be the closed linear span in X of $\{x\} \cup \{y_k : k \in N\}$, $\widetilde{W}_n := S_n^{-1}(\widetilde{X})$, $\widetilde{S}_n$ the restriction of S_n to $\widetilde{W}_n$ and $\widetilde{Y} := Y \cap \widetilde{X}$. We work now with the Banach spaces $\widetilde{W}_n$ and the sequence $\widetilde{S}_n : \widetilde{W}_n \to \widetilde{X}$ of continuous operators with $\widetilde{Y} = \bigcup_{n=1}^{\infty} \widetilde{S}_n(\widetilde{W}_n)$, that is an increasing union of operator ranges. Of course $\widetilde{X}$ is separable and Fréchet. We know that $\widetilde{Y}$ is not closed because of the sequence (y_k) and so, by the argument given above for Y, the vector space $\widetilde{Y}$ must have infinite codimension in $\widetilde{X}$.

If there are infinitely many $n \in \mathbb{N}$ such that $\widetilde{S}_n(\widetilde{W}_n)$ is closed in $\widetilde{X}$, say $n_1 < n_2 < \cdots$, then we can write each $\widetilde{S}_{n_j}(\widetilde{W}_{n_j}) \setminus \{0\}$ as a countable union $\bigcup_{k=1}^{\infty} C_{j,k}$ of closed convex subsets of $\widetilde{X}$: indeed, use translates of open convex neighborhoods of the origin in the subspace to cover the set, taking care that 0 is not in the closures of any of these translates. Then invoke the Lindelöf property of the separable metric spaces to get a countable subcover. Thus

$$\widetilde{Y} \setminus \{0\} = \bigcup_{j=1}^{\infty} \widetilde{S}_{n_j}(\widetilde{W}_{n_j}) \setminus \{0\} = \bigcup_{j,k=1}^{\infty} C_{j,k}$$

is a countable union of closed convex subsets of $\widetilde{X}$. Since $\widetilde{Y}$ is also of infinite codimension in $\widetilde{X}$, a result of Drewnowski [192, Corollary 5.5] implies that $\widetilde{X} \setminus \widetilde{Y}$ is spaceable. Hence, using $\widetilde{X} \setminus \widetilde{Y} = \widetilde{X} \cap (X \setminus Y)$, we get that $X \setminus Y$ is spaceable in this case.

Thus we are left with the case where there are at most finitely many $n \in \mathbb{N}$ such that $\widetilde{S}_n(\widetilde{W}_n)$ is closed. By discarding that finite number of n and renumbering, we may assume that all fail to be closed. Consider the unit ball U_n in $\widetilde{W}_n$. Then

$$Y_n = \mathrm{span}(\overline{\widetilde{S}_n(U_n)}) = \bigcup_{r>0} r\,\overline{\widetilde{S}_n(U_n)}$$

is not barrelled in the induced topology from $\widetilde{X}$. This follows by an argument of Drewnowski [192, p. 388]. Since $\overline{\widetilde{S}_n(U_n)}$ is a closed absorbing balanced subset of Y_n, if Y_n were barrelled, $\overline{\widetilde{S}_n(U_n)}$ would contain a neighborhood of 0 in Y_n. Now, by completeness of $\widetilde{W}_n$, we get $\widetilde{S}_n(\widetilde{W}_n) = Y_n$ and $\widetilde{S}_n$ is open (see, e.g., [359, Lemma III.2.1]). Hence $\widetilde{S}_n$ induces a linear isomorphism from $\widetilde{W}_n/\mathrm{Ker}\,\widetilde{S}_n$ onto Y_n. Therefore $Y_n = \widetilde{S}_n(\widetilde{W}_n)$ is completely metrizable, hence closed in $\widetilde{X}$, which is a contradiction.

We observe next that $Y_n \setminus \{0\}$ has a countable cover by closed convex subsets of $\widetilde{X}$. Via the Lindelof argument mentioned above, we can write $\widetilde{X} \setminus \{0\} = \bigcup_{k=1}^{\infty} A_k$, where the A_k's are convex open sets in $\widetilde{X}$ with $0 \notin A_k$ for all $k \in \mathbb{N}$. Then

$$Y_n \setminus \{0\} = \bigcup_{j,k=1}^{\infty} (\overline{A_k} \cap (j\,\widetilde{S}_n(U_n))).$$

By construction, it is easy to verify that $\widetilde{S}_n(U_n) \subset \widetilde{S}_{n+1}(U_{n+1})$ and so $Y_n \subset Y_{n+1}$ for all $n \in \mathbb{N}$. Since each Y_n is a vector subspace, we get that their (increasing) union $Y_\infty := \bigcup_{n=1}^{\infty} Y_n$ is also a vector subspace of $\widetilde{X}$. As each Y_n is not barrelled, each one has infinite codimension in the space $\widetilde{X}$. We claim that Y_∞ has also infinite codimension. If it were not the case, consider a finite dimensional subspace $F \subset \widetilde{X}$ with $F \cap Y_\infty = \{0\}$, $F + Y_\infty = \widetilde{X}$ and take the quotient map $q : \widetilde{X} \to \widetilde{X}/F$. Then, by the Baire category theorem, there must be $n \in \mathbb{N}$ so that $q(Y_n)$ is a subspace of second category in $\widetilde{X}/F$. As

$$q(Y_n) = \bigcup_{k \in \mathbb{N}} k\, q(\overline{\widetilde{S}_n(U_n)}),$$

the set $q(\overline{\widetilde{S}_n(U_n)})$ must have nonempty interior in $\widetilde{X}/F$ and so $q \circ \widetilde{S}_n$ is surjective by [358, Lemma III.2.1]. Thus $Y_n = Y_\infty$, but this contradicts the fact that Y_n has infinite codimension.

Since Y_∞ has infinite codimension, there is an infinite dimensional subspace of $\widetilde{X}$ intersecting Y_∞ only in 0. As $Y_\infty \setminus \{0\} = \bigcup_{n \in \mathbb{N}}(Y_n \setminus \{0\})$ can be expressed as a countable union of closed convex subsets of $\widetilde{X}$, we can invoke [192, Corollary 5.5] to conclude that $\widetilde{X} \setminus Y_\infty$ is spaceable. Since

$$\widetilde{X} \setminus Y_\infty \subset \widetilde{X} \setminus \widetilde{Y} = \widetilde{X} \cap (X \setminus Y),$$

we derive that $X \setminus Y$ must be spaceable. $\qquad\square$

Among other applications, the last result is used in [275] to show spaceability of the set of non-absolutely convergent power series in the disc algebra $A(\mathbb{D})$; see Exercise 7.1. As mentioned in Section 3.2, Theorem 7.4.2 also yields that $\ell_p \setminus \ell_q$ (if $p > q \geq 1$) is spaceable in ℓ_p and even that $\ell_p \setminus \bigcup_{1 \leq p < q} \ell_p$ in ℓ_p for each $p \geq 1$.

In the more specific setting of spaces of functions, we have at our disposal the following theorem. By $\mathcal{P}(\Omega)$ we represent, as usual, the family of subsets of a set Ω, while $\sigma(f)$ will denote the support of a function $f : \Omega \to \mathbb{K}$, that is, the set

$$\sigma(f) = \{x \in \Omega : f(x) \neq 0\}.$$

Theorem 7.4.3. *Let Ω be a nonempty set and Z be a topological vector space on $\mathbb{K}$. Assume that X is an F-space on $\mathbb{K}$ consisting of Z-valued functions on Ω and that $\|\cdot\|$ is an F-norm defining the topology of X. Suppose, in* *addition, that S is a nonempty subset of X and that $\mathcal{S} : \mathcal{P}(\Omega) \to \mathcal{P}(\Omega)$ is a set function with $\mathcal{S}(A) \supset A$ for all $A \in \mathcal{P}(\Omega)$ satisfying the following properties:*

(i) *If $(g_n) \subset X$ satisfies $g_n \to g$ in X then there is a subsequence $(n_k) \subset \mathbb{N}$ such that, for every $x \in \Omega$, $g_{n_k}(x) \to g(x)$.*

(ii) *There is a constant $C \in (0, \infty)$ such that $\|f + g\| \geq C\|f\|$ for all $f, g \in X$ with $\sigma(f) \cap \sigma(g) = \varnothing$.*

(iii) *$\alpha f \in S$ for all $\alpha \in \mathbb{K}$ and all $f \in S$.*

(iv) *If $f, g \in X$ are such that $f + g \in S$ and $\mathcal{S}(\sigma(f)) \cap \sigma(g) = \varnothing$, then $f \in S$.*

(v) *There is a sequence of functions $\{f_n\}_{n \geq 1} \subset X \setminus S$ such that $\mathcal{S}(\sigma(f_m)) \cap \sigma(f_n) = \varnothing$ for all m, n with $m \neq n$.*

Then the set $X \setminus S$ is spaceable in X.

A weaker version of Theorem 7.4.3 was established in Theorem 3.5.2 and used to prove Theorem 3.5.5 characterizing spaceability of families of Lebesgue integrable functions of a given strict order p.

Proof of Theorem 7.4.3. Let us show that (f_n) is a basic sequence. Indeed, by (iii) one derives that $0 \in S$, so from (v) we get $f_n \neq 0$ for all n; moreover, for every pair $r, s \in \mathbb{N}$ with $s \geq r$ and any scalars $a_1, \ldots, a_s$ it follows from (ii) and (v) [and the fact $\mathcal{S}(\sigma(f_n)) \supset \sigma(f_n)$ for all n] that

$$\left\| \sum_{n=1}^{s} a_n f_n \right\| = \left\| \sum_{n=1}^{r} a_n f_n + \sum_{n=r+1}^{s} a_n f_n \right\| \geq C \left\| \sum_{n=1}^{r} a_n x_n \right\|,$$

because the supports of $\sum_{n=1}^{r} a_n f_n$ and $\sum_{n=r+1}^{s} a_n f_n$ have empty intersection, since $\sigma(\sum_{n \in F} a_n f_n) \subset \bigcup_{n \in F} \sigma(f_n)$ for every finite set $F \subset \mathbb{N}$. According to Nikolskii's theorem, (f_n) is a basic sequence (with basic constant $\alpha = 1/C$).

In particular, the functions f_n $(n \geq 1)$ are linearly independent. Consider the set

$$M := \overline{\text{span}} \{f_n : n \in \mathbb{N}\}.$$

It is plain that M is a closed infinite dimensional vector subspace of X. It is enough to show that $M \setminus \{0\} \subset X \setminus S$. To this end, fix a function $F \in M \setminus \{0\}$. Then there is a uniquely determined sequence $(c_n) \subset \mathbb{K}$ such that

$$F = \sum_{n=1}^{\infty} c_n f_n = \|\cdot\| - \lim_{n \to \infty} \sum_{k=1}^{n} c_k f_k.$$

Let $N = \min\{n \in \mathbb{N} : c_n \neq 0\}$. Then $F = c_N f_N + h$, with $h = \|\cdot\| - \lim_{n \to \infty} h_n$ and $h_n := \sum_{k=N+1}^{n} c_k f_k$ $(n \geq N+1)$. Note that $\sigma(f_N) = \sigma(c_N f_N)$ as $c_N \neq 0$. If $x \in \mathcal{S}(\sigma(c_N f_N)) = \mathcal{S}(\sigma(f_N))$ then, by (v), $x \notin \sigma(f_k)$ for all $k > N$. Hence

$h_n(x) = 0$ for all $n > N$. But, from (i), there is a subsequence $(n_k) \subset \mathbb{N}$ with $h_{n_k} \longrightarrow h$ pointwise. Thus $h(x) = 0$ or, that is the same, $x \notin \sigma(h)$. Therefore

$$\mathcal{S}(\sigma(c_N f_N)) \cap \sigma(h) = \varnothing.$$

By way of contradiction, assume that $F \in S$. Since $F = c_N f_N + h$, we obtain from (iv) that $c_N f_N \in S$. By applying (iii) we get $f_N = c_N^{-1}(c_N f_N) \in S$, which contradicts (v). Consequently, $F \in X \setminus S$, as required. $\qquad\square$

The last theorem can be used to give alternative, shorter proofs of some spaceability results presented in this book; see again the Exercises section.

7.5   An algebrability criterion

As in the properties of dense-lineability and spaceability, there are not many algebrability criteria in the related literature to date. An attempt has been given, in the specific setting of hypercyclity, by Proposition 4.6.2. The criterium given in Theorem 7.5.1 below is an extension of the one provided by Balcerzak, Bartoszewicz and Filipczak in [39, Proposition 7], who worked with subfamlies of $\mathbb{R}^{[0,1]}$. In fact, strong algebrability is obtained.

By $\mathcal{E}$ we denote the family of *exponential-like functions* $\mathbb{C} \to \mathbb{C}$, that is, the functions of the form

$$\varphi(z) = \sum_{j=1}^{m} a_j e^{b_j z}$$

for some $m \in \mathbb{N}$, some $a_1, \dots, a_m \in \mathbb{C} \setminus \{0\}$ and some distinct $b_1, \dots, b_m \in \mathbb{C} \setminus \{0\}$. The number m is called the rank of m.

Theorem 7.5.1. *Let Ω be a nonempty set and $\mathcal{F}$ be a family of functions $\Omega \to \mathbb{K}$, where $\mathbb{K} = \mathbb{R}$ or $\mathbb{C}$. Assume that there exists a function $f \in \mathcal{F}$ such that $f(\Omega)$ is uncountable and $\varphi \circ f \in \mathcal{F}$ for every $\varphi \in \mathcal{E}$. Then $\mathcal{F}$ is strongly $\mathfrak{c}$-algebrable. More precisely, if $H \subset (0, \infty)$ is a set with $\mathrm{card}(H) = \mathfrak{c}$ and linearly independent over the field $\mathbb{Q}$ of rational numbers, then*

$$\{\exp \circ (rf) : r \in H\}$$

is a free system of generators of an algebra contained in $\mathcal{F} \cup \{0\}$.

Proof. Clearly, we may suppose that $\mathbb{K} = \mathbb{C}$. Firstly, each function $\varphi(z) = \sum_{j=1}^{m} a_j e^{b_j z}$ in $\mathcal{E}$ (with $a_1, \dots, b_m \in \mathbb{C} \setminus \{0\}$ and $b_1, \dots, b_m$ distinct) has at most countably many zeros. Indeed, we can assume $|b_1| = \cdots = |b_p| > |b_j|$ $(j = p+1, \cdots, m)$. Then $b_j = |b_1| e^{i\theta_j}$ $(j = 1, \dots, p)$ with $|\theta_j - \theta_1| \in (0, \pi]$ for $j = 2, \dots, p$ (so $c_j := \cos(\theta_j - \theta_1) < 1$ for $j = 2, ..., p$). Hence we have for all $r > 0$ that

$$|\varphi(re^{-i\theta_1})| \geq |a_1|e^{|b_1|r} - \sum_{j=2}^{p}|a_j|e^{|b_1|c_j r} - \sum_{j=p+1}^{m}|a_j|e^{|b_j|r} \longrightarrow \infty \text{ as } r \to +\infty.$$

Therefore φ is a nonconstant entire function, so $\varphi^{-1}(\{0\})$ is countable because the zero-points of a nonconstant analytic function are isolated.

Now, consider a nonzero polynomial P in N complex variables without constant term, as well as numbers $r_1, \ldots, r_N \in H$. The function $\Phi : \Omega \to \mathbb{C}$ given by $\Phi = P(\exp \circ (r_1 f), \ldots, \exp \circ (r_N f))$ is of the form

$$\sum_{j=1}^{m} a_i (e^{r_1 f(x)})^{k(j,1)} \cdots (e^{r_N f(x)})^{k(j,N)} = \sum_{j=1}^{m} a_i \exp\left(f(x)\sum_{l=1}^{N} r_l\, k(j,l)\right),$$

where $a_1, \ldots, a_m \in \mathbb{C} \setminus \{0\}$ and the matrix $[k(j,l)]_{\substack{j=1,\ldots,m \\ l=1,\ldots,N}}$ of nonnegative integers has distinct nonzero rows. Thus, the numbers $b_j := \sum_{l=1}^{N} r_l k(j,l)$ $(j = 1, \ldots, m)$ are distinct and nonzero, due to the $\mathbb{Q}$-independence of the r_l's. Hence the function $\varphi(z) := \sum_{j=1}^{m} a_j e^{b_j z}$ belongs to $\mathcal{E}$. But $\Phi = \varphi \circ f$, so if $\Phi \equiv 0$ then we would have $\varphi|_{f(\Omega)} \equiv 0$, which contradicts the fact that $f(\Omega)$ is uncountable. Consequently, $\Phi \neq 0$ and, by hypothesis, $\Phi \in \mathcal{F}$. This proves the theorem. $\qquad\square$

Next, in order to illustrate the last result, we are going to give two examples: one in the setting of smooth functions (see Chapter 1) and the other one in the setting of holomorphic functions (cf. Chapter 2). Recall that by $\mathcal{S}(\mathbb{R})$ and $H_e(G)$ we denoted, respectively, the family of nowhere analytic (or singular) C^∞-functions on $\mathbb{R}$ and the family of all holomorphic functions $f : G \to \mathbb{C}$ –where G is a domain of $\mathbb{C}$– that are not holomorphically extendable beyond the boundary of G. Both families have been proved to be algebrable; see Theorems 1.4.3 and 2.3.1, respectively. The next two theorems improve these results.

Theorem 7.5.2. *The set $\mathcal{S}(\mathbb{R})$ is strongly $\mathfrak{c}$-algebrable.*

Proof. Let us first prove that for every positive integer m, any exponential-like function $\varphi : \mathbb{R} \to \mathbb{R}$ of rank m and each $c \in \mathbb{R}$, the preimage $\varphi^{-1}(\{c\})$ has at most m elements.

For this, we proceed by induction. If $m = 1$, the function φ is of the form $\varphi(x) = a\, e^{\beta x}$ $(x \in \mathbb{R})$, with $a \neq 0 \neq \beta$. Hence φ is strictly monotone and the property is obvious. Assume that the property holds for all exponential-like functions of range m. Let $\varphi(x) = \sum_{i=1}^{m+1} a_i\, e^{\beta_i x}$, for some distinct nonzero real numbers $\beta_1, \ldots, \beta_{m+1}$ and some nonzero real numbers $a_1, \ldots, a_{m+1}$. Consider the derivative

$$\varphi'(x) = \sum_{i=1}^{m+1} \beta_i a_i e^{\beta_i x} = e^{\beta_1 x}\left(\beta_1 a_1 + \sum_{i=2}^{m+1} \beta_i a_i e^{(\beta_i - \beta_1)x}\right) \quad (x \in \mathbb{R}).$$

Note that $\gamma_i := \beta_i - \beta_1$ $(i = 2, \ldots, m+1)$ are nonzero distinct real numbers. So, we may apply the induction hypothesis to $\psi(x) := \sum_{i=2}^{m+1} \beta_i a_i e^{\gamma_i x}$ and $c := -\beta_1 a_1$. This shows that $\varphi^{-1}(\{0\})$ has at most m elements. Hence f has at most m local extrema on $\mathbb{R}$. This implies that for each $c \in \mathbb{R}$, the preimage $\varphi^{-1}(\{c\})$ has at most $m+1$ elements, as desired.

Consequently, each function φ as before is not constant in every subinterval of $\mathbb{R}$. In particular there exists a finite decomposition of $\mathbb{R}$ into intervals, such that φ is strictly monotone on each of them.

Now, fix any function $F \in \mathcal{S}(\mathbb{R})$, for instance, the Lerch function $F(x) = \sum_{n=1}^{\infty} \frac{\cos(3^n x)}{n!}$ [286]. Fix also an exponential-like function φ. Denote $g := \varphi \circ F$. Clearly, $g \in \mathcal{C}^{\infty}$. We want to show that $g \in \mathcal{S}(\mathbb{R})$. Suppose, on the contrary, that g is analytic at a point $x_0 \in \mathbb{R}$. Then there is an open set $V \ni x_0$ such that g is analytic on V (see, e.g., [279]). By the result proved at the beginning of the proof, there is an open set $U \subset F(V)$ on which φ is invertible. Hence $F = \varphi^{-1} \circ g$ is analytic (as a composition of analytic functions; see [279]) on the set $F^{-1}(U) \cap V$, that leads us to a contradiction. By an application of Theorem 7.5.1, the proof is finished. $\qquad \square$

Recall that $A^{\infty}(G)$ denotes the vector space of all functions $f \in H(G)$, all of whose derivatives $f^{(n)}$ $(n \geq 0)$ can be continuously extended to ∂G.

Theorem 7.5.3. *If G is a Jordan domain in the complex plane $\mathbb{C}$ then the set $H_e(G) \cap A^{\infty}(G)$ is strongly $\mathfrak{c}$-algebrable.*

Proof. Since G is a Jordan domain, a function $f \in H(G)$ belongs to $H_e(G)$ if and only if there is no domain containing G strictly to which f can be holomorphically extended.

Since G is a regular domain, a result of Chmielowski ([162]; see Chapter 2) stated that

$$H_e(G) \cap A^{\infty}(G) \neq \varnothing.$$

Pick a function $f \in H_e(G) \cap A^{\infty}(G)$ and let

$$\Omega := G \quad \text{and} \quad \mathcal{F} := H_e(G) \cap A^{\infty}(G).$$

Since f is nonconstant, the set $f(\Omega)$ is open, so uncountable. According to Theorem 7.5.1, it is enough to prove that $\varphi \circ f \in \mathcal{F}$ for every $\varphi \in \mathcal{E}$, the family of exponential-like functions. Since such member of $\mathcal{E}$ is nonconstant, we would be done as soon as we demonstrate the following assertion: If φ is a nonconstant entire function then $\varphi \circ f \in H_e(G)$ (observe that, of course, $\varphi \circ f \in A^{\infty}(G)$ for every entire function φ).

With this aim, assume, by way of contradiction, that $F := \varphi \circ f \notin H_e(G)$. Then there is an open ball B centered at some point $z_0 \in \partial G$ as well as a function $\widetilde{F} \in H(G \cup B)$ such that $\widetilde{F} = F$ on G. Let $B_1 \subset B$ be any closed ball centered at z_0. If $\widetilde{F}'(z) = 0$ for all $z \in B \cap \partial G$ then, since ∂G has no isolated points, we would have $\widetilde{F}' = 0$ on a subset of B having some accumulation point in B (namely, on $B_1 \cap \partial G$). By the Analytic Continuation Principle, $\widetilde{F}' = 0$ on B, hence $\widetilde{F}$ is constant on B. By the same Principle, and since $\widetilde{F} = F$ on G, we get $F = \text{constant}$ in G. Since $f \in H_e(G)$, f is not constant, so $f(G)$ is an open subset of $\mathbb{C}$ by the Open Mapping theorem for analytic functions (see, e.g., [2]). Therefore φ is constant on the open set $f(G)$, and a third application of the Analytic Continuation Principle yields $\varphi = \text{constant}$, which is absurd. Then there must be $z_1 \in B \cap \partial G$ with

$$\widetilde{F}'(z_1) \neq 0.$$

From the Local Representation theorem (see [2]) we derive the existence of an open ball $B_2 \subset B$ centered at z_1 and of a domain W with $W \ni \widetilde{F}(z_1) = \varphi(f(z_1))$ (recall that f extends continuously to ∂G) such that $\widetilde{F} : B_2 \to W$ is bijective. In particular,

$$0 \neq \widetilde{F}'(z_1) = \varphi'(f(z_1)) f'(z_1),$$

where in the last equality the fact that f' has continuous extension to ∂G has been used. Thus $\varphi'(f(z_1)) \neq 0$ and, again by the Local Representation theorem, there are an open ball B_0 centered at $f(z_1)$ and a domain V with $\varphi(f(z_1)) \in V \subset W$ such that $\varphi : B_0 \to V$ is bijective. Then $(\widetilde{F}|_{B_2})^{-1}(V)$ is a domain satisfying $z_1 \in (\widetilde{F}|_{B_2})^{-1}(V) \subset B$. Consequently, $G \cup (\widetilde{F}|_{B_2})^{-1}(V)$ is a domain containing G strictly and the function

$$\widetilde{f} : G \cup (\widetilde{F}|_{B_2})^{-1}(V) \to \mathbb{C}$$

given by

$$\widetilde{f}(z) = \begin{cases} f(z) & \text{if } z \in G \\ \varphi^{-1}(\widetilde{F}(z)) & \text{if } z \in (\widetilde{F}|_{B_2})^{-1}(V) \end{cases}$$

is well defined, holomorphic and extends f. This contradicts the fact that $f \in H_e(G)$ and the proof is finished. □

We conclude this section with a special case of Theorem 7.5.1 in which one can derive dense strong algebrability if some condition on the generating function f is imposed.

Theorem 7.5.4. *Let $\mathcal{F}$ be a family of functions in $\mathcal{C}[0,1]$. Assume that there exists a strictly monotonic function $f \in \mathcal{C}[0,1]$ such that $\varphi \circ f \in \mathcal{F}$ for every $\varphi \in \mathcal{E}$. Then $\mathcal{F}$ is densely strongly $\mathfrak{c}$-algebrable in $\mathcal{C}[0,1]$.*

Proof. Firstly, observe that if $\Omega = [0,1]$ then $f(\Omega)$ is a non-degenerate interval, so it is an uncountable set. Then, it is sufficient to show that the algebra $\mathcal{A}$ generated by the system

$$\{\exp \circ (rf) : r \in H\}$$

given in Theorem 7.5.1 is dense. For this, we invoke the Weierstrass-Stone theorem. Take any $\alpha \in H \subset (0, +\infty)$. Given $x_0 \in [0, 1]$, the function $F(x) :=$ $e^{\alpha f(x)}$ belongs to $\mathcal{A}$ and satisfies $F(x_0) \neq 0$. Moreover, for prescribed distinct points $x_0, x_1 \in [0, 1]$, the same function F fulfills $F(x_0) \neq F(x_1)$, because both functions f and $x \mapsto e^{\alpha x}$ are one-to-one. As a conclusion, $\mathcal{A}$ is dense in $\mathcal{C}[0, 1]$. $\qquad\square$

7.6 Additivity and cardinal invariants: a brief account

One conclusion that we can infer from what we have been presenting in the previous chapters is that one of the most common strategies to obtain lineability is the following: once a mathematical object enjoying a particular property is found, one tries to modify it in order to construct a basis for the *potential* candidate to linear space.

Thus, almost all techniques are, to date, constructive. Only a handful of *existence techniques* (although not very general ones) have been obtained in the recent years. In this section, we are interested in giving a brief account of a recent approach to these lineability questions via set theory and cardinal invariants. It is not our purpose to give a thorough study of this approach but some motivation for future study.

It was not until very recently that some *existence results* to guarantee the lineability of certain sets were given. These mentioned results, although not giving the *specific* linear space, supply the positive answer we look for in this area. Due to the lack of existence results in this topic, we believe that it is interesting to have this kind of *machine-proving result* that, right away, guarantees the lineability of a given family of functions. In order to do that, we relate the notion of lineability with that of *additivity,* introduced by T. Natkaniec in [321, 322] and thoroughly studied by K.C. Ciesielski [166, 168–170, 173] and F.E. Jordan [261] (see, also, [210]). The last notion is defined as follows.

Definition 7.6.1. Let $\mathcal{F} \subset \mathbb{R}^{\mathbb{R}}$. The *additivity* of $\mathcal{F}$ is defined as the following cardinal number:

$$\mathcal{A}(\mathcal{F}) = \min\left(\{\, \text{card}(F) : F \subset \mathbb{R}^{\mathbb{R}},\ \varphi + F \not\subset \mathcal{F},\ \text{for all}\ \varphi \in \mathbb{R}^{\mathbb{R}} \,\} \cup \{(2^{\mathfrak{c}})^+\}\right),$$

where $(2^{\mathfrak{c}})^+$ stands for the successor cardinal of $2^{\mathfrak{c}}$.

The above definition gives us, roughly, the *biggest* cardinal number κ for which every family $\mathcal{G}$, with $\text{card}(\mathcal{G}) < \kappa$, can be translated into $\mathcal{F}$. Let us remark that for some families the additivity is relatively "easy" to compute, whereas lineability is, in general, "hard" to calculate.

Additivity and lineability have not been related until now. Although it may seem like this concept has nothing to do with the concept of lineability, it actually has a lot to do with it, as the following assertion by Gámez, Muñoz and Seoane [210] shows.

Theorem 7.6.2. *Let $\mathcal{F} \subset \mathbb{R}^\mathbb{R}$ be star-like, that is, $\alpha \mathcal{F} \subset \mathcal{F}$ for all $\alpha \in \mathbb{R}$. If $\mathfrak{c} < \mathcal{A}(\mathcal{F}) \leq 2^\mathfrak{c}$, then $\mathcal{F}$ is $\mathcal{A}(\mathcal{F})$-lineable.*

Proof. We first prove the following result: Let $\mathcal{F}, F$ be proper subsets of $\mathbb{R}^\mathbb{R}$ such that $F - F \subset \mathcal{F}$ and $\aleph_0 < \text{card}(F) < \mathcal{A}(F)$. Then there exists $g \in \mathcal{F} \backslash F$ such that $g + F \subset \mathcal{F}$.

Indeed, let $h \in \mathbb{R}^\mathbb{R} \setminus F$, and define $F_h = (h + F) \cup F$. Then $\text{card}(F_h) = \text{card}(F) < \mathcal{A}(\mathcal{F})$. Therefore, there exists $g \in \mathbb{R}^\mathbb{R}$ such that $g + F_h \subset \mathcal{F}$. Consequently, we have that $g + F \subset \mathcal{F}$ and $(g + h) + F \subset \mathcal{F}$. Since $0 \in F - F \subset \mathcal{F}$, it is obvious that both functions $g, g + h \in \mathcal{F}$. Assuming that $g \in \mathcal{F}$, it must be $g + h \notin \mathcal{F}$, since otherwise we would have

$$h = (g + h) - g \in F - F \subset F,$$

and this is a contradiction.

Now, let $\mathcal{F}$ be as in the statement of the theorem. Obviously, $\mathcal{F}$ contains a linear space (namely, $\{0\}$) and, using Zorn's Lemma, we obtain that there exists a maximal linear space X contained in $\mathcal{F}$. Clearly, we have $\text{card}(X) = \max\{\dim X, \mathfrak{c}\}$. If the statement does not hold, we shall have $\text{card}(X) < \mathcal{A}(\mathcal{F})$ and, by the result established at the beginning of this proof, there exists $g \in \mathcal{F} \setminus X$ such that $g + X \subset \mathcal{F}$. Define $Y = [g] + X$, where $[g]$ denotes the linear span of g. Using that $\mathcal{F}$ is star-like, it is easy to see that $Y \subset \mathcal{F}$, in plain contradiction with the maximality of X. This contradiction proves the theorem. $\square$

As an immediate consequence, we obtain the following.

Corollary 7.6.3. *Let $\mathcal{F} \subset \mathbb{R}^\mathbb{R}$ be star-like with $\mathcal{A}(\mathcal{F}) \geq \mathfrak{c}$. Then $\mathcal{F}$ is lineable.*

Theorem 7.6.2 can be applied to many classes of functions, such as those presented in Definition 1.5.1 from Section 1.5 (see, also, [210]).

The result provides a technique (that can only be used or applied within the framework of $\mathbb{R}^\mathbb{R}$).

After this concept of additivity was linked to lineability questions, several authors delved into this theory and considered alternatives to the original lineability definition. For instance, the following concept was introduced in [48]:

Definition 7.6.4. Let M be a subset of some vector space W. The *lineability cardinal number* of M is defined as

$$\mathcal{L}(M) = \min\{\kappa \colon M \cup \{0\} \text{ contains no vector space of dimension } \kappa\}.$$

Notice that $M \subset W$ is μ-lineable if, and only if, $\mu < \mathcal{L}(M)$. In particular, μ is the maximal dimension of a subspace of $M \cup \{0\}$ if, and only if, $\mathcal{L}(M) = \mu^+$. The number $\mathcal{L}(M)$ need not be a cardinal successor (see, e.g., [28] or [105]). Thus, a vector subspace of $M \cup \{0\}$ with maximal dimension does not necessarily exist.

If W is a vector space over the field $\mathbb{K}$ and $M \subset W$, let

$$\mathrm{st}(M) = \{w \in W : (\mathbb{K} \setminus \{0\})\, w \subset M\}.$$

The following property is immediate from the definitions.

Proposition 7.6.5. *Let W be a vector space and $M \subset W$. The following holds:*

(a) *If V is a subspace of W, then $V \subset M \cup \{0\}$ if, and only if, $V \subset \mathrm{st}\,(M) \cup \{0\}$.*

(b) *In particular, we have $\mathcal{L}(M) = \mathcal{L}(\mathrm{st}(M))$.*

Recall (see, e.g., [209]) that a family $M \subset W$ is said to be *star-like* provided that $\mathrm{st}(M) = M$. Proposition 7.6.5 explains why the assumption that M is star-like appears in quite a few results on lineability. Also, a simple use of Zorn's lemma shows that any linear subspace V_0 of $M \cup \{0\}$ can be extended to a maximal linear subspace V of $M \cup \{0\}$. Thus, one can consider the following concept (see, e.g., [167]).

Definition 7.6.6. Let M be any arbitrary subset of a vector space W. We define the *maximal lineability cardinal number* of M as

$$\mathrm{m}\mathcal{L}(M) = \min\{\dim(V) : V \text{ is a maximal linear subspace of } M \cup \{0\}\}.$$

Although this notion might seem similar to that of maximal-lineability and maximal-spaceability (introduced by Bernal in [83]) they are, in general, not related. In any case, Proposition 7.6.5 implies that $\mathrm{m}\mathcal{L}(M) = \mathrm{m}\mathcal{L}(\mathrm{st}(M))$.

Remark 7.6.7. *It is easy to see that $\mathcal{HL}(M) = \mathrm{m}\mathcal{L}(M)^+$, where $\mathcal{HL}(M)$ is the homogeneous lineability number, defined in [48], given by*

$$\mathcal{HL}(M) = \min\left(\{\lambda : \text{there is a linear space } Y \subset M \cup \{0\} \text{ with } \dim(Y) < \lambda \text{ which cannot be extended to a linear space } X \subset M \cup \{0\} \text{ with } \dim(X) = \lambda\} \cup \{(\mathrm{card}(V))^+\}\right).$$

Observe that $\mathcal{HL}(M)$ is always a successor cardinal, as shown in [48]. Clearly we have

$$\mathcal{HL}(M) \leq \mathcal{L}(M).$$

The inequality may be strict, as proved in [48].

Also, besides the above notions, which were all attempts to generalize the concept to additivity within the framework of lineability, we can (for

$M \subset W$) consider the following *additivity* number (compare [48]), which is a generalization of the notion of additivity (Definition 7.6.1) for $V = \mathbb{R}^{\mathbb{R}}$:

$$A(M, W) = \min(\{\mathrm{card}\,(F) \colon F \subset W \quad \text{and} \quad (\forall w \in W)\,(w + F \not\subset M)\}$$
$$\cup \{(\mathrm{card}(W))^+\}),$$

where $w + F = \{w + f \colon f \in F\}$. Most of the times the space W, usually $W = \mathbb{R}^{\mathbb{R}}$, will be clear by the context.

However, this definition above is of interest whenever we are working with W being a topological vector space. We shall say that $M \subset W$ is μ-*spaceable* with respect to a topology τ on W, provided there exists a τ-closed vector space $V \subset M \cup \{0\}$ of dimension μ. Most of the times the space W, usually $W = \mathbb{R}^{\mathbb{R}}$, will be clear by the context. In such cases one can write $A(M)$ instead of $A(M, W)$.

In [167, Proposition 2.1] the authors provided the following relation between additivity and lineability numbers:

Theorem 7.6.8. *Let W be a vector space over a field K and let $\varnothing \neq M \subsetneq W$. Then*

1. *$2 \leq A(M) \leq |W|$ and $\mathrm{m}\mathcal{L}(M) < \mathcal{L}(M) \leq \dim(W)^+$;*

2. *if $A(\mathrm{st}(M)) > |K|$, then $A(\mathrm{st}(M)) \leq \mathrm{m}\mathcal{L}(M)$.*

In particular, if M is star-like, then $A(M) > |K|$ implies that

3. *$A(M) \leq \mathrm{m}\mathcal{L}(M) < \mathcal{L}(M) \leq \dim(W)^+$.*

Proof. The inequalities from item *1* are easy to see. As for item *2*, this can be proved by an easy transfinite induction. Alternatively, notice that Bartoszewicz and Głąb proved, in [48, Corollary 2.3], that if $M \subset W$ is star-like and $A(M) > |K|$, then $A(M) < \mathcal{HL}(M)$. Hence, $A(\mathrm{st}(M)) > |K|$ implies that

$$A(\mathrm{st}(M)) < \mathcal{HL}(\mathrm{st}(M)) = \mathrm{m}\mathcal{L}(\mathrm{st}(M))^+ = \mathrm{m}\mathcal{L}(M)^+.$$

Therefore, $A(\mathrm{st}(M)) \leq \mathrm{m}\mathcal{L}(M)$. $\qquad\square$

In particular, we can consider also the following *spaceability* cardinal number:

$$\mathcal{L}_\tau(M) = \min\{\kappa \colon M \cup \{0\} \text{ contains no } \tau\text{-closed subspace of dimension } \kappa\}.$$

Notice that $\mathcal{L}(M) = \mathcal{L}_\tau(M)$ when τ is the discrete topology. Of course, there might be some other topological properties distinguishing between the families M with the same value $\mathcal{L}_\tau(M)$. For example, in [39] it is shown that if M is the family of strongly singular functions (a continuous function $[0, 1] \to \mathbb{R}$ is called *strongly singular* provided it is constant on no subinterval and its derivative exists and equals 0 almost everywhere in $[0, 1]$) in the space

CBV$[0,1]$ of continuous bounded variation in $[0,1]$, then $\mathcal{L}_u(M) = \mathfrak{c}^+$ and M contains a linear subspace generated by a discrete set of the cardinality $\mathfrak{c}$. Similarly, if M is the family of all nowhere differentiable functions in $\mathcal{C}([0,1])$, then $\mathcal{L}_u(M) = \mathfrak{c}^+$, as proven in [345]. However, the linear subspace of M constructed in [345] is only separable.

Let us give a final example. Fix $N \in \mathbb{N}$ and the space $W = \mathbb{R}^X$ of all functions from $X = \mathbb{R}^N$ to $\mathbb{R}$, and consider the topologies τ_u and τ_p of uniform and pointwise convergence, respectively. Then we have, clearly, the inequalities:

$$\mathcal{L}_{\tau_p}(M) \leq \mathcal{L}_{\tau_u}(M) \leq \mathcal{L}(M).$$

As we can see, since the concept of additivity was linked to that of lineability, many papers have been appearing and there is very active ongoing research in this direction. It is very likely that, by the time this monograph appears in print, there will be even more ongoing works in the direction of additivity-lineability. However, it is not the purpose of this monograph to deepen into this highly technical study but provide a brief account and motivation to the reader for future lines of research. At the moment we would like to cite [48, 167, 210, 336] as the most modern references in this direction.

7.7 Exercises

Exercise 7.1. Prove that the family of power series $\sum_{n=0}^{\infty} a_n z^n$ with radius of convergence 1 such that $\sum_{n=0}^{\infty} |a_n|$ diverges is spaceable and dense-lineable in the disc algebra $A(\mathbb{D}) := \{f \in H(\mathbb{D}) : f \text{ extends continuously on } \overline{\mathbb{D}}\}$, endowed with the supremum norm $\|f\| = \sup\{|f(z)| : z \in \mathbb{D}\}$.
Hint: Apply Theorems 7.4.2 and 7.3.3.

Exercise 7.2. Show that the set $\mathcal{F}$ described in Theorem 4.5.11 is maximal dense-lineable in $\mathcal{C}(\mathbb{T})$. That is, we are asking for solving again Exercise 4.19, but this time by a direct application of Theorem 7.3.1.
Hint: The set of complex trigonometrical polynomials is dense in $\mathcal{C}(\mathbb{T})$.

Exercise 7.3. Give an alternative, quicker proof of the fact (given in Theorem 1.8.1) that the set $CS(\mathbb{R}, \mathbb{R}^2)$ of continuous surjections $\mathbb{R} \to \mathbb{R}^2$ is dense-lineable and spaceable in $\mathcal{C}(\mathbb{R}, \mathbb{R}^2)$. As a matter of fact, it is even maximal dense-lineable.
Hint: Use Theorems 7.4.3 and 7.3.1, in this order, and take into account that the set of functions in $\mathcal{C}(\mathbb{R}, \mathbb{R}^2)$ having bounded support is dense in this space.

Exercise 7.4. Prove that the set of strongly singular functions $[0,1] \to \mathbb{R}$ is strongly $\mathfrak{c}$-algebrable.
Hint: Fix any strongly singular function F, as for instance a strictly increasing

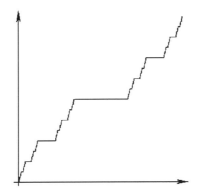

FIGURE 7.1: Cantor's staircase.

version of the so-called "Cantor's staircase" (see, e.g., [376, p. 100] and Figure 7.1), and follow the approach of the proof of Theorem 7.5.2.

Exercise 7.5. Let $I = [0, \infty)$ and denote by $C_0(I)$ and $R(I)$, respectively, the vector space of continuous functions $f : I \to \mathbb{R}$ with $\lim_{x \to \infty} f(x) = 0$ and the vector space of Riemann integrable functions on I.

(a) Show that the expression $\|f\| := \sup_{x \geq 0} |f(x)| + \sup_{x \geq 0} \left| \int_0^x f(t)\, dt \right|$ defines a norm on the space $C_0(I) \cap R(I)$ which makes it a separable Banach space, and that the set B of continuous functions $I \to \mathbb{R}$ with bounded support is a dense vector subspace of $(C_0(I) \cap R(I), \|\cdot\|)$.

(b) Prove that the set $A := C_0(I) \cap R(I) \setminus \bigcup_{0 < p < \infty} L_p(I)$ is spaceable and maximal dense-lineable in $(C_0(I) \cap R(I), \|\cdot\|)$.

Hint: To prove (b), set $X := C_0(I) \cap R(I)$, $Z_n := C_0(I) \cap R(I) \cap L_n(I)$ and $T_n :=$ the inclusion $Z_n \hookrightarrow X$ ($n \in \mathbb{N}$). Then take into account that $L_p(I) \subset L_q(I)$ if $p \leq q$ and apply Theorem 7.4.2 to get spaceability. Derive maximal dense-lineability from (a) and Theorem 7.3.1.

Exercise 7.6. Consider the space $CBV([0, 1])$ of continuous bounded variation functions $[0, 1] \to \mathbb{R}$, endowed with the norm $\|f\| := |f(0)| + V_f([0, 1])$, where $V_f([0, 1])$ is the variation of f on $[0, 1]$, given by

$$V_f([0, 1]) = \sup \left\{ \sum_{i=1}^n |f(t_i) - f(t_{i-1})| : 0 = t_0 < t_1 < \cdots < t_N = 1, \ N \in \mathbb{N} \right\}.$$

Let $AC([0, 1])$ be the subset of absolutely continuous functions, that is, the set of functions f which are differentiable almost everywhere on $[0, 1]$ and such that Barrow's rule $f(b) - f(a) = \int_a^b f'(t)\, dt$ holds for every interval $[a, b] \subset [0, 1]$. Demonstrate that the set $CBV([0, 1]) \setminus AC([0, 1])$ is spaceable in $CBV([0, 1])$.

Hint: Use Theorem 7.4.3 by constructing a sequence (f_n) of appropriate translation-dilations of the function F considered in Exercise 7.4. Alternatively, prove that $AC([0,1])$ is a closed subspace of $CBV([0,1])$ with infinite codimension, and then apply Theorem 7.4.1.

Exercise 7.7. By using Theorem 7.3.1, give a rapid proof of the dense lineability in $\mathcal{C}^\infty(\mathbb{R})$ of the family $\mathcal{A}$ of smooth functions $f : \mathbb{R} \to \mathbb{R}$ for which the origin is a Pringsheim-singularity (see Section 1.4), that is, the radius of convergence of the associated Taylor series at the origin is 0.

Hint: Prove that $\mathcal{A}$ is lineable by choosing a smooth function φ with $\varphi^{(n)}(0) = n! \, n^n$ for all $n \in \mathbb{N}$ (whose existence is guaranteed by Borel's theorem; see, e.g., [190, pp. 50–51]) and show that $\mathcal{A}$ is lineable by proving that $P \cdot f \in \mathcal{A}$ for each nonzero polynomial P. This can be done with the help of Leibnitz's rule for the nth derivative of a product. Finally, use the denseness in $\mathcal{C}^\infty(\mathbb{R})$ of the set of polynomials.

Exercise 7.8. This exercise shows that the conclusion of the Lebesgue dominated convergence fails for many "tamed" sequences of functions. Consider the vector space $(\mathbb{R}^\mathbb{R})^\mathbb{N}$ of sequences $(f_k)_{k \geq 1}$ of functions $\mathbb{R} \to \mathbb{R}$, as well as the subspace given by

$CBL_s := \{(f_k) \in (\mathbb{R}^\mathbb{R})^\mathbb{N} :$ each f_k is continuous, bounded and integrable,

$$\|f_k\|_\infty \underset{k \to \infty}{\longrightarrow} 0 \text{ and } \sup_k \|f_k\|_1 < \infty\}$$

as well as the family

$$\mathcal{F} := \{(f_k) \in CBL_s : \|f_k\|_1 \not\to 0 \text{ as } k \to \infty\}.$$

(a) Verify that CBL_s becomes a Banach space when endowed with the norm $\|(f_k)\| = \sup_k \|f_k\|_\infty + \sup_k \|f_k\|_1$.

(b) Prove that $\mathcal{F}$ is spaceable in CBL_s.

Hint: For (b), make use of Wilansky–Kalton's criterion (Theorem 7.4.1) together with the countable family $\{\Phi_n : n \geq 1\}$ of sequences given by $\Phi_n = \{f_{n,k}\}_{k \geq 1}$, where $f_{n,k}(x) := \frac{k^n}{k^{2n}+x^2}$.

Exercise 7.9. Let $\varphi : [0,\infty) \to (0,\infty)$ be an increasing function. Demonstrate that the set

$$\mathcal{E}_\varphi := \left\{ f \in \mathcal{E} : \limsup_{r \to +\infty} \frac{\max\{|f(z)| : |z| = r\}}{\varphi(r)} = \infty \right\}$$

is maximal dense-lineable and spaceable in the space $\mathcal{E} = H(\mathbb{C})$ of entire functions.

Hint: Firstly, observe that it is enough to show the result for $\mathcal{E}_\psi$, where $\psi(r) = e^r + \varphi(r)$. Then the (dense) set of polynomials is contained in $Y := \mathcal{E} \setminus \mathcal{E}_\psi$. Use the classical Weierstrass interpolation theorem for entire functions to prove that the set Y is not the whole $\mathcal{E}$. Hence Y is not closed in $\mathcal{E}$. Apply Theorem 7.4.2 (with only one Z_n) and, subsequently, Theorem 7.3.1.

Exercise 7.10. The weighted Dirichlet spaces of the unit disc $\mathbb{D}$ are defined by

$$\mathcal{S}_\nu := \left\{ f(z) = \sum_{n=0}^\infty a_n z^n \in H(\mathbb{D}) : \sum_{n=0}^\infty |a_n|^2 (n+1)^{2\nu} < \infty \right\},$$

where $\nu \in \mathbb{R}$. For instance, if $\nu = 0, -1/2, 1/2$, then $\mathcal{S}_\nu$ is, respectively, the classical Hardy space $H^2(\mathbb{D})$, the Bergman space $A^2(\mathbb{D})$ and the Dirichlet space $\mathcal{D}$.

(a) Demonstrate that each $\mathcal{S}_\nu$ becomes a Hilbert space under the inner product
$\langle \sum_{n=0}^\infty a_n z^n, \sum_{n=0}^\infty b_n z^n \rangle = \sum_{n=0}^\infty a_n \overline{b_n} (n+1)^{2\nu}.$

(b) Show that $\mathcal{S}_\alpha \supsetneq \mathcal{S}_\beta$ if $\beta > \alpha$.

(c) Define the set $\mathcal{S}_{\nu,\text{strict}} := \mathcal{S}_\nu \setminus \bigcup_{a>\nu} \mathcal{S}_a$. Prove that this set is maximal dense-lineable and spaceable in $\mathcal{S}_\nu$.

Hint: For part (c) invoke, again, Theorems 7.4.2 and 7.3.1, and use that each polynomial is in $Y := \bigcup_{a>\nu} \mathcal{S}_a$ and that $Y \neq \mathcal{S}_\nu$. To this end, verify that the function $f(z) = \sum_{k=0}^\infty \frac{z^k}{(k+1)^{\nu+\frac{1}{2}} \cdot \log(k+1)}$ belongs to $\mathcal{S}_\nu \setminus Y$.

7.8 Notes and remarks

Section 7.2. Another remarkable example of a set that is linearly large in one sense but linearly not too large in another sense is the following one due to Cariello and Seoane [153]. Let X denote any of the sequence spaces $X = c_0$ or ℓ_p $(1 \leq p \leq \infty)$. Then the subset

$$Z(X) := \{x = (x_n) \in X : x_n = 0 \text{ only for finitely many } n \in \mathbb{N}\}$$

is $\mathfrak{c}$-lineable (so maximal lineable) but not spaceable (see Section 6.5). Note that, if we take $A = Z(X)$ and $B = c_{00} =$ the space of sequences with only finitely many nonzero entries, then $A+B \subset A$, and Theorem 7.3.1 yields that $Z(X)$ is even maximal dense-lineable.

Section 7.3. Theorem 7.3.1 and Corollary 7.3.2 contains a number of results due to Aron, Bernal, García, Pérez, Ordóñez and Seoane that have evolved along the papers [81], [26], [83] and [104]. Theorem 7.3.3 can be found in [104].

Section 7.4. Kitson–Timoney's Theorem 7.4.2 can be found in [275, Theorem 3.3]. It is an open problem whether it can be extended to allow the Z_n's to be Fréchet spaces. In 2009, Botelho, Diniz and Pellegrino [142] gave sufficient conditions in order to ensure $\mathfrak{c}$-lineability of the family $K(X,Y) \setminus \Pi_p(X,Y)$

consisting of those compact operators that are not absolutely p-summing, where X and Y are Banach spaces and $1 \leq p < \infty$. Recall that a linear operator $T : X \to Y$ is called *compact* if the image $T(B)$ of the unit ball of X is relatively compact in Y, and that T is called *absolutely p-summing* provided that there is a constant $C \in (0, \infty)$ such that

$$\left(\sum_{i=1}^n \|Tx_i\|^p \right)^{1/p} \leq C \cdot \sup \left\{ \left(\sum_{i=1}^n |\varphi(x_i)|^p \right)^{1/p} : \varphi \in X^*, \|\varphi\| = 1 \right\}.$$

Theorem 7.4.2 is used in [275] to improve the results in [142]: namely, under conditions milder than those in [142], it is established that $K(X,Y) \setminus \bigcup_{1 \leq p < \infty} \Pi_p(X,Y)$ is spaceable in $K(X,Y)$. It is also used in [247] to show spaceability of $I_1(E,F) \setminus I_2(E,F)$ in $I_1(E,F)$, where E, F are Banach spaces and I_1, I_2 are operator ideals in the sense of Pietsch satisfying appropriate conditions.

Theorem 7.4.3 is due to Bernal and Ordóñez [104], being an extension of one of their results in [103]. The usual choices for the mapping $\mathcal{S}$ in this theorem are $\mathcal{S}(A) = A$ and $\mathcal{S}(A) = \overline{A}$ (in this case, if Ω is a topological space). The result is employed in [103, 104] to get spaceability in various settings. Exercise 7.5 is borrowed from [104, section 4.6], and its content have already been mentioned in the Notes and Remarks section of Chapter 3. Exercises 7.8, 7.9 and 7.10 also contain some results from [104].

Section 7.5. The real version of Theorem 7.5.1 is given in Proposition 7 of [39]. Theorems 7.5.2 and 7.5.3 are due, respectively, to Balcerzak, Bienias, Filipczak and Głąb [45] and Bernal [85]. The criterium given in Theorem 7.5.1 is exploited in [39] and [45] to obtain (strong) algebrability in many further families of functions, such as the set of strongly singular functions in $CBV([0,1])$ (see Exercise 7.4), the family of functions of class $\mathcal{C}^N$ ($N \in \mathbb{N}$) that do not have derivative of order $N + 1$ somewhere in any open subinterval of $[0,1]$, the set of Sierpinski–Zygmund functions (see Section 6.6), the family $\mathcal{C}_G$ of all functions $[0,1] \to \mathbb{R}$ for which the set of points of continuity is exactly a prescribed G_δ subset G of $[0,1]$ and the set of approximately continuous functions $[0,1] \to \mathbb{R}$ (i.e. functions f such that, for every $x \in [0,1]$, there is a Lebesgue measurable set $E \subset [0,1]$ such that

$$\lim_{h \to 0} \frac{\lambda(E \cap [x - h, x + h])}{2h} = 1$$

and $f|_{E \cup \{x\}}$ is a continuous function; see also Section 1.4) that are discontinuous almost everywhere, among others.

Theorem 7.5.4 has been extracted from [45, Section 6]. Criteria for dense algebrability that do not need monotonicity of any member of $\mathcal{F}$ can be found (with applications) in the paper [49] by Bartoszewicz and Głąb. This paper contains, in addition, a number of results about closed and non-closed algebrability. For instance, it is proved there that if X is a nonempty set and

$\mathcal{A} \subset \mathbb{K}^X$ is an algebra consisting of finite range functions, then the following holds: if $\mathcal{A}$ is finitely generated (not finitely generated) then it is τ_p-closed (not τ_u-closed, respectively; in particular, $\mathcal{A}$ is not τ_p-closed). Here we have denoted by τ_p and τ_u, respectively, the topology of pointwise convergence and the topology of uniform convergence on $\mathbb{K}^X$.

Section 7.6. Theorem 7.6.2 carries a number of interesting consequences; see [210]. One of them is that the set $J(\mathbb{R})$ of Jones functions (recall that a function belongs to $J(\mathbb{R})$ provided that its graph intersects every closed subset of $\mathbb{R}^2$ with uncountable projection on the x-axis) is $e_{\mathfrak{c}}$-lineable, where

$$e_{\mathfrak{c}} = \min\{\operatorname{card}(F) : F \subset \mathbb{R}^{\mathbb{R}} \text{ and for every } \varphi \in \mathbb{R}^{\mathbb{R}} \text{ there exists}$$
$$f \in F \text{ such that } \operatorname{card}(f \cap \varphi) < \mathfrak{c}\}.$$

It would be very interesting to know whether a dual result of Theorem 7.6.2 could also be found for algebrability, that is: is there any cardinal invariant that relates to the concept of algebrability as additivity does with lineability?

Bibliography

[1] M. D. Acosta, A. Aizpuru, R. M. Aron, and F. J. García-Pacheco, *Functionals that do not attain their norm*, Bull. Belg. Math. Soc. Simon Stevin **14** (2007), no. 3, 407–418.

[2] L. V. Ahlfors, *Complex Analysis*, 3rd ed., McGraw-Hill, London, 1979.

[3] A. Aizpuru, C. Pérez-Eslava, F. J. García-Pacheco, and J. B. Seoane-Sepúlveda, *Lineability and coneability of discontinuous functions on* ℝ, Publ. Math. Debrecen **72** (2008), no. 1-2, 129–139.

[4] I. Akbarbaglu and S. Maghsoudi, *Large structures in certain subsets of Orlicz spaces*, Linear Algebra Appl. **438** (2013), no. 11, 4363–4373.

[5] I. Akbarbaglu, S. Maghsoudi, and J. B. Seoane-Sepúlveda, *Porous sets and lineability of continuous functions on locally compact groups*, J. Math. Anal. Appl. **406** (2013), no. 1, 211–218, DOI 10.1016/j.jmaa.2013.04.051.

[6] A. Albanese, X. Barrachina, E. M. Mangino, and A. Peris, *Distributional chaos for strongly continuous semigroups of operators*, Comm. Pure Appl. Analysis **12** (2013), no. 5, 2069–2082.

[7] F. Albiac and N. J. Kalton, *Topics in Banach space theory*, Graduate Texts in Mathematics, vol. 233, Springer, New York, 2006.

[8] N. Albuquerque, *Maximal lineability of the set of continuous surjections*, Bull. Belg. Math. Soc. Simon Stevin **21** (2014), no. 1, 83–87.

[9] N. Albuquerque, L. Bernal-González, D. Pellegrino, and J. B. Seoane-Sepúlveda, *Peano curves on topological vector spaces*, Linear Algebra Appl. **460** (2014), 81–96.

[10] P. C. Allaart and K. Kawamura, *The improper infinite derivatives of Takagi's nowhere-differentiable function*, J. Math. Anal. Appl. **372** (2010), no. 2, 656–665.

[11] T. R. Alves, *Lineability and algebrability of the set of holomorphic functions with given domain of existence*, Studia Math. **220** (2014), no. 2, 157–167.

[12] ———, *Spaceability and algebrability in the theory of domains of existence in Banach spaces*, RACSAM. Rev. R. Acad. Cienc. Exactas Fís. Nat. Ser. A Mat., accepted for publication.

[13] A. I. Ansari, *Existence of hypercyclic operators on topological vector spaces*, J. Funct. Anal. **148** (1997), no. 2, 384–390.

[14] D. H. Armitage, *A non-constant entire function on the plane whose integral on every line is zero*, Amer. Math. Monthly **101** (1994), no. 9, 892–894.

[15] ———, *Dense vector spaces of universal harmonic functions*, Advances in multivariate approximation (Witten-Bemmerholz, 1998), 33–42, Math. Res. **107**, Wiley-VCH, Berlin, 1999.

[16] ———, *Entire functions decaying rapidly on strips*, Quaestiones Math. **23** (2000), 417–424.

[17] ———, *Harmonic monsters*, J. Approx. Theory **123** (2003), 110–116.

[18] R. M. Aron, *Linearity in nonlinear situations*, Advanced courses of mathematical analysis. II, World Sci. Publ., Hackensack, NJ, 2007, pp. 1–15.

[19] R. M. Aron and J. Bès, *Hypercyclic differentiation operators*, Contemp. Math. **232** (1999), 39–46.

[20] R. M. Aron, J. Bès, F. León, and A. Peris, *Operators with common hypercyclic subspaces*, J. Operator Theory **54** (2005), 251–260.

[21] R. M. Aron, C. Boyd, R. A. Ryan, and I. Zalduendo, *Zeros of polynomials on Banach spaces: the real story*, Positivity **7** (2003), no. 4, 285–295.

[22] R. M. Aron, J. A. Conejero, A. Peris, and J. B. Seoane-Sepúlveda, *Sums and products of bad functions*, Function spaces, Contemp. Math., vol. 435, Amer. Math. Soc., Providence, RI, 2007, pp. 47–52.

[23] _____, *Powers of hypercyclic functions for some classical hypercyclic operators*, Integr. Equ. Oper. Theory **58** (2007), no. 4, 591–596.

[24] _____, *Uncountably generated algebras of everywhere surjective functions*, Bull. Belg. Math. Soc. Simon Stevin **17** (2010), no. 3, 571–575.

[25] R. Aron, D. García, and M. Maestre, *Linearity in nonlinear problems*, RACSAM. Rev. R. Acad. Cienc. Exactas Fís. Nat. Ser. A Mat. **95** (2001), no. 1, 7–12 (English, with English and Spanish summaries).

[26] R. M. Aron, F. J. García-Pacheco, D. Pérez-García, and J. B. Seoane-Sepúlveda, *On dense-lineability of sets of functions on* ℝ, Topology **48** (2009), no. 2–4, 149–156.

[27] R. Aron, R. Gonzalo, and A. Zagorodnyuk, *Zeros of real polynomials*, Linear and Multilinear Algebra **48** (2000), no. 2, 107–115.

[28] R. M. Aron, V. I. Gurariy, and J. B. Seoane-Sepúlveda, *Lineability and spaceability of sets of functions on* ℝ, Proc. Amer. Math. Soc. **133** (2005), no. 3, 795–803.

[29] R. M. Aron and P. Hájek, *Zero sets of polynomials in several variables*, Arch. Math. (Basel) **86** (2006), no. 6, 561–568.

[30] _____, *Odd degree polynomials on real Banach spaces*, Positivity **11** (2007), no. 1, 143–153.

[31] R. M. Aron, J. A. Jaramillo, and T. Ransford, *Smooth surjections without surjective restrictions*, J. Geom. Anal. **23** (2013), 2081–2090.

[32] R. M. Aron and D. Markose, *On universal functions*, J. Korean Math. Soc. **41** (2004), 65–76.

[33] R. M. Aron, D. Pérez-García, and J. B. Seoane-Sepúlveda, *Algebrability of the set of nonconvergent Fourier series*, Studia Math. **175** (2006), no. 1, 83–90.

[34] R. M. Aron and M. P. Rueda, *A problem concerning zero-subspaces of homogeneous polynomials*, Linear Topol. Spaces Complex Anal. **3** (1997), 20–23. Dedicated to Professor Vyacheslav Pavlovich Zahariuta.

[35] R. M. Aron and J. B. Seoane-Sepúlveda, *Algebrability of the set of everywhere surjective functions on* ℂ, Bull. Belg. Math. Soc. Simon Stevin **14** (2007), no. 1, 25–31.

[36] A. Avilés and S. Todorcevic, *Zero subspaces of polynomials on* $l_1(\Gamma)$, J. Math. Anal. Appl. **350** (2009), no. 2, 427–435.

[37] D. Azagra, G. A. Muñoz-Fernández, V. M. Sánchez, and J. B. Seoane-Sepúlveda, *Riemann integrability and Lebesgue measurability of the composite function*, J. Math. Anal. Appl. **354** (2009), no. 1, 229–233.

[38] A. G. Bacharoglu, *Universal Taylor series on doubly connected domains*, Results in Maths. **53** (2009), no. 1-2, 9–18.

[39] M. Balcerzak, A. Bartoszewicz, and M. Filipczac, *Nonseparable spaceability and strong algebrability of sets of continuous singular functions*, J. Math. Anal. Appl. **407** (2013), 263–269.

[40] M. Balcerzak, K. Ciesielski, and T. Natkaniec, *Sierpiński-Zygmund functions that are Darboux, almost continuous, or have a perfect road*, Arch. Math. Logic **37** (1997), no. 1, 29–35.

[41] S Banach, *Théorie des opérations linéaires*, Chelsea Publishing Co., New York, 1955 (French).

[42] T. Banakh, A. Bartoszewicz, S. Głąb, and E. Szymonik, *Algebrability, lineability and the subsums of series*, Colloq. Math. **129** (2012), no. 3, 75–85.

[43] P. Bandyopadhyay and G. Godefroy, *Linear structures in the set of norm-attaining functionals on a Banach space*, J. Convex Anal. **13** (2006), no. 3-4, 489–497.

[44] C. S. Barroso, G. Botelho, V. V. Fávaro, and D. Pellegrino, *Lineability and spaceability for the weak form of Peano's theorem and vector-valued sequence spaces*, Proc. Amer. Math. Soc. **141** (2013), no. 6, 1913–1923, DOI 10.1090/S0002-9939-2012-11466-2.

[45] A. Bartoszewicz, M. Bienias, M. Filipczak, and S. Głąb, *Strong c-algebrability of strong Sierpiński-Zygmund, smooth nowhere analytic and other sets of functions*, J. Math. Anal. Appl. **412** (2014), no. 2, 620–630, DOI 10.1016/j.jmaa.2013.10.075.

[46] A. Bartoszewicz, M. Bienias, and S. Głąb, *Independent Bernstein sets and algebraic constructions*, J. Math. Anal. Appl. **393** (2012), no. 1, 138–143.

[47] A. Bartoszewicz and S. Głąb, *Strong algebrability of sets of sequences of functions*, Proc. Amer. Math. Soc. **141** (2013), 827–835.

[48] ———, *Additivity and lineability in vector spaces*, Linear Algebra Appl. **439** (2013), no. 7, 2123–2130, DOI 10.1016/j.laa.2013.06.007.

[49] ———, *Large function algebras with certain topological properties*, J. Function Spaces **2015** (2015), Article ID 761924, 7 pages.

[50] A. Bartoszewicz, S. Głąb, D. Pellegrino, and J. B. Seoane-Sepúlveda, *Algebrability, non-linear properties, and special functions*, Proc. Amer. Math. Soc. **141** (2013), no. 10, 3391–3402.

[51] F. Bastin, J. A. Conejero, C. Esser, and J. B. Seoane-Sepúlveda, *Algebrability and nowhere Gevrey differentiability*, Israel J. Math. **205** (2015), no. 1, 127–143, DOI 10.1007/s11856-014-1104-1.

[52] F. Bastin, C. Esser, and S. Nikolay, *Prevalence of "nowhere analyticity"*, Studia Math. **210** (2012), no. 3, 239–246.

[53] F. Bayart, *Linearity of sets of strange functions*, Michigan Math. J. **53** (2005), no. 2, 291–303.

[54] ———, *Topological and algebraic genericity of divergence and universality*, Studia Math. **167** (2005), no. 2, 161–181.

[55] ———, *Common hypercyclic subspaces*, Integr. Equ. Oper. Theory **53** (2005), no. 4, 467–476.

[56] F. Bayart and S. Grivaux, *Frequently hypercyclic operators*, Trans. Amer. Math. Soc. **358** (2006), no. 11, 5083–5117.

[57] ———, *Invariant gaussian measures for operators on Banach spaces and linear dynamics*, Proc. London Math. Soc. **94** (2007), no. 1, 181–210.

[58] F. Bayart, K.-G. Grosse-Erdmann, V. Nestoridis, and C. Papadimitropoulos, *Abstract theory of universal series and applications*, Proc. London Math. Soc. **96** (2008), no. 2, 417–463.

[59] F. Bayart, S. V. Konyagin, and H. Queffélec, *Convergence almost everywhere and divergence everywhere of Taylor and Dirichlet series*, Real Anal. Exchange **29** (2003/04), no. 2, 557–587.

[60] F. Bayart and E. Matheron, *Hypercyclic operators failing the Hypercyclic Criterion on classical Banach spaces*, J. Funct. Anal. **250** (2007), no. 2, 426–441.

[61] ———, *Dynamics of Linear Operators*, Cambridge University Press, 2009. Cambridge Tracts in Mathematics.

[62] F. Bayart and V. Nestoridis, *Universal Taylor series have a strong form of universality*, J. Anal. Math. **104** (2008), no. 1, 69–82.

[63] F. Bayart and L. Quarta, *Algebras in sets of queer functions*, Israel J. Math. **158** (2007), no. 1, 285–296.

[64] F. Bayart and I. Z. Ruzsa, *Difference sets and frequently hypercyclic weighted shifts*, Ergodic Theory Dynamical Systems, accepted for publication.

[65] B. Beauzamy, *Introduction to Banach spaces and their geometry*, North-Holland Mathematics Studies, vol. 68, North-Holland Publishing Co., Amsterdam-New York, 1982. Notas de Matemática [Mathematical Notes], 86.

[66] B. Beauzamy, *Un opérateur, sur l'espace de Hilbert, dont tous les polinômes sont hypercycliques*, C. R. Acad. Sci. Paris **303** (1986), no. 18, Série I, 923–925.

[67] _____, *Introduction to Operator Theory and Invariant Subspaces*, Cambridge University Press, Amsterdam, 1988.

[68] E. I. Berezhnoi, *The subspace of $C[0, 1]$ consisting of functions having no finite one-sided derivatives at any points*, Mat. Zametki **73** (2003), no. 3, 348–354.

[69] T. Bermúdez, A. Bonilla, F. Martínez-Giménez, and A. Peris, *Li-Yorke and distributionally chaotic operators*, J. Math. Anal. Appl. **373** (2011), no. 1, 83–93.

[70] L. Bernal-González, *Funciones con derivadas sucesivas grandes y pequeñas por doquier*, Collect. Math. **38** (1987), 117–122.

[71] _____, *A lot of "counterexamples" to Liouville's theorem*, J. Math. Anal. Appl. **201** (1996), no. 3, 1002–1009.

[72] _____, *Small entire functions with extremely fast growth*, J. Math. Anal. Appl. **207** (1997), no. 2, 541–548.

[73] _____, *On hypercyclic operators on Banach spaces*, Proc. Amer. Math. Soc. **127** (1999), no. 4, 1003–1010.

[74] _____, *Densely hereditarily hypercyclic sequences and large hypercyclic manifolds*, Proc. Amer. Math. Soc. **127** (1999), no. 11, 3279–3285.

[75] _____, *Hypercyclic sequences of differential and antidifferential operators*, J. Approx. Theory **96** (1999), no. 2, 323–337.

[76] _____, *Universal images of universal elements*, Studia Math. **138** (2000), no. 3, 241–250.

[77] _____, *Linear Kierst-Szpilrajn theorems*, Studia Math. **166** (2005), no. 1, 55–69.

[78] _____, *Hypercyclic subspaces in Fréchet spaces*, Proc. Amer. Math. Soc. **134** (2006), no. 7, 1955–1961.

[79] _____, *Linear structure of weighted holomorphic nonextendibility*, Bull. Austral. Math. Soc. **73** (2006), no. 3, 335–344.

[80] _____, *Disjoint hypercyclic operators*, Studia Math. **182** (2007), 113–131.

[81] _____, *Dense-lineability in spaces of continuous functions*, Proc. Amer. Math. Soc. **136** (2008), no. 9, 3163–3169.

[82] _____, *Lineability of sets of nowhere analytic functions*, J. Math. Anal. Appl. **340** (2008), no. 2, 1284–1295.

[83] _____, *Algebraic genericity of strict-order integrability*, Studia Math. **199** (2010), no. 3, 279–293.

[84] _____, *Lineability of universal divergence of Fourier series*, Integr. Equ. Oper. Theory **74** (2012), no. 2, 271–279.

[85] _____, *Vector spaces of non-extendable holomorphic functions*, J. Analyse Math. (2015), accepted for publication.

[86] L. Bernal-González and A. Bonilla, *Families of strongly annular functions: linear structure*, Rev. Mat. Complut. **26** (2013), 283–297.

[87] L. Bernal-González, A. Bonilla, M. C. Calderón-Moreno, and J. A. Prado-Bassas, *Maximal cluster sets of L-analytic functions on arbitrary curves*, Constr. Approx. **25** (2007), no. 2, 211–219.

[88] ———, *Universal Taylor series with maximal cluster sets*, Rev. Mat. Iberoam. **25** (2009), no. 2, 757–780.

[89] L. Bernal-González and M. C. Calderón-Moreno, *Holomorphic T-monsters and strongly omnipresent operators*, J. Approx. Theory **104** (2000), no. 2, 204–219.

[90] ———, *Dense linear manifolds of monsters*, J. Approx. Theory **119** (2002), no. 2, 156–180.

[91] L. Bernal-González, M. C. Calderón-Moreno, and K. G. Grosse-Erdmann, *Strongly omnipresent operators: general conditions and applications to composition operators*, J. Austral. Math. Soc. (Series A) **72** (2002), 335–348.

[92] ———, *Strongly omnipresent integral operators*, Integr. Equ. Oper. Theory **44** (2002), 397–409.

[93] L. Bernal-González, M. C. Calderón-Moreno, and W. Luh, *Large linear manifolds of noncontinuable boundary-regular holomorphic functions*, J. Math. Anal. Appl. **341** (2008), no. 1, 337–345.

[94] ———, *Dense-lineability of sets of Birkhoff-universal functions with rapid decay*, J. Math. Anal. Appl. **363** (2010), no. 1, 327–335.

[95] L. Bernal-González, M. C. Calderón-Moreno, and J. A. Prado-Bassas, *Maximal cluster sets along arbitrary curves*, J. Approx. Theory **129** (2004), no. 2, 207–216.

[96] ———, *Cyclicity of coefficient multipliers: linear structure*, Acta Math. Hungar. **114** (2007), no. 4, 287–300.

[97] ———, *Holomorphic operators generating dense images*, Integr. Equ. Oper. Theory **60** (2008), no. 1, 1–11.

[98] ———, *Large subspaces of compositionally universal functions with maximal cluster sets*, J. Approx. Theory **164** (2012), no. 2, 253-267.

[99] ———, *The set of space-filling curves: topological and algebraic structure*, Linear Algebra Appl. **467** (2015), 57–74.

[100] L. Bernal-González and K.-G. Grosse-Erdmann, *The Hypercyclicity Criterion for sequences of operators*, Studia Math. **157** (2003), no. 1, 17–32.

[101] L. Bernal-González and A. Montes-Rodríguez, *Universal functions for composition operators*, Complex Variables Theory Appl. **27** (1995), no. 1, 47–56.

[102] ———, *Non-finite dimensional closed vector spaces of universal functions for composition operators*, J. Approx. Theory **82** (1995), no. 3, 375–391.

[103] L. Bernal-González and M. Ordóñez Cabrera, *Spaceability of strict order integrability*, J. Math. Anal. Appl. **385** (2012), no. 1, 303–309.

[104] ———, *Lineability criteria, with applications*, J. Funct. Anal. **266** (2014), no. 6, 3997–4025, DOI 10.1016/j.jfa.2013.11.014.

[105] L. Bernal-González, D. Pellegrino, and J. B. Seoane-Sepúlveda, *Linear subsets of nonlinear sets in topological vector spaces*, Bull. Amer. Math. Soc. (N.S.) **51** (2014), no. 1, 71–130, DOI 10.1090/S0273-0979-2013-01421-6.

[106] L. Bernal-González and J. A. Prado-Tendero, *U-operators*, J. Austral. Math. Soc. (Series A) **78** (2005), no. 1, 59–89.

[107] N. Bernardes, A. Bonilla, V. Müller, and A. Peris, *Distributional chaos for linear operators*, J. Funct. Anal. **265** (2013), 2143–2163.

[108] ———, *Li-Yorke chaos in Linear Dynamics*, Ergod. Th. Dynam. Sys., DOI 10.1017/etds.2014.20.

[109] F. J. Bertoloto, G. Botelho, V. V. Fávaro, and A. M. Jatobá, *Hypercyclicity of convolution operators on spaces of entire functions*, Ann. Inst. Fourier Grenoble **63** (2013), no. 4, 1263–1283.

[110] J. P. Bès, *Invariant manifolds of hypercyclic vectors for the real scalar case*, Proc. Amer. Math. Soc. **127** (1999), no. 6, 1801–1804.

[111] ———, *Dynamics of composition operators with holomorphic symbol*, Rev. Real Acad. Cien. Ser. A Mat. **107** (2013), 437–449.

[112] J. P. Bès and J. A. Conejero, *Hypercyclic subspaces in omega*, J. Math. Anal. Appl. **316** (2006), no. 1, 16–23.

[113] J. P. Bès and O. Martin, *Geometric characteristics and common hypercyclic subspaces*, Top. Proc. **33** (2009), 197-209.

[114] J. P. Bès, O. Martin, and R. Sanders, *Weighted shifts and disjoint hypercyclicity*, J. Operator Theory, to appear.

[115] J. P. Bès and Q. Menet, *Existence of common and upper frequently hypercyclic subspaces*, arXiv:1409.0995v1 [math.DS], Preprint (2014).

[116] J. P. Bès and A. Peris, *Hereditarily hypercyclic operators*, J. Funct. Anal. **167** (1999), no. 1, 94–113.

[117] ———, *Disjointness in hypercyclicity*, J. Math. Anal. Appl. **336** (2007), no. 1, 297–315.

[118] A. S. Besicovitch, *Diskussion der stetigen Funktionen im Zusammenhang mit der Frage über ihre Differentierbarkeit*, Bull. Acad. Sci. de Russie **19** (1925), 527–540.

[119] C. Bessaga and A. Pełczyński, *On bases and unconditional convergence of series in Banach spaces*, Studia Math. **17** (1958), 151–164.

[120] B. J. Birch, *Homogeneous forms of odd degree in a large number of variables*, Mathematika **4** (1957), 102–105.

[121] G. D. Birkhoff, *Démonstration d'un théorème élémentaire sur les fonctions entières*, C. R. Acad. Sci. Paris **189** (1929), 473–475.

[122] E. Bishop and R. R. Phelps, *A proof that every Banach space is subreflexive*, Bull. Amer. Math. Soc. **67** (1961), 97–98.

[123] H. Blumberg, *New properties of all real functions*, Trans. Amer. Math. Soc. **82** (1922), 53–61.

[124] R. P. Boas Jr., *Entire functions*, Academic Press, Inc., 1954.

[125] J. Bobok, *Infinite dimensional Banach space of Besicovitch Functions*, Real Anal. Exchange **32** (2007), 319–333.

[126] ———, *On spaceability of Besicovitch functions*, Preprint (2014).

[127] D. D. Bonar and F. Carroll, *Annular functions form a residual set*, J. Reine Angew. Math. **272** (1975), 23–24.

[128] J. Bonet, F. Martínez-Giménez, and A. Peris, *Linear chaos on Fréchet spaces*, Internat. J. Bifur. Chaos Appl. Sci. Engrg. **13** (2003), no. 7, 1649–1655.

[129] ———, *Universal and chaotic multipliers on spaces of operators*, J. Math. Anal. Appl. **297** (2004), no. 2, 599–611.

[130] J. Bonet and A. Peris, *Hypercyclic operators on nonnormable Fréchet spaces*, J. Funct. Anal. **159** (1998), 587–596.

[131] B. Bongiorno, U. B. Darji, and L. Di Piazza, *Lineability of non-differentiable Pettis primitives*, Monatsh. Math., DOI 10.1007/s00605-014-0703-6.

[132] A. Bonilla, *"Counterexamples" to harmonic Liouville's theorem and harmonic functions with zero nontangential limits*, Colloq. Math. **83** (2000), no. 2, 155–160.

[133] _____, *Small entire functions with infinite growth index*, J. Math. Anal. Appl. **267** (2002), no. 1, 400–404.

[134] _____, *Universal harmonic functions*, Quaestiones Math. **25** (2002), 527–530.

[135] A. Bonilla and K.-G. Grosse-Erdmann, *On a theorem of Godefroy and Shapiro*, Integr. Equ. Oper. Theory **56** (2006), no. 2, 151–162.

[136] _____, *Frequently hypercyclic operators and vectors*, Ergod. Th. Dynam. Sys. **27** (2007), no. 2, 383–404.

[137] _____, *Frequently hypercyclic subspaces*, Monatsh. Math. **168** (2012), no. 3-4, 305–320.

[138] G. Botelho, D. Cariello, V. V. Fávaro, and D. Pellegrino, *Maximal spaceability in sequence spaces*, Linear Algebra Appl. **437** (2012), no. 12, 2978–2985.

[139] G. Botelho, D. Cariello, V. V. Fávaro, D. Pellegrino, and J. B. Seoane-Sepúlveda, *Distinguished subspaces of L_p of maximal dimension*, Studia Math. **215** (2013), no. 3, 261–280.

[140] _____, *On very non-linear subsets of continuous functions*, Q. J. Math. **65** (2014), 841–850.

[141] G. Botelho, D. Diniz, V. V. Fávaro, and D. Pellegrino, *Spaceability in Banach and quasi-Banach sequence spaces*, Linear Algebra Appl. **434** (2011), no. 5, 1255–1260.

[142] G. Botelho, D. Diniz, and D. Pellegrino, *Lineability of the set of bounded linear non-absolutely summing operators*, J. Math. Anal. Appl. **357** (2009), no. 1, 171–175.

[143] G. Botelho and V. V. Fávaro, *Constructing Banach spaces of vector-valued sequences with special properties*, Michigan Math. J., accepted for publication.

[144] P. S. Bourdon, *Invariant manifolds of hypercyclic operators*, Proc. Amer. Math. Soc. **118** (1993), no. 3, 845–847.

[145] P. S. Bourdon and J. H. Shapiro, *The role of the spectrum in the cyclic behavior of composition operators*, Memoirs Amer. Math. Soc. 596, AMS, Providence, RI, 1997.

[146] A. M. Bruckner, *Differentiation of real functions*, Lecture Notes in Mathematics, vol. 659, Springer-Verlag, Berlin, 1978.

[147] Z. Buczolich, *Solution to the gradient problem of C. E. Weil*, Rev. Mat. Iberoam. **21** (2005), no. 3, 889–910, DOI 10.4171/RMI/439.

[148] D. Burdick and F. D. Lesley, *Some uniqueness theorems for analytic functions*, Amer. Math. Monthly **82** (1975), no. 2, 152–155.

[149] M. C. Calderón-Moreno, *Universal functions with small derivatives and extremely fast growth*, Analysis **22** (2002), 57–66.

[150] M. C. Calderón-Moreno and S. Müller, *Universal holomorphic functions and harmonic functions with additional properties*, Acta Math. Hungar. **105** (2004), 1–16.

[151] M. C. Calderón-Moreno and J. A. Prado-Bassas, *A dynamical characterization of sub-Arakelian subsets*, J. Math. Anal. Appl. **405** (2013), 499–506.

[152] D. Carando, V. Dimant, and S. Muro, *Hypercyclic convolution operators on Fréchet spaces of analytic functions*, J. Math. Anal. Appl. **336** (2007), 1324–1340.

[153] D. Cariello and J. B. Seoane-Sepúlveda, *Basic sequences and spaceability in ℓ_p spaces*, J. Funct. Anal. **266** (2014), no. 6, 3797–3814, DOI 10.1016/j.jfa.2013.12.011.

[154] N. L. Carothers, *A short course on Banach space theory*, London Mathematical Society Student Texts, vol. 64, Cambridge University Press, Cambridge, 2005.

[155] F. S. Cater, *Differentiable, nowhere analytic functions*, Amer. Math. Monthly **91** (1984), no. 10, 618–624.

[156] K. C. Chan, *Hypercyclicity of the operator algebra for a separable Hilbert space*, J. Operator Theory **42** (1999), 231–244.

[157] K. C. Chan and R. D. Taylor Jr., *Hypercyclic spaces of a Banach spaces*, Integr. Equ. Oper. Theory **41** (2001), 381–388.

[158] S. Charpentier, *On the closed subspaces of universal series in Banach spaces and Fréchet spaces*, Studia Math. **198** (2010), no. 2, 121–145.

[159] S. Charpentier, Q. Menet, and A. Mouze, *Closed universal subspaces of spaces of infinitely differentiable functions*, Preprint (2014).

[160] S. Charpentier and A. Mouze, *Universal Taylor series and summability*, Rev. Mat. Complut. **28** (2015), 153–167.

[161] S. Charpentier, V. Nestoridis, and F. Wielonsky, *Generic properties of Padé approximants and Padé universal series*, arXiv:1404.3409v1 [math.FA], Preprint (2014).

[162] J. Chmielowski, *Domains of holomorphy of type A^k*, Proc. Royal Irish Acad. **80-A** (1980), 97–101.

[163] J. P. R. Christensen, *A topological analogue of the Fubini theorem and some applications*, in: Papers from the Open House for Probabilists, Mat. Inst. Aarhus Univ., Aarhus (1971), pp. 26–31.

[164] C. Chui and M. N. Parnes, *Approximation by overconvergence of power series*, J. Math. Anal. Appl. **36** (1971), 693–696.

[165] S.-Y. Chung and J. Chung, *There exist no gaps between Gevrey differentiable and nowhere Gevrey differentiable*, Proc. Amer. Math. Soc. **133** (2005), no. 3, 859–863.

[166] K. C. Ciesielski, *Set-theoretic real analysis*, J. Appl. Anal. **3** (1997), no. 2, 143–190.

[167] K. C. Ciesielski, J. L. Gámez-Merino, D. Pellegrino, and J. B. Seoane-Sepúlveda, *Lineability, spaceability, and additivity cardinals for Darboux-like functions*, Linear Algebra Appl. **440** (2014), 307–317, DOI 10.1016/j.laa.2013.10.033.

[168] K. C. Ciesielski and J. Jastrzębski, *Darboux-like functions within the classes of Baire one, Baire two, and additive functions*, Topology Appl. **103** (2000), no. 2, 203–219.

[169] K. C. Ciesielski and A. W. Miller, *Cardinal invariants concerning functions whose sum is almost continuous*, Real Anal. Exchange **20** (1994/95), no. 2, 657–672.

[170] K. C. Ciesielski and T. Natkaniec, *Algebraic properties of the class of Sierpiński-Zygmund functions*, Topology Appl. **79** (1997), no. 1, 75–99.

[171] ———, *On Sierpiński-Zygmund bijections and their inverses*, Topology Proc. **22** (1997), no. Spring, 155–164.

[172] K. C. Ciesielski and J. Pawlikowski, *The covering property axiom, CPA*, Cambridge Tracts in Mathematics, vol. 164, Cambridge University Press, Cambridge, 2004. A combinatorial core of the iterated perfect set model.

[173] K. C. Ciesielski and I. Recław, *Cardinal invariants concerning extendable and peripherally continuous functions*, Real Anal. Exchange **21** (1995/96), no. 2, 459–472.

[174] J. A. Conejero, P. Jiménez-Rodríguez, G. A. Muñoz-Fernández, and J. B. Seoane-Sepúlveda, *When the identity theorem "seems" to fail*, Amer. Math. Monthly **121** (2014), no. 1, 60–68, DOI 10.4169/amer.math.monthly.121.01.060.

[175] J. A. Conejero, V. Müller, and A. Peris, *Hypercyclic behaviour of operators in a hypercyclic C_0-semigroup*, J. Funct. Anal. **244** (2007), no. 1, 342–348.

[176] J. A. Conejero, G. A. Muñoz-Fernández, M. Murillo Arcila, and J. B. Seoane-Sepúlveda, *Smooth functions with uncountably many zeros*, Bull. Belg. Math. Soc. Simon Stevin **22** (2015), no. 1, 71–75.

[177] J. B. Conway, *Functions of one complex variable*, Springer-Verlag, New York, 1986.

[178] ———, *A course in functional analysis*, 2nd ed., Graduate Texts in Mathematics, vol. 96, Springer-Verlag, New York, 1990.

[179] S. D'Alessandro and P. Hájek, *Polynomial algebras and smooth functions in Banach spaces*, J. Funct. Anal. **266** (2014), no. 3, 1627–1646, DOI 10.1016/j.jfa.2013.11.017.

[180] R. B. Darst, *Most infinitely differentiable functions are nowhere analytic*, Canad. Math. Bull. **16** (1973), 597–598.

[181] M. M. De La Rosa and C. Read, *A hypercyclic operator whose direct sum $T \oplus T$ is not hypercyclic*, J. Operator Theory **61** (2009), no. 2, 369–380.

[182] R. Deville and É. Matheron, *Infinite games, Banach space geometry and the eikonal equation*, Proc. Lond. Math. Soc. (3) **95** (2007), no. 1, 49–68.

[183] E. Diamantopoulos, C. Kariofillis, and C. Mouratides, *Universal Laurent series in finitely connected domains*, Arch. Math. **91** (2008), no. 2, 145–154.

[184] J. Diestel, *Sequences and series in Banach spaces*, Graduate Texts in Mathematics, vol. 92, Springer-Verlag, New York, 1984.

[185] J. Diestel, J. H. Fourie, and J. Swart, *The metric theory of tensor products*, American Mathematical Society, Providence, RI, 2008. Grothendieck's résumé revisited.

[186] J. Diestel, H. Jarchow, and A. Tonge, *Absolutely summing operators*, Cambridge Studies in Advanced Mathematics, vol. 43, Cambridge University Press, Cambridge, 1995.

[187] J. Diestel and J. J. Uhl, *Vector measures*, Mathematical Surveys **15**, American Mathematical Society, Providence, RI, 1977.

[188] S. Dineen, *Complex analysis on infinite-dimensional spaces*, Springer Monographs in Mathematics, Springer-Verlag London Ltd., London, 1999.

[189] A. E. Djrbashian and F. A. Shamoian, *Topics in the theory of A_a^p spaces*, Teubner-Texte zur Mathematik, vol. 105, 1988.

[190] W. F. Donoghe, *Distributions and Fourier transforms*, Academic Press, New York, 1969.

[191] H. R. Dowson, *Spectral Theory of Linear Operators*, Academic Press, London, 1978.

[192] L. Drewnowski, *Quasicomplements in F-spaces*, Studia Math. **77** (1984), no. 4, 373–391.

[193] V. Drobot and M. Morayne, *Continuous functions with a dense set of proper local maxima*, Amer. Math. Monthly **92** (1985), no. 3, 209–211.

[194] P. L. Duren, *Theory of H^p spaces*, Dover, New York, 2000.

[195] A. Dvoretzky and C. A. Rogers, *Absolute and unconditional convergence in normed linear spaces*, Proc. Nat. Acad. Sci. U.S.A. **36** (1950), 192–197.

[196] P. H. Enflo, V. I. Gurariy, and J. B. Seoane-Sepúlveda, *Some results and open questions on spaceability in function spaces*, Trans. Amer. Math. Soc. **366** (2014), no. 2, 611–625, DOI 10.1090/S0002-9947-2013-05747-9.

[197] ———, *On Montgomery's conjecture and the distribution of Dirichlet sums*, J. Funct. Anal. **267** (2014), no. 4, 1241–1255, DOI 10.1016/j.jfa.2014.04.001.

[198] S. Episkoposian and J. Müller, *Universality properties of Walsh–Fourier series*, Monats. Math. **175** (2014), 511–518.

[199] C. Esser, *Generic results in classes of ultradifferentiable functions*, J. Math. Anal. Appl. **413** (2014), no. 1, 378–391.

[200] M. Fenoy and J. B. Seoane-Sepúlveda, *Lineability within Probability Theory settings* (2015), preprint.

[201] G Fernández and A. A. Hallack, *Remarks on a result about hypercyclic non-convolution operators*, J. Math. Anal. Appl. **309** (2005), 52–55.

[202] M. Fernández-Unzueta, *Zeroes of polynomials on ℓ_∞*, J. Math. Anal. Appl. **324** (2006), no. 2, 1115–1124.

[203] V. P. Fonf, V. I. Gurariy, and M. I. Kadets, *An infinite dimensional subspace of C[0,1] consisting of nowhere differentiable functions*, C. R. Acad. Bulgare Sci. **52** (1999), no. 11-12, 13–16.

[204] J. Foran, *Fundamentals of real analysis*, Monographs and Textbooks in Pure and Applied Mathematics, vol. 144, Marcel Dekker Inc., New York, 1991.

[205] D. Gaier, *Lectures on complex aproximation*, Birkhäuser, Basel-London-Stuttgart, 1987.

[206] _____, *The Faber operator and its boundedness*, J. Approx. Theory **101** (1999), 265–277.

[207] E. A. Gallardo-Gutiérrez and A. Montes-Rodríguez, *The role of the spectrum in the cyclic behavior of composition operators*, Memoirs Amer. Math. Soc. 791, AMS, Providence, RI, 2004.

[208] J. L. Gámez-Merino, *Large algebraic structures inside the set of surjective functions*, Bull. Belg. Math. Soc. Simon Stevin **18** (2011), no. 2, 297–300.

[209] J. L. Gámez-Merino, G. A. Muñoz-Fernández, V. M. Sánchez, and J. B. Seoane-Sepúlveda, *Sierpiński-Zygmund functions and other problems on lineability*, Proc. Amer. Math. Soc. **138** (2010), no. 11, 3863–3876.

[210] J. L. Gámez-Merino, G. A. Muñoz-Fernández, and J. B. Seoane-Sepúlveda, *Lineability and additivity in* $\mathbb{R}^{\mathbb{R}}$, J. Math. Anal. Appl. **369** (2010), no. 1, 265–272.

[211] J. L. Gámez-Merino, G. A. Muñoz-Fernández, and J. B. Seoane-Sepúlveda, *A characterization of continuity revisited*, Amer. Math. Monthly **118** (2011), no. 2, 167–170.

[212] J. L. Gámez-Merino and J. B. Seoane-Sepúlveda, *An undecidable case of lineability in* $\mathbb{R}^{\mathbb{R}}$, J. Math. Anal. Appl. **401** (2013), no. 2, 959–962.

[213] D. García, B. C. Grecu, M. Maestre, and J. B. Seoane-Sepúlveda, *Infinite dimensional Banach spaces of functions with nonlinear properties*, Math. Nachr. **283** (2010), no. 5, 712–720.

[214] F. J. García-Pacheco, *Vector subspaces of the set of non-norm-attaining functionals*, Bull. Aust. Math. Soc. **77** (2008), no. 3, 425–432, DOI 10.1017/S0004972708000348.

[215] F. J. García-Pacheco, M. Martín, and J. B. Seoane-Sepúlveda, *Lineability, spaceability, and algebrability of certain subsets of function spaces*, Taiwanese J. Math. **13** (2009), no. 4, 1257–1269.

[216] F. J. García-Pacheco, N. Palmberg, and J. B. Seoane-Sepúlveda, *Lineability and algebrability of pathological phenomena in analysis*, J. Math. Anal. Appl. **326** (2007), no. 2, 929–939.

[217] F. J. García-Pacheco, C. Pérez-Eslava, and J. B. Seoane-Sepúlveda, *Moduleability, algebraic structures, and nonlinear properties*, J. Math. Anal. Appl. **370** (2010), no. 1, 159–167.

[218] F. J. García-Pacheco and D. Puglisi, *Lineability of functionals and operators*, Studia Math. **201** (2010), no. 1, 37–47.

[219] F. J. García-Pacheco, F. Rambla-Barreno, and J. B. Seoane-Sepúlveda, *Q-linear functions, functions with dense graph, and everywhere surjectivity*, Math. Scand. **102** (2008), no. 1, 156–160.

[220] F. J. García-Pacheco and M. A. Sofi, *Lineability issues including vector-valued measurable and integrable functions*, J. Convex Anal. **19** (2012), no. 2, 393–402.

[221] B. R. Gelbaum and J. M. H. Olmsted, *Counterexamples in analysis*, The Mathesis Series, Holden-Day Inc., San Francisco, CA, 1964.

[222] _____, *Counterexamples in analysis*, Dover Publications Inc., Mineola, NY, 2003. Corrected reprint of the second (1965) edition.

[223] S. Givant and P. Halmos, *Introduction to Boolean algebras*, Undergraduate Texts in Mathematics, Springer, New York, 2009.

[224] S. Głąb, P. Kaufmann, and L. Pellegrini, *Large structures made of nowhere L^q functions*, Studia Math. **221** (2014), no. 1, 13–34.

[225] J. Glovebnik, *The range of vector-valued analytic functions*, Ark. Mat. **14** (1976), 113–118.

[226] G. Godefroy, *The Banach space c_0*, Extracta Math. **16** (2001), no. 1, 1–25.

[227] G. Godefroy and J. H. Shapiro, *Operators with dense, invariant, cyclic vectors manifolds*, J. Funct. Anal. **98** (1991), no. 2, 229–269.

[228] M. Goliński, *Invariant subspace problem for classical spaces of functions*, J. Funct. Anal. **262** (2012), no. 3, 1251–1273.

[229] M. González, F. León-Saavedra, and A. Montes-Rodríguez, *Semi-Fredholm Theory: hypercyclic and supercyclic subspaces*, Proc. London Math. Soc. **81** (2000), no. 3, 169–189.

[230] K. Grandjot, *Über Grenzwerte ganzer transzendenter Funktionen*, Math. Ann. **91** (1924), no. 3-4, 316–320.

[231] S. Grivaux, *Construction of operators with a prescribed behaviour*, Arch. Math. (Basel) **81** (2003), no. 3, 291–299.

[232] S. Grivaux and E. Matheron, *Invariant measures for frequently hypercyclic operators*, Adv. Math. **265** (2015), 371–427.

[233] K.-G. Grosse-Erdmann, *Holomorphe Monster und universelle Funktionen*, Vol. 176, Mit. Math. Sem. Giessen, 1987.

[234] _____, *Universal families and hypercyclic operators*, Bull. Amer. Math. Soc. **36** (1999), no. 3, 345–381.

[235] _____, *Recent developments in hypercyclicity*, Rev. R. Acad. Cien. Serie A. Mat. **97** (2003), 273–286.

[236] K.-G. Grosse-Erdmann and R. Mortini, *Universal functions for composition operators with nonautomorphic symbol*, J. Anal. Math. **107** (2009), no. 1, 355–376.

[237] K.-G. Grosse-Erdmann and A. Peris, *Linear Chaos*, Springer, London, 2011.

[238] R. C. Gunning and H. Rossi, *Analytic functions of several complex variables*, AMS Chelsea Publishing, Providence, RI, 2009. Reprint of the 1965 original.

[239] V. I. Gurariǐ, *Subspaces and bases in spaces of continuous functions*, Dokl. Akad. Nauk SSSR **167** (1966), 971–973 (Russian).

[240] _____, *Linear spaces composed of everywhere nondifferentiable functions*, C. R. Acad. Bulgare Sci. **44** (1991), no. 5, 13–16 (Russian).

[241] V. I. Gurariy and L. Quarta, *On lineability of sets of continuous functions*, J. Math. Anal. Appl. **294** (2004), no. 1, 62–72.

[242] T. R. Hamlett, *Compact maps, connected maps and continuity*, J. London Math. Soc. (2) **10** (1975), 25–26.

[243] H. Hedenmalm, *Topics in the theory of Dirichlet series*, Visn. Khark. Univ., Ser. Mat. Prykl. Mat. Mekh. **475** (2000), 195–203.

[244] H. Hedenmalm, B. Korenblum, and K. Zhu, *Theory of Bergman spaces*, Springer, New York, 2000.

[245] S. Hencl, *Isometrical embeddings of separable Banach spaces into the set of nowhere approximatively differentiable and nowhere Hölder functions*, Proc. Amer. Math. Soc. **128** (2000), no. 12, 3505–3511.

[246] P. Henrici, *Applied and computational complex analysis*, Vol. 3, J. Wiley, New York, 1986.

[247] F. L. Hernández, C. Ruiz, and V. M. Sánchez, *Spaceability and operators ideals*, J. Math. Anal. Appl. (2015), accepted for publication.

[248] D. Herrero, *Limits of hypercyclic and supercyclic operators*, J. Funct. Anal. **99** (1991), no. 1, 179–190.

[249] ———, *Hypercyclic operators and chaos*, J. Operator Theory **28** (1992), 93–103.

[250] H. M. Hilden and L. J. Wallen, *Some cyclic and noncyclic vectors of certain operators*, Indiana Univ. Math. **23** (1974), 557–565.

[251] J. G. Hocking and G. S. Young, *Topology*, Dover, New York, 1988.

[252] K. Hoffman, *Banach spaces of analytic functions*, Dover Pub., New York, 1962.

[253] R. A. Horn, *Editor's Endnotes*, Amer. Math. Monthly **107** (2000), no. 10, 968–969.

[254] J. Horváth, *Topological Vector Spaces and Distributions* (Vol. I), Addison-Wesley, Reading, MA, 1966.

[255] B. R. Hunt, T. Sauer, and J. A. Yorke, *Prevalence: a translation-invariant "almost every" on infinite-dimensional spaces*, Bull. Amer. Math. Soc. (N.S.) **27** (1992), no. 2, 217–238.

[256] P. Jiménez-Rodríguez, c_0 *is isometrically isomorphic to a subspace of Cantor-Lebesgue functions*, J. Math. Anal. Appl. **407** (2013), no. 2, 567–570, DOI 10.1016/j.jmaa.2013.05.033.

[257] P. Jiménez-Rodríguez, G. A. Muñoz-Fernández, and J. B. Seoane-Sepúlveda, *Non-Lipschitz functions with bounded gradient and related problems*, Linear Algebra Appl. **437** (2012), no. 4, 1174–1181.

[258] P. Jiménez-Rodríguez, G. A. Muñoz-Fernández, and J. B. Seoane-Sepúlveda, *On Weierstrass' Monsters and lineability*, Bull. Belg. Math. Soc. Simon Stevin **20** (2013), no. 4, 577–586.

[259] E. Johnston, *The Faber transform and analytic continuation*, Proc. Amer. Math. Soc. **103** (1988), 237–243.

[260] F. B. Jones, *Connected and disconnected plane sets and the functional equation* $f(x) + f(y) = f(x + y)$, Bull. Amer. Math. Soc. **48** (1942), 115–120.

[261] F. E. Jordan, *Cardinal numbers connected with adding Darboux-like functions*, Ph. D. dissertation, West Virginia University, USA, 1998.

[262] J. P. Kahane, *Baire's category theorem and trigonometric series*, J. Analyse Math. **80** (2000), no. 1, 143–182.

[263] N. J. Kalton, *Basic sequences in F-spaces and their applications*, Proc. Edinb. Math. Soc. **(2) 19** (1974/1975), 151–167.

[264] ———, *The basic sequence problem*, Studia Math. **116** (1995), no. 2, 167–187.

[265] N. Kalton, *Quasi-Banach spaces*, Handbook of the geometry of Banach spaces, Vol. 2, North-Holland, Amsterdam, 2003, pp. 1099–1130, DOI 10.1016/S1874-5849(03)80032-3.

[266] N. J. Kalton, N. T. Peck, and J. W. Roberts, *An F-Space Sampler*, Cambridge University Press, Cambridge, 1984.

[267] P. K. Kamtan and M. Gupta, *Theory of bases and cones*, Pitman, Boston, 1985.

[268] E. Katsoprinakis, V. Nestoridis, and I. I. Papadoperakis, *Universal Faber Series*, Analysis **21** (2001), 339–363.

[269] Y. Katznelson, *An introduction to harmonic analysis*, John Wiley & Sons Inc., New York, 1968.

[270] Y. Katznelson and K. Stromberg, *Everywhere differentiable, nowhere monotone, functions*, Amer. Math. Monthly **81** (1974), no. 4, 349–354.

[271] L. Kaup and K. Kaup, *Holomorphic functions of several variables*, Walter de Gruyter, Berlin and New York, 1983.

[272] A. B. Kharazishvili, *Strange functions in real analysis*, 2nd ed., Pure and Applied Mathematics, vol. 272, Chapman & Hall/CRC, Boca Raton, FL, 2006.

[273] St. St. Kierst and D. Szpilrajn, *Sur certaines singularités des fonctions analytiques uniformes*, Fundamenta Math. **21** (1933), 276–294.

[274] S. S. Kim and K. H. Kwon, *Smooth (C^∞) but nowhere analytic functions*, Amer. Math. Monthly **107** (2000), no. 3, 264–266.

[275] D. Kitson and R. M. Timoney, *Operator ranges and spaceability*, J. Math. Anal. Appl. **378** (2011), no. 2, 680–686.

[276] S. Kolyada and L. Snoha, *Some aspects of topological transitivity – A survey*, Grazer Math. Ber., Bericht Nr. **334** (1997), 3–35.

[277] T. W. Körner, *Fourier analysis*, Cambridge University Press, Cambridge, 1988.

[278] S. Koumandos, V. Nestoridis, Y. S. Smyrlis, and V. Stefanopoulos, *Universal series in $\bigcap_{p>1} l^p$*, Bull. London Math. Soc. **42** (2010), no. 1, 119–129.

[279] S. G. Krantz and H. R. Parks, *A primer of real analytic functions,* 2nd edition, Birkähuser Advanced Texts: Basler Lehrbcher, Birkhäuser Boston, Inc., Boston, MA, 2002.

[280] K. Kunen, *Set theory: An introduction to independence proofs*, Studies in Logic and the Foundations of Mathematics, vol. 102, North-Holland Publishing Co., Amsterdam, 1983. Reprint of the 1980 original.

[281] H. Lebesgue, *Leçons sur l'intégration et la recherche des fonctions primitives*, Gauthier-Willars, 1904.

[282] F. León-Saavedra and A. Montes-Rodríguez, *Linear structure of hypercyclic vectors*, J. Funct. Anal. **148** (1997), no. 2, 524–545.

[283] ———, *Spectral theory and hypercyclic subspaces*, Trans. Amer. Math. Soc. **353** (2000), no. 1, 247–267.

[284] F. León-Saavedra and V. Müller, *Hypercyclic sequences of operators*, Studia Math. **175** (2006), no. 1, 1–18.

[285] F. León-Saavedra and P. Romero de la Rosa, *Fixed points and orbits of non-convolution operators*, Fixed Point Theory Appl. **2014**, no. 2014:221.

[286] M. Lerch, *Ueber die Nichtdifferentiirbarkeit bewisser Funktionen*, J. Reine angew. Math. **103** (1888), 126–138.

[287] B. Levine and D. Milman, *On linear sets in space C consisting of functions of bounded variation*, Comm. Inst. Sci. Math. Méc. Univ. Kharkoff [Zapiski Inst. Mat. Mech.] (4) **16** (1940), 102–105 (Russian, with English summary).

[288] A. Lévy, *Basic Set Theory*, Springer-Verlag, Berlin, 1979.

[289] J. Lindenstrauss, *On subspaces of Banach spaces without quasi-complements*, Israel J. Math. **6** (1968), no. 1, 36–38.

[290] J. Lindenstrauss and L. Tzafriri, *Classical Banach spaces*, Lecture Notes in Mathematics, Vol. 338, Springer-Verlag, Berlin-New York, 1973.

[291] N. G. Lloyd, *Degree theory*, Cambridge University Press, UK, 1978.

[292] J. López-Salazar, *Vector spaces of entire functions of unbounded type*, Proc. Amer. Math. Soc. **139** (2011), no. 4, 1347–1360.

[293] ———, *Lineability of the set of holomorphic mappings with dense range*, Studia Math. **210** (2012), 177–188.

[294] W. Luh, *Approximation analytischer Funktionen durch überkonvergente Potenzreihen und deren Matrix-Transformierten*, Vol. 88, Mit. Math. Sem. Giessen, 1970.

[295] ———, *Holomorphic monsters*, J. Approx. Theory **53** (1988), no. 2, 128–144.

[296] G. R. MacLane, *Sequences of derivatives and normal families*, J. Anal. Math. **2** (1952), no. 1, 72–87.

[297] J. Malý, *The Darboux property for gradients*, Real Anal. Exchange **22** (1996/97), no. 1, 167–173.

[298] F. Martínez-Giménez, P. Oprocha, and A. Peris, *Distributional chaos for operators with full scrambled sets*, Math. Z. **274** (2013), no. 1-2, 603–612.

[299] F. Martínez-Giménez and A. Peris, *Universality and chaos for tensor products of operators*, J. Approx. Theory **124** (2003), 7–24.

[300] R. D. Mauldin (ed.), *The Scottish Book*, Birkhäuser, Boston, Mass., 1981. Mathematics from the Scottish Café; Including selected papers presented at the Scottish Book Conference held at North Texas State University, Denton, TX, May 1979.

[301] Q. Menet, *Sous-espaces fermés de séries universelles sur un espace de Fréchet*, Studia Math. **207** (2011), no. 2, 181–195.

[302] ———, *Hypercyclic subspaces on Fréchet spaces without continuous norm*, Integr. Equ. Oper. Theory **77** (2013), 489–520.

[303] ———, *Hypercyclic subspaces and weighted shifts*, Adv. Math. **255** (2014), 305–337.

[304] ———, *Existence and non-existence of frequently hypercyclic subspaces for weighted shifts*, Proc. Amer. Math. Soc., DOI http://dx.doi.org/10.1090/S0002-9939-2015-12444-6.

[305] ———, *Hereditarily hypercyclic subspaces*, arXiv:1312.5786v1 [math.DS], Preprint (2013).

[306] G. Mittag-Leffler, *Sur la représentation analytique d'une branche uniforme d'une fonction monogène (Sixième note)*, Acta Math. **42** (1920), 285–308.

[307] A. Montes-Rodríguez, *Banach spaces of hypercyclic vectors*, Michigan Math. J. **43** (1996), no. 3, 419–436.

[308] ———, *A Birkhoff theorem for Riemann surfaces*, Rocky Mount. J. Math. **28** (1998), no. 2, 663–693.

[309] A. Montes-Rodríguez and M. C. Romero-Moreno, *Supercyclicity in the operator algebra*, Studia Math. **150** (2002), 201–213.

[310] A. Montes-Rodríguez and H. N. Salas, *Supercyclic subspaces: spectral theory and weighted shifts*, Adv. Math. **163** (2001), no. 1, 74–134.

[311] ———, *Supercyclic subspaces*, Bull. London Math. Soc. **35** (2003), no. 6, 721–737.

[312] T. K. S. Moothathu, *Two remarks on frequent hypercyclicity*, J. Math. Anal. Appl. **408** (2013), no. 2, 843–845.

[313] G. Mora and J. A. Mira, *Alpha-dense curves in infinite dimensional spaces*, Int. J. Pure Appl. Math. **5(4)** (2003), 437–449.

[314] D. Morgenstern, *Unendlich oft differenzierbare nicht-analytische Funktionen*, Math. Nachr. **12** (1954), 74 (German).

[315] J. Müller, *Continuous functions with universally divergent Fourier series on small subsets of the circle*, C. R. Acad. Sci. Paris Ser. I **348** (2010), no. 21-22, 1155–1158.

[316] J. Müller, V. Vlachou, and A. Yavrian, *Overconvergent series of rational functions and universal Laurent series*, J. Anal. Math. **104** (2008), no. 1, 235-245.

[317] J. Mujica, *Complex analysis in Banach spaces*, North-Holland Mathematics Studies, vol. 120, North-Holland Publishing Co., Amsterdam, 1986. Holomorphic functions and domains of holomorphy in finite and infinite dimensions. Mathematical Notes, 107.

[318] ———, *Complex analysis in Banach spaces*, Dover Publ. Inc., New York, 2010.

[319] J. R. Munkres, *Topology: a First Course*, 2nd ed., Prentice-Hall Inc., Upper Saddle River, NJ, 2000.

[320] G. A. Muñoz-Fernández, N. Palmberg, D. Puglisi, and J. B. Seoane-Sepúlveda, *Lineability in subsets of measure and function spaces*, Linear Algebra Appl. **428** (2008), no. 11-12, 2805–2812.

[321] T. Natkaniec, *Almost continuity*, Real Anal. Exchange **17** (1991/92), no. 2, 462–520.

[322] ———, *New cardinal invariants in real analysis*, Bull. Polish Acad. Sci. Math. **44** (1996), no. 2, 251–256.

[323] T. Natkaniec, *Algebrability of some families of Darboux-like functions*, Linear Algebra Appl. **439** (2013), no. 10, 3256–3263, DOI 10.1016/j.laa.2013.08.040.

[324] V. Nestoridis, *Universal Taylor series*, Ann. Inst. Fourier (Grenoble) **46** (1996), no. 5, 1293–1306.

[325] V. Nestoridis and C. Papadimitropoulos, *Abstract theory of universal series*, C. R. Math. Acad. Sci. Paris **341** (2005), 539–543.

[326] D. J. Newman, *An entire function bounded in every direction*, Amer. Math. Monthly **83** (1976), no. 3, 192–193.

[327] O. A. Nielsen, *An Introduction to Integration and Measure Theory*, Canadian Mathematical Society Series of Monographs and Advanced Texts, John Wiley & Sons, Inc., New York, 1997. A Wiley-Interscience Publication.

[328] T. K. Nogueira, *Lineabilidade em conjuntos de funções reais que atingem o máximo em um único ponto*, 2014. Master's Thesis – UFPB (Brazil).

[329] W. F. Osgood, *A Jordan curve of positive area*, Trans. Amer. Math. Soc. **4** (1903), 107–112.

[330] J. C. Oxtoby, *Measure and Category*, Springer-Verlag, New York, 1980. 2nd edition.

[331] G. Peano, *Sur une courbe, qui remplit toute une aire plane*, Math. Ann. **36** (1890), 157–160 (French).

[332] D. Pellegrino and E. V. Teixeira, *Norm optimization problem for linear operators in classical Banach spaces*, Bull. Braz. Math. Soc. (N.S.) **40** (2009), no. 3, 417–431.

[333] H. Petersson, *Hypercyclic subspaces for Fréchet spaces operators*, J. Math. Anal. Appl. **319** (2006), no. 2, 764–782.

[334] A. Plichko and A. Zagorodnyuk, *On automatic continuity and three problems of The Scottish book concerning the boundedness of polynomial functionals*, J. Math. Anal. Appl. **220** (1998), no. 2, 477–494.

[335] K. Płotka, *Sum of Sierpiński-Zygmund and Darboux like functions*, Topology Appl. **122** (2002), no. 3, 547–564.

[336] ———, *On lineability and additivity of real functions with finite preimages*, J. Math. Anal. Appl. **421** (2015), no. 2, 1396–1404.

[337] E. E. Posey and J. E. Vaughan, *Functions with a proper local maximum in each interval*, Amer. Math. Monthly **90** (1983), no. 4, 281–282.

[338] H. Queffélec and M. Queffélec, *Diophantine approximation and Dirichlet series*, Harish-Chandra Research Institute Lecture Notes, vol. 2, Hindustan Book Agency, New Delhi, 2013.

[339] T. I. Ramsamujh, *Nowhere analytic C^∞ functions*, J. Math. Anal. Appl. **160** (1991), 263–266.

[340] C. J. Read, *The invariant subspace problem for a class of Banach spaces, 2: hypercyclic operators*, Israel J. Math. **63** (1988), no. 1, 1–40.

[341] D. Redett, *Strongly annular functions in Bergman space*, Comput. Methods Funct. Theory **7** (2007), no. 2, 429–432.

[342] B. Reznick, *Sums of even powers of real linear forms*, Mem. Amer. Math. Soc. **96** (1992), no. 463, viii+155, DOI 10.1090/memo/0463.

[343] M. Rmoutil, *Norm-attaining functionals and proximinal subspaces* (2015), arXiv:1503.06112v1.

[344] J. Rodríguez, *On lineability on vector integration*, Mediterr. J. Math. **10** (2013), no. 1, 425–438.

[345] L. Rodríguez-Piazza, *Every separable Banach space is isometric to a space of continuous nowhere differentiable functions*, Proc. Amer. Math. Soc. **123** (1995), no. 12, 3649–3654.

[346] S. Rolewicz, *On orbits of elements*, Studia Math. **32** (1969), 17–22.

[347] J. L. Romero, *When is $L^p(\mu)$ contained in $L^q(\mu)$?*, Amer. Math. Monthly **90** (1983), no. 3, 203–206, DOI 10.2307/2975553.

[348] H. P. Rosenthal, *On quasi-complemented subspaces of Banach spaces*, Proc. Nat. Acad. Sci. U.S.A. **59** (1968), 361–364.

[349] ———, *On quasi-complemented subspaces of Banach spaces, with an appendix on compactness of operators from $L^p(\mu)$ to $L^r(\nu)$*, J. Functional Analysis **4** (1969), 176–214.

[350] W. Rudin, *Holomorphic maps of discs into F-spaces*, in: Complex Analysis (Lexington, KY, 1976), Lecture Notes in Math. 599, Springer, 1977, 104–108.

[351] ———, *Real and Complex Analysis*, 3rd ed., McGraw-Hill Book Co., New York, 1987.

[352] ———, *Functional Analysis*, 2nd ed., McGraw-Hill Book Co., New York, 1991.

[353] C. Ruiz and V. M. Sánchez, *Nonlinear subsets of function spaces and spaceability*, Linear Algebra Appl. **463** (2014), 56–67.

[354] H. Sagan, *Space-filling Curves*, Universitext, Springer-Verlag, New York, 1994.

[355] H. N. Salas, *Supercyclicity and weighted shifts*, Studia Math. **135** (1999), no. 1, 55–74.

[356] ———, *Eigenvalues and hypercyclicity in omega*, RACSAM. Rev. R. Acad. Cienc. Exactas Fís. Nat. Ser. A Mat. **105** (2011), 379–388.

[357] H. Salzmann and K. Zeller, *Singularitäten unendlich oft differenzierbarer Funktionen*, Math. Z. **62** (1955), 354–367 (German).

[358] S. Saxon and M. Levin, *Every countable-codimensional subspace of a barrelled space is barrelled*, Proc. Amer. Math. Soc. **29** (1971), 91–96.

[359] H. H. Schaeffer and M. P. Wolff, *Topological Vector Spaces*, 2nd ed., Grad. Texts in Math., vol. 3, Springer-Verlag, New York, 1999.

[360] J. B. Seoane-Sepúlveda, *Chaos and lineability of pathological phenomena in analysis*, ProQuest LLC, Ann Arbor, MI, 2006. Thesis (Ph.D.)–Kent State University.

[361] J. B. Seoane-Sepúlveda, *Explicit constructions of dense common hypercyclic subspaces*, Publ. Res. Inst. Math. Sci. **43** (2007), no. 2, 373–384.

[362] J. H. Shapiro, *Composition Operators and Classical Function Theory*, Universitext, Springer-Verlag, New York, 1993.

[363] G. Schoeber, *Univalent functions*, Lecture Notes in Math., vol. 478, Springer-Verlag, Berlin and New York, 1975.

[364] B. Schweizer and J. Smítal, *Measures of chaos and a spectral decomposition of dynamical systems on the interval*, Trans. Amer. Math. Soc. **344** (1994), no. 2, 737–754.

[365] J. Shinoda, *Some consequences of Martin's axiom and the negation of the continuum hypothesis*, Nagoya Math. J. **49** (1973), 117–125.

[366] S. Shkarin, *On the set of hypercyclic vectors for the differentiation operators*, Israel J. Math. **180** (2010), no. 1, 271–283.

[367] _____, *Hypercyclic operators on topological vector spaces*, J. London Math. Soc. (2) **86** (2012), no. 1, 195–213.

[368] W. Sierpiński and A. Zygmund, *Sur une fonction qui est discontinue sur tout ensemble de puissance du continu*, Fund. Math. **4** (1923), 316–318.

[369] J. Stallings, *Fixed point theorems for connectivity maps*, Fund. Math. **47** (1959), 249–263.

[370] S. Z. Sternberg, *Lectures on Differential Geometry*, Prentice Hall, Englewood Cliffs, NJ, 1964.

[371] B. Subramanian, *On the inclusion $L^p(\mu) \subset L^q(\mu)$*, Amer. Math. Monthly **85** (1978), no. 6, 479–481.

[372] J. Thim, *Continuous nowhere differentiable functions*, Luleå University of Technology, 2003. Master Thesis.

[373] J. I. Úbeda García, *Aspectos geométricos y topológicos de las curvas α-densas*, Ph. D. Thesis, Universidad de Alicante, Spain, 2006 (Spanish).

[374] M. Valdivia, *The space $\mathcal{H}(\Omega, (z_j))$ of holomorphic functions*, J. Math. Anal. Appl. **337** (2008), no. 2, 821–839.

[375] _____, *Spaces of holomorphic functions in regular domains*, J. Math. Anal. Appl. **350** (2009), no. 2, 651–662.

[376] A. C. M. Van Rooij and W. H. Schikhof, *A Second Course on Real Functions*, Cambridge University Press, Cambridge, 1982.

[377] D. J. Velleman, *Characterizing continuity*, Amer. Math. Monthly **104** (1997), no. 4, 318–322.

[378] C. E. Weil, *On properties of derivatives*, Trans. Amer. Math. Soc. **114** (1965), 363–376.

[379] J. Wengenroth, *Hypercyclic operators on nonlocally convex spaces*, Proc. Amer. Math. Soc. **131** (2003), no. 6, 1759–1761.

[380] D. J. White, *Functions preserving compactness and connectedness are continuous*, J. London Math. Soc. **43** (1968), 714–716.

[381] A. Wilansky, *SemiFredholm maps in FK spaces*, Math. Z. **144** (1975), 9–12.

[382] S. Willard, *General Topology*, Dover, New York, 2004.

[383] G. L. Wise and E. B. Hall, *Counterexamples in Probability and Real Analysis*, The Clarendon Press Oxford University Press, New York, 1993.

[384] T. Yamanaka, *A new higher order chain rule and Gevrey class*, Ann. Global Anal. Geom. **7** (1989), no. 3, 179–203.

[385] Z. Zahorski, *Supplément au mémoire "Sur l'ensemble des points singuliers d'une fonction d'une variable réelle admettant les dérivées de tous les ordres"*, Fund. Math. **36** (1949), 319–320 (French).

[386] L. Zalcman, *Uniqueness and nonuniqueness for the Radon transform*, Bull. London Math. Soc. **14** (1982), 241–245.

[387] L. Zhang and Z-H. Zhou, *Notes about the structure of common supercyclic vectors*, J. Math. Anal. Appl. **418** (2014), 336–343.

Index

Printed and bound by CPI Group (UK) Ltd, Croydon, CR0 4YY

22/10/2024

01777620-0019